X-Ray Scattering from Semiconductors and other Materials

3rd Edition

X-Ray Scattering from Semiconductors and other Materials

3rd Edition

Paul F Fewster

PANalytical Research Centre, UK

World Scientific

NEW JERSEY • LONDON • SINGAPORE • BEIJING • SHANGHAI • HONG KONG • TAIPEI • CHENNAI

Published by

World Scientific Publishing Co. Pte. Ltd.
5 Toh Tuck Link, Singapore 596224
USA office: 27 Warren Street, Suite 401-402, Hackensack, NJ 07601
UK office: 57 Shelton Street, Covent Garden, London WC2H 9HE

Library of Congress Cataloging-in-Publication Data
Fewster, Paul F., author.
X-ray scattering from semiconductors and other materials / Paul F. Fewster (PANalytical Research, UK). -- 3rd edition.
pages cm
Includes bibliographical references and index.
ISBN 978-9814436922 (alk. paper)
1. X-rays--Scattering. 2. Semiconductors. I. Title.
QC482.S3F49 2015
537.5'35--dc23
2014031060

British Library Cataloguing-in-Publication Data
A catalogue record for this book is available from the British Library.

Typeset by Stallion Press
Email: enquiries@stallionpress.com

In-house Editor: Rhaimie Wahap

Printed in Singapore

Preface

This third edition includes many changes to the second edition, which in itself was a version that included the corrections supposed to appear in the first edition. This third edition is much more comprehensive with a fuller account of the different scattering theories, including; the very latest explanation that is especially relevant to powder diffraction, and a more complete dynamical scattering theory for perfect materials. These approaches are compared with the conventional theories so that they can be put into context. The elasticity theory is dealt with in more detail for modelling mixtures of structures and orientations that are strained to accommodate the bonding across interfaces. A larger range of diffractometers is discussed in detail; as individual components and their combination. The sample, theory and diffractometer are all brought together with illustrated examples, to give a practical guide to analysis. The emphasis is still primarily on semiconductors, since they offer the greatest range of possible analyses, but some of these techniques are applicable to other materials, hence the small change in the title. For example results are given for metallic multilayers and even biological macro-molecules. This basically reflects some of the work done in my laboratory, so it is in a way a personal view from many years in this field.

Since the last edition I have considered new more rapid ways of obtaining high-resolution data both for perfect single crystal multilayer semiconductors to powder samples. This has been possible by building up a better understanding of the scattering and modelling the whole diffraction process from source to detector. Some of the anomalies encountered have made me rethink aspects of scattering theory, which I now believe gives a better account of the experimental evidence. The content is less concerned about the average atom or molecular structure, since this is generally known in the examples given, but concentrates on aspects of the microstructure, layer thickness and composition in these increasing complex multilayer structures. Semiconductors range from the most perfect through to heavily distorted materials; this makes them a fascinating material system to test the limits in X-ray scattering, also they have a clear technological importance in the world we live. The concentration is on small laboratory experiments, although much is applicable to synchrotron work, so there is no attempt in bringing in the exploitation of very intense sources for specialised experiments. The challenge of making the most of what is available in small laboratories is fascinating and requires considerable innovation.

I would like to thank all those I have interacted with, from work colleagues, those with whom I have collaborated, and those who I have listened to at meetings. It is inevitable that these interactions all influence my thinking that is essentially what is contained in this edition. I also thank my wife Penelope for her patience during the course of this work.

Contents

CHAPTER 1

AN INTRODUCTION TO SEMICONDUCTOR MATERIALS

This chapter discusses the intentions of the book, the general properties and generation of X-rays, and the structural and chemical information that can be extracted through diffraction. Much of the emphasis is on composite layer structures, because of their technological importance. These layer structures create interesting challenges because of the stresses generated, which cause strains that can be very anisotropic. A detailed explanation is presented of the elasticity for all combinations of orientation and symmetry. Some of the basics of epitaxy are discussed and how with certain combinations of materials, complex microstructures can exist.

1.1. General outline

Semiconductor materials exist in many structural forms and therefore require a large range of experimental techniques for their analysis. The advantage of concentrating on this class of materials is that the amount of detailed structural information is impressive, however the application of the methods described is not exclusive to these material types. Scattering from powder samples will be covered theoretically and how a greater understanding of the scattering can give rise to extra information. The overall intention of this book though, is to allow the reader to obtain a good working knowledge of X-ray diffraction theory, techniques and analysis, so that he or she is fully aware of the possibilities, assumptions and limitations with this form of structural investigation.

The molecular structure of semiconductors is in general well known. Most materials of interest have been manufactured in some way; therefore an approximate knowledge of the elements and layer thicknesses and sequence is assumed, and is the starting point for many of the approaches used. Most samples of interest, however, are not of a simple molecular form but are composite structures, commonly consisting of multiple thin layers with different compositional phases. There are many important structural parameters that can modify semiconductor device performance. These parameters include phase composition, micro-structural or layer dimensions and imperfections, etc. A description of how the properties are categorised and material form is given, since this largely determines the X-ray scattering experiment for the analysis, Fewster (1996).

The theory of X-ray scattering is presented from a physical basis and therefore naturally starts with dynamical theory and its extensions before describing the more approximate kinematical theory. The theories are largely considered at the single photon level. In reality the experiment collects many photons that are divergent and occupy a range of energies. These effects influence the experimental results and are covered in the description of the instruments. In general the scattering is three-dimensional because the sample is three-dimensional and this is born in mind throughout. The mapping of the scattering in three-dimensions is the most general experiment and all other approaches are obvious projections. The assumptions on moving to two-dimensional reciprocal space mapping and ultimately one-dimensional X-ray scattering, e.g. rocking curves will then appear more obvious and understandable. The assumptions associated with interpreting data collected in various ways will be discussed. This will then allow the reader to understand the subject conceptually and extend the techniques to his or her particular problem.

The concentration is on near-perfect semiconductor materials, whose elements are defined by their valence in the Periodic Table, that is group IV, III-V, II-VI, etc., since the available information is large as well as being commercially very important. The concentration on near-perfect materials is to contain the range of techniques and available information. Although the techniques will be considered in general terms so as not to limit the book to specific materials. These near-perfect materials are essentially extended single crystals, however as the perfection declines they become more mosaic until eventually polycrystalline. Since this edition is more comprehensive and concentrates on the whole process of collecting data, including the diffraction process and the instrumental aberrations, this is extended to include polycrystalline material analysis and powder diffractometers.

Interpretation requires a good understanding of the instrument aberrations, the assumptions in the methods and presumed details concerning the sample. Aspects of quality control with X-ray diffraction methods based on comparative measurements requires an understanding of the sensitivity of the conditions to the parameters of interest, these aspects will also be covered.

1.2. Semiconductors

Semiconductors range from the most perfect crystals available to amorphous materials. The sample dimensions can range from 12 inch (although 18 inch is possible) diameter ingots of bulk Si, down to layers of partial-atomic coverage with nanometre scale lateral dimensions embedded in a multi-layer structure. The molecular structure of most common semiconductors is of a high symmetry extended lattice, for example Si has the space group (this represents the relative relationships between atoms) of $Fd3m$ and the space group for GaAs and InP is $F\bar{4}3m$, whereas GaN in hexagonal form has the space group $P6_3mc$. The two former space groups are simple face-centred cubic lattices, figure 1.2.1, which leads to relatively simple crystallographic relationships, whereas the latter space group is more complex and requires a deeper understanding of the crystallography.

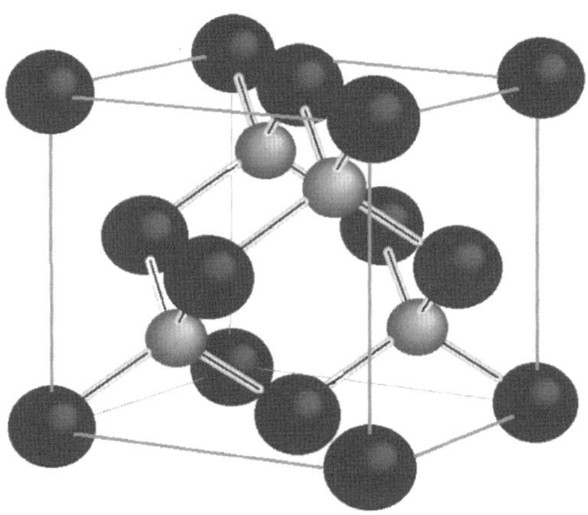

Figure 1.2.1 The arrangement of atoms in typical III-V semiconductor structures (e.g. GaAs, InP).

The degree of complexity in semiconductor samples is increasing with the manufacture of laterally patterned or phase separated structures having dimensions sufficiently small to create "zero" dimensional quantum size effects. The list of structural parameters required to define these materials is increasing, although the basic definitions can be described quite simply, Table 1.2.1. Since these structures are manufactured the analysis has a good starting point in that an approximate understanding of the structure will exist. The challenge therefore is to determine these parameters more precisely, not only to aid the manufacturing process but also to analyse them for defects and structural quality, etc.

All physical properties, be they electronic or optical, rely in some way on the structural properties of the material. As the degree of device sophistication increases the structural tolerances are reduced. Therefore the necessity for accurate determination of the composition, thickness and defects, etc., becomes very important. X-ray analysis techniques are very well developed for obtaining this information and mature enough to recognise its weaknesses. Of course no technique should be used in isolation but compared and complimented with other methods. Most techniques measure something that does not compare directly with any other; this in itself gives valuable information not only on the assumptions in deriving the information, but also on the sensitivity of the technique.

Amorphous semiconductors are important in the areas of large area electronics, although because of the low free-carrier mobility some of the associated circuitry requires re-crystallisation. This change of structural form improves the mobility and device speed. The sizes of the crystallites, the location of crystallite boundaries with respect to the device active region are all-important parameters. As we increase the degree of perfection the long-range structural order gives rise to well-defined electronic band structures. This leads into the possibilities of band structure engineering. The variables that the device engineer has are composition, shape and orientation to create structures with the required optical or electronic responses. Control over the growth of these structural parameters at the required level, sometimes at the atomic level, is not a trivial task. The growth of these structures is dominated by epitaxy either from the liquid phase (LPE), chemical vapour deposition (CVD), molecular beam epitaxy (MBE) or extended forms of these; metal organic CVD (MOCVD), metal organic MBE (MOMBE), etc.

These growth methods require very careful control and therefore careful analysis (in-situ and ex-situ) to ensure the structural parameters are those wanted. Also different compositional phases have different interatomic spacings and therefore these must be accommodated either by elastic strains or plastic deformation. Plastic deformation exists in the form of cracks and dislocations, which can act as charge carrier recombination centres and alter the device performance. It is therefore very necessary to have knowledge of the defects in the active region of the device that controls their behaviour. Again very careful structural analysis is required. All these properties can depend strongly on the quality of the substrate material, i.e. its defect density, orientation and surface strains, etc.

It should be clear that these semiconductors cannot be grown by pressing a few buttons and achieving the ultimate performance. The growth machine has to be characterized for growth-rate that could relate to temperature stability, which can influence the alloy composition in a layer, etc. A very good and thorough understanding of the materials and the growth method are required. In-situ analysis methods used to monitor the growth are developing but generally the most thorough analyses are performed ex-situ that can be very exhaustive. The in-situ methods generally rely on a detailed understanding from post-growth analysis. X-ray diffraction methods are sometimes used in-situ but in general contribute to improving yield by analysing material at various stages in manufacture, help in controlling the growth process and for detailed materials analysis ex-situ.

The X-ray analysis technique to apply depends on the material quality, the level of detail and the precision required. This book will describe all the levels of precision and assumptions made to carry out certain types of analysis. Because the crystalline quality of many semiconductors is very high the diffraction process cannot be treated in a simple way. Most analyses require the application of the dynamical diffraction theory and therefore an understanding of this and the assumptions involved are important and described in Chapter 2. The development of instrumentation for collecting the scattered X-rays has also created new possibilities in analysis that make X-ray methods a very versatile tool in probing the structure of materials.

Table 1.2.1. Definition of the structural properties of materials.

Type of property	General property	Specific property
Macroscopic	*Shape*	Layer thickness Lateral dimensions
	Composition	Structural phase Elements present Phase extent
	Form	Amorphous Polycrystalline Single Crystal
	Orientation	General preferred texture Layer tilt
	Distortion	Layer strain tensor Lattice relaxation Warping
	Homogeneity	Between analysed regions
	Interfaces	Interface spreading
	Density	Porosity Coverage
Microscopic	*Shape*	Average crystallite size Crystallite size distribution
	Composition	Local chemistry
	Orientation	Crystallite tilt distribution
	Distortion	Crystallite inter-strain distribution Crystallite intra-strain distribution Dislocation strain fields Point defects Cracks Strain from precipitates
	Interface	Roughness laterally
	Homogeneity	Distribution within region of sample studied

1.3. Method

In this section a brief description of the accessible information to X-ray diffraction techniques will be given. How and why this following information is possible to extract will become clear in later chapters. Table 1.2.1 presents the definition of various structural parameters used to define a material. The first subdivision of the structural properties is into macroscopic and microscopic. These are X-ray definitions and can be considered respectively as aspects that define the major features of the diffraction pattern (peak position and intensity) and those that alter the pattern in a more subtle way (peak shape and weak diffuse scattering).

Table 1.3.1. Definition for structural types.

Structural type	Definition
Nearly perfect epitaxial	A single extended crystal having near perfect registry with the same orientation as the underlayer, which is also nearly perfect.
Textured epitaxial	The layer orientation is close to registry with the underlayer, both normal and parallel to the surface plane. The layer is composed of mosaic blocks.
Textured polycrystalline	Crystallites preferentially orientated normal to the surface, but random in the plane. They have a distribution in sizes.
Nearly perfect polycrystalline	Random orientated crystallites of similar size and shape.
Amorphous extended lattice	Similar strength interatomic bonds but no length scale correlation greater than this.
Random molecules	Essentially amorphous structure with weak interlinking between molecules, possibly giving some ordering.

X-ray diffraction is a very sensitive structural analysis tool and the extent to which detailed information can be obtained depends on the sample itself. Suppose that the sample is poorly defined and contains numerous crystallites, with a distribution of structural phases, sizes, orientations and strains, then, separating the various contributions is not trivial. However if certain properties can be determined rather precisely then others can be determined by extending the range of experiments. Clearly therefore, the initial assumptions concerning the sample will define which structural details can be obtained readily. At this stage we should define the sample since these will define the likely information that can be determined by X-ray methods, and the applicable information to use, Table 1.3.1.

Orientation in this context refers to the alignment of low index atomic planes (these are planes separated by distances of about one unit cell spacing) to some other reference, e.g. the surface. These definitions concentrate on samples with laterally extended homogeneity and therefore can be expanded to include patterned structures and random structural variations in the lateral plane by considering them as columns.

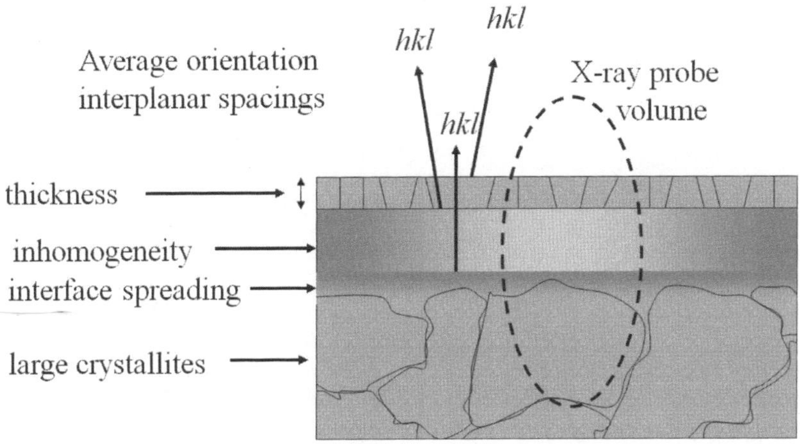

Figure 1.3.1. The main macroscopic parameters that characterise a layered structure.

Since any structural probe will determine an average of a region or analyse an unrepresentative region of an inhomogeneous sample, it is clear to see that the useful information may be limited to some average parameter and its variation. For structural types that are highly inhomogeneous, e.g. random molecules and textured polycrystalline materials, then X-ray diffraction will average some long-range order, orientation distribution and their variations. This is where it is important to link the physical or chemical property of the material to the structural property, for example is it the macroscopic average or the microscopic details that determines the property of interest? The next question may well be the scale of the variation; is it homogeneous at the micron or nanometre scale? X-ray diffraction averages in several ways, within a coherently diffracting volume and the X-ray beam dimensions on the sample. Controlling the beam divergence can modify the former and the latter can be subdivided by analysing the scattered beam with an area detector as in X-ray topography. The range of X-ray analysis techniques therefore cannot be simply categorised into finite bounds of applicability but depend upon the material, the property of the material of interest, the versatility of the diffractometer, the X-ray wavelength, etc. Understanding the details of diffraction process, the nature of X-rays and assumptions concerning the sample are all-important to making a good and reliable analysis.

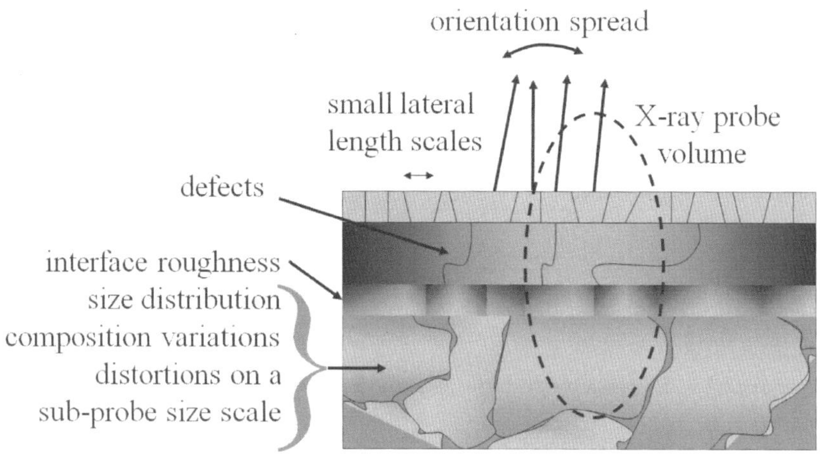

Figure 1.3.2. The main microscopic parameters that characterise a layered structure.

Some typical macroscopic and microscopic properties are given in figures 1.3.1 and 1.3.2 respectively. The important aspect here is the X-ray probe dimension with respect to the properties. Clearly the probe is not simply defined two dimensionally but also has some depth into the figures, consequently we must be aware how this probe brings all this information together to create a signal which is then interpreted. Having defined some basics concerning the sample we shall consider some basic information about the X-rays used to extract this information.

1.4. Properties of X-rays

X-ray wavelengths compare with the energy transitions of inner electron orbitals in atoms. It is this property that is used to create laboratory monochromatic X-rays. High energy decelerating electrons will also emit X-radiation and this is the reason for the continuum of radiation from laboratory sources. A laboratory source is shown diagrammatically in figure 1.4.1 with an accompanying spectrum. The radiation from a laboratory source is not very uniformly distributed and in general much of the radiation is not used. However the intense characteristic lines act as a good internal standard and it is these lines that are used in the majority of laboratory experiments.

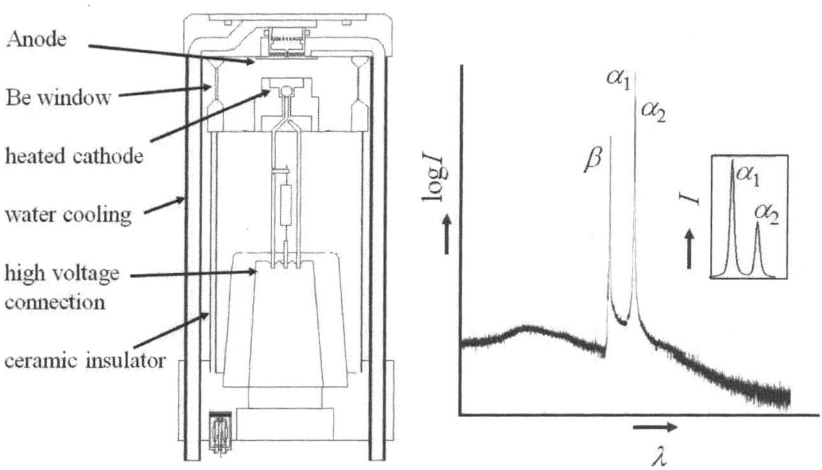

Figure 1.4.1. The interior of a modern sealed source X-ray tube and an indication of the spectral variation with intensity for a typical anode material

Synchrotron radiation sources work on a very different principle. A synchrotron is really a storage ring for electrons, which are contained by magnetic fields to prevent excessive divergence and consequent energy loss. When the electrons are deviated from a straight line using so-called bending magnets, wigglers and undulators, the consequent acceleration towards the centre of the curve creates an energy orbital jump thus producing electromagnetic radiation. If this energy change is large (i.e. high speed electrons and small bending radius from intense magnetic fields) then X-rays can be produced. The X-rays from the synchrotron are emitted tangentially from the radius and are concentrated into a narrow cone with the electric field vector predominately confined to the plane of the orbit; i.e. the beam is horizontally polarised. However it is possible to rotate this plane of polarisation but in general this aspect does restrict most experiments to scattering in the vertical plane. This can lead to extra tolerances required for mechanical movements of diffractometers because of the gravitational pull. Another aspect to consider is the wavelength calibration; this has to be done before any experiment since the emission is smooth and there are few reference lines except at absorption edges. Laboratory sources are rather less efficient at producing X-rays. The emerging X-rays are randomly polarised and almost radially symmetric, yet only a small percentage of this divergent source can be used.

Because of the method of injecting electrons into a synchrotron they are arranged in bunches and consequently the X-ray emission will have a time structure. This can prove useful for some experiments especially when only a single bunch is injected, if the X-ray pulse can be synchronised with some dynamic experiment, (Barrington-Leigh and Rosenbaum, 1976, Whatmore, Goddard, Tanner and Clark, 1982). The third generation sources have a very high brilliance level, giving rise to very small source sizes that can create some phase coherence across the whole source. This phase coherence can lead to observable interference effects when the beam travels along different optical paths (phase contrast topography and tomography, Cloetens *et al*, 1999). This coherence over the source can also create routes to reconstructing the scattering object (Miao, Charalambous, Kirz and Sayre, 1999). Laboratory X-rays from similar sized sources have very low power output. However there are methods of effectively moving the source close to infinity with crystal optics: the phase front formed from an object containing several different optical paths can then be separated with an analyser crystal (Davis, Gao, Gureyev, Stevenson and Wilkins,

1995). It is clear that the developments and possibilities continue and this is far from a static subject. These developments will then lead to new possibilities in analysis.

Another source that concentrates on the phase coherence is the "Free Electron Laser." This source utilizes the alternating magnets of an undulator that deviate the electrons from side to side, producing X-rays. Now if the X-rays produced at each sideways movement are in phase with each other, i.e. being separated by an integral number of wavelengths, then the output will be very strongly coherent. This makes the coherent scattering more useful, rather then the synchrotron that is not fully coherent. The disadvantage though, is that this necessitates a pulsed output rather than a more continuous source, because this is a linear device and not a storage ring as in the case with a synchrotron. The early free electron lasers are predominately soft X-rays (long wavelengths or low energy), although harder X-ray sources with wavelengths closer to those created at synchrotrons and sealed laboratory sources are close to completion.

The highly directional aspects of the synchrotron generated X-rays leads to very intense sources compared with laboratory sources. However the convenience and improvements in intensity output makes the laboratory sources suitable for most experiments. One of the earliest methods for increasing the intensity in the laboratory was achieved by rapidly rotating the anode (rotating anode source) to distribute the heat. This has lead to increases in intensities by almost an order of magnitude for 15 kW sources; 60 kW sources are also available. However there are many other ways of improving the intensity and whatever method is used it has to be related to the problem to be solved, since the intensity output should be qualified with flux, divergence, wavelength distribution, etc.

Since X-rays are primarily generated from inner atom core transitions the photon wavelengths are in the region of 0.1 nm, which is of the order of the interatomic spacings in materials. Bragg's equation (derived in Chapter 2) indicates that the difference in the scattering angle of two interatomic spacings of 0.14 and 0.15 nm determined with a 0.15 nm

X-ray wavelength is ~2.4^0. As will be seen later the peak widths of diffraction maxima can be located within about 0.0002^0. This gives X-rays the high strain sensitivity at the part per million level and sensitivity to atomic scale spatial resolutions.

1.5. Instrumentation

There have been considerable developments in new instrumentation. The power of laboratory X-ray sources have increased and various focusing mirrors and X-ray lenses can recover the divergence of laboratory X-ray sources with considerable intensity enhancements. The degree of sophistication is increasing with the various components recognising each other (i.e. exchanging the X-ray tube will be recognised by the system, thus limiting the power delivered, etc.). Computer automation has made very significant improvements in time and freed user involvement and this will continue. This has considerably helped in the thinking to doing ratio.

The mechanical stability has also improved with optical encoding on the axes, allowing fast movement to very high precision. Interchangeable components (monochromators, X-ray mirrors, slits, etc.) increase the versatility of diffractometers and can be pre-aligned so that several very different experiments can be performed on one instrument with a simple change. The experiment can now be fitted to the sample and property of interest instead of the former more established approach of having an array of instruments for each experimental technique. The choice of instrumental configuration and its consequential influence on the information required from the sample will be covered in Chapter 4.

1.6. Sample definition

The various sample types have been described in section 1.2 and 1.3, but here the description will be defined more closely to that required for X-ray diffraction. These definitions also indicate the information of importance in analysis, Table 1.6.1. Basically any structure will be an arrangement of atoms. However a crystal is defined as "any structure having essentially a discrete diffraction pattern." This is the accepted definition (Acta Crystallographica **A48** 928, 1992). To have a diffraction pattern that is observable with X-rays in the simplest case requires some form of periodicity or repeat unit cell.

A semiconductor, for example GaAs, Si, GaN, consists of an extended periodic array and would fit into the above categories of perfect epitaxy, textured epitaxy and possibly textured polycrystalline in thin layer form. Although material with no dislocations (missing lines of atoms) can be grown, most do have dislocations threading through them. The generation of convection currents during growth can create mosaic blocks (crystallites surrounded by defects) that can be tilted with respect to each other. These are all fairly typical features found in bulk material and thin films. One of the most fundamental problems in thin films is that the atomic spacing of the layer differs from that of the underlying material. This will cause either elastic distortion or, if the internal stress exceeds that which can be accommodated by elastic strains, plastic deformation occurs and misfit dislocations are generated. Misfit dislocations can be formed from the high stress levels and imperfections at the growing surface nucleating dislocation loops that glide to the interface, or by turning a threading dislocation to lie in the interface plane. Knowledge of the state of strain, the number of defects, etc., can be very important for device performance and X-ray diffraction methods are very sensitive to these effects.

Figure 1.6.1 gives a three dimensional view of the structural properties of a thin film. Basically we have a unit cell repeat that can vary laterally and in depth, having parameters a, b, c, α, β and γ. Within this there are relative rotations between regions and layers, defects (dislocations and point defects (atomic site errors, e.g. interstitials, vacancies and impurity atoms)). These features all influence the diffraction pattern of X-rays. The creation of the scattering pattern from X-rays is one thing but to interpret the features is quite another and a reasonable understanding of the sample in question is necessary. The important aspects will now be considered here.

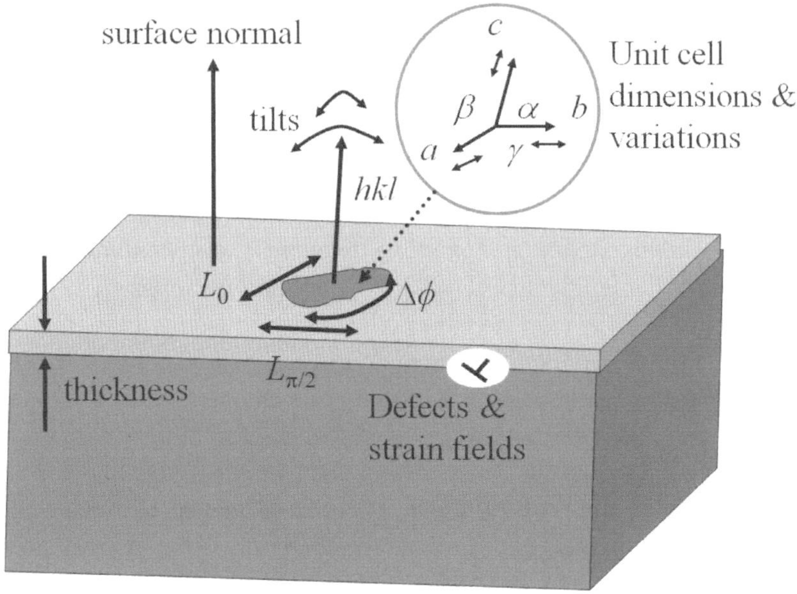

Figure 1.6.1. The range of structural parameters accessible to X-ray methods for imperfect (real) samples.

The very high strain sensitivity can allow measurement of x <1% (absolute) composition variations in $Al_xGa_{1-x}As$ alloys, <0.1% (absolute) composition variations in $In_xGa_{1-x}As$, etc., for peak shifts of 0.001° or 3.6 seconds of arc. However to achieve this, the strain has to be related to the composition using some assumptions. As discussed above a thin layer grown epitaxially on a substrate will distort either elastically or plastically. In both cases we need to determine the unit cell parameters of the layer of interest and calculate how this would change if it was free standing. Clearly we have to include the influence of elastic parameters.

Table 1.6.1. A broad overview of the structural parameters that characterise various material types. Those parameters that have meaning in the various materials are given with filled diamonds, those that could have meaning are given by open squares.

	Thickness	Composition	Relaxation	Distortion	Crystallite size	Orientation	Defects
Perfect Epitaxy	♦	♦				♦	
Nearly perfect epitaxy	♦	♦	□	□	□	♦	♦
Textured epitaxy	♦	♦	♦	♦	♦	♦	♦
Textured polycrystalline	♦	♦	□	♦	♦	♦	□
Perfect polycrystalline	♦	♦		♦	♦		□
Amorphous layers	♦	♦					

1.6.1. The influence of elastic distortions

The arrangement of atoms in silicon is similar to that given in figure 1.2.1, except that all the atoms are identical. If we try and compress the structure along the bonds, then the [*111*] type directions will be much more difficult to compress than along a [*100*] direction, where we would distort angles, for example. So although the structure is of high symmetry, its elastic properties are very anisotropic. Compression along one direction will necessitate an expansion in another. This can be characterised by examining the relationship between stress and strain. Initially we can suppose that the strain is elastic until the internal stress is too large and plastic deformation occurs. The plastic deformation will occur as cracks or dislocations, however the strain parallel to any interface will be related to the degree of alignment of the atoms in a layer with that underneath. Hooke's law gives the relationship of stress to strain, but because we are considering an anisotropic medium we have to generalise the problem and the elastic stiffness to fourth rank tensors, Nye (1985).

$$\sigma_{ij} = a_{im} a_{jn} a_{ko} a_{lp} T_{mnop} \varepsilon_{ij} \qquad 1.6.1$$

The a_{ij} are the direction cosines of the direction associated with i to that of j, σ_{ij} and ε_{ij} are the stresses and strains along the various directions related by i and j. The parameters σ_{zz} ($=\sigma_{33}$) and ε_{zz} ($=\varepsilon_{33}$) are the stress and strain directions normal to the surface. T_{mnop} is a fourth rank tensor with 81 coefficients, where m, n, o and p run from 1 to 3. When equivalent coefficients are considered, this can be simplified to a 6×6 matrix, c_{ij}, where i and j run from 1 to 6. The conversion from one to another is obtained by combining the suffixes m and n, and o and p, using the following rules:

Table 1.6.2. The rules for converting the 4th rank stiffness tensor to a more convenient matrix form.

$m,o=1$, $n,p=1$	$m,o=2$, $n,p=2$	$m,o=3$, $n,p=3$	$m,o=2$, $n,p=3$	$m,o=1$, $n,p=3$	$m,o=1$, $n,p=2$
$i,j=1$	$i,j=2$	$i,j=3$	$i,j=4$	$i,j=5$	$i,j=6$

T_{ij} represents the stiffness coefficients, whereas the c_{ij} that are tabulated for most semiconductor materials. This simplified form is given by

$$\begin{pmatrix} \sigma_{xx} \\ \sigma_{yy} \\ \sigma_{zz} \\ \sigma_{yz} \\ \sigma_{xz} \\ \sigma_{xy} \end{pmatrix} = \begin{pmatrix} c_{11} c_{12} c_{13} c_{14} c_{15} c_{16} \\ c_{21} c_{22} c_{23} c_{24} c_{25} c_{26} \\ c_{31} c_{32} c_{33} c_{34} c_{35} c_{36} \\ c_{41} c_{42} c_{43} c_{44} c_{45} c_{46} \\ c_{51} c_{52} c_{53} c_{54} c_{55} c_{56} \\ c_{61} c_{62} c_{63} c_{64} c_{65} c_{66} \end{pmatrix} \begin{pmatrix} \varepsilon_{xx} \\ \varepsilon_{yy} \\ \varepsilon_{zz} \\ \varepsilon_{yz} \\ \varepsilon_{xz} \\ \varepsilon_{xy} \end{pmatrix} \qquad 1.6.2$$

σ_{ij} and ε_{ij} represent the stresses and strains along various directions. The convention is that; c_{11} is the stiffness coefficient for the *a* axis [*100*] direction for cubic system, whereas for other symmetries this can vary, for example the hexagonal symmetries have c_{11} along [*2-1-10*], c_{22} along [*01-10*] and c_{33} along [*0001*]. This full array of coefficients is the general case for triclinic structures, although there are further equivalences, $c_{ij} = c_{ji}$, hence for the triclinic case there are 21 unique stiffness coefficients required. For higher symmetry many of these coefficients become zero and some become equivalent.

Suppose we consider the growth along the <*001*> direction then this corresponds to the *zz* indices and the layer will be constrained in the plane of the interface and the stress will be zero normal to this direction, i.e. the top surface is unconstrained then from equation 1.6.2

$$\sigma_{zz} = 0 = c_{31}\varepsilon_{xx} + c_{32}\varepsilon_{yy} + c_{33}\varepsilon_{zz} + c_{34}\varepsilon_{yz} + c_{35}\varepsilon_{xz} + c_{36}\varepsilon_{xy} \qquad 1.6.3$$

Now if we take the example of a cubic system, the high symmetry leads to the following equivalents $c_{31} = c_{32} = c_{13} = c_{23} = c_{12} = c_{21}$, $c_{22} = c_{33} = c_{11}$ and $c_{34} = c_{35} = c_{36} = 0$, therefore, equation 1.6.3 can be rearranged to yield the strains along three orthogonal directions to be related rather simply:

$$\varepsilon_{zz} = -\frac{c_{31}}{c_{33}}\varepsilon_{xx} - \frac{c_{32}}{c_{33}}\varepsilon_{yy} \equiv -\frac{c_{12}}{c_{11}}\{\varepsilon_{xx} + \varepsilon_{yy}\} \qquad 1.6.4$$

Hornstra and Bartels (1978) have solved the condition for several typical orientations of cubic systems, giving some examples for III-V compounds. Any solutions of the cubic system will just include combinations of the coefficients c_{11}, c_{12} and c_{44}. However it is most useful with the increasing range of materials and orientations that a completely general solution is required. So working with equation 1.6.1, the coefficients need to be transformed back into their 4th rank form, e.g. c_{46} becomes T_{2312}, etc., following Table 1.6.2. In this tensor form the "crystal" can be rotated by invoking the relevant direction cosines and then related to sample stresses and strains.

For the case of an extended thin layer, there will be no overall shear for the probed volume, and again it is possible to extract the σ_{zz} line, but now the number of terms increases (depending on the orientation). So the general relationship for an extended thin layer with no shear is given by

$$\varepsilon_{zz} = \frac{F_1}{F_3}\varepsilon_{xx} + \frac{F_2}{F_3}\varepsilon_{yy} \qquad 1.6.5$$

where

$$F_i = \sum_{m=1}^{3}\sum_{n=1}^{3}\sum_{o=1}^{3}\sum_{p=1}^{3} a_{3m}a_{3n}a_{io}a_{ip}T_{mnop} \qquad 1.6.6$$

The T_{mnop} is now the full fourth rank tensor representation and can be converted according to Table 1.6.2 to relate them to the more familiar tabulated form, c_{ij}. Each summation has 81 terms, however this is reduced by symmetry since some of the T_{mnop}/c_{ij} values are zero.

If we now consider the cubic system for a sample, with a surface direction that is not isotropic in the surface plane, e.g. *<110>*, then $F_1 \neq F_2$. Equation 1.6.5 then becomes:

$$\varepsilon_{zz} = \frac{0.5c_{11} + 0.5c_{12} - c_{44}}{0.5c_{11} + 0.5c_{12} + c_{44}}\varepsilon_{xx} + \frac{c_{12}}{0.5c_{11} + 0.5c_{12} + c_{44}}\varepsilon_{yy} \qquad 1.6.7$$

This formula is based on the calculation of the strains, ε_{xx} and ε_{yy} along *<110>* and *<001>*, and therefore these parameters F_1 and F_2 are only correct if the strains in these direction are known. Clearly the elastic parameters and therefore the distortion is a function of the direction. This leads to significant asymmetry in the elastic parameters in GaAs for example:

$$\varepsilon_{zz} = 0.1843\varepsilon_{xx} + 0.3687\varepsilon_{yy} \qquad 1.6.8$$

So provided the layer symmetry (cubic in this example) matches that of the substrate and both have the same orientation and the layer is strained to match that of the substrate then equation 1.6.7 is valid for obtaining the state of strain in this *<110>* layer. However if there is some strain relaxation, the strains need to be measured along the directions chosen for ε_{xx} and ε_{yy}. Similarly if the layer and substrate are of a different symmetry or orientation (e.g. GaN on sapphire not grown on *0001*, i.e. non isotropic in the basal plane) then the elastic parameters need to be calculated along the directions of measurement.

Layer growth is sometimes conducted on vicinal planes to improve surface morphology or to grow quantum wires, or the substrate is not perfectly on orientation. This will also influence the parameters F_1 and F_2. By way of example a cubic GaAs layer grown on a cubic substrate that has a surface orientation of 0^0, 0.25^0, 4^0 and 10^0 away from the *<001>* towards *<101>* will have differing parameters F_1 and F_2 given in Table 1.6.3:

Table 1.6.3. The variation in the principal elastic parameters as a function of misorientation with respect to the *001* direction, the % error in ε_{zz} is independent of strain magnitude.

Vicinal angle	F_1/F_3	F_2/F_3	% error in ε_{zz} introduced
0^0	0.45202	0.45202	0%
0.25^0	0.45201	0.45199	0.004%
4^0	0.45005	0.44569	0.91%
10^0	0.44038	0.41462	5.4%

From Table 1.6.3, it is clear that minor misorientations, i.e. those within typical nominal values, have little effect on the strain estimations, however for larger misorientations this can be significant. The X-ray diffraction experiment will give an approximate one to one relationship of composition to the strain and therefore this should be considered for any precision measurement.

To further illustrate the variation in the relationship between the stiffness coefficients and the strain, a few examples for hexagonal GaN are given, since several growth surfaces are used, e.g. c-plane (*0001*), a-plane (*2-1-10*), m-plane (*01-10*) and r-plane (*10-12*). These planes have relevance as an orientation option since the a- and m-planes are non-polar surfaces and r-plane is semi-polar. The a, m and r orientations will remove or reduce the piezo-electric effect brought about by the strain of epitaxy, which can be quite a significant effect in c-plane material. For c-plane GaN, equation 1.6.5 simplifies to

$$\varepsilon_{zz} = \frac{c_{13}}{c_{33}}\varepsilon_{xx} + \frac{c_{23}}{c_{33}}\varepsilon_{yy} = 0.2543\varepsilon_{xx} + 0.2543\varepsilon_{yy} \qquad 1.6.9$$

From symmetry $c_{23} = c_{13}$ and hence any stress applied in the surface will have a similar strain response, i.e. it is isotropic in the surface plane. However for a-plane GaN

$$\varepsilon_{zz} = \frac{c_{13}}{c_{11}}\varepsilon_{xx} + \frac{c_{12}}{c_{11}}\varepsilon_{yy} = 0.2806\varepsilon_{xx} + 0.3678\varepsilon_{yy} \qquad 1.6.10$$

Also in this case of m-plane GaN

$$\varepsilon_{zz} = \frac{c_{23}}{c_{22}}\varepsilon_{xx} + \frac{c_{12}}{c_{22}}\varepsilon_{yy} = 0.2806\varepsilon_{xx} + 0.3678\varepsilon_{yy} \qquad 1.6.11$$

Because the stiffness coefficients $c_{11} = c_{22}$ for this symmetry the response to in-plane stress for these two orientations is identical. Whereas for r-plane GaN

$$\varepsilon_{zz} = 0.4460\varepsilon_{xx} + 0.5502\varepsilon_{yy} \qquad 1.6.12$$

The combination of coefficents runs to 8 non-zero terms to evalute F_1 and F_3, and 4 non-zero terms to evalute F_2, equations 1.6.5 and 1.6.6.

The orientations within the surface plane, for c-plane, a-plane and m-plane are simple low index planes, whereas for the r-plane surface (*10-12*), the two orthogonal directions for determining these coefficients are [*-0.568,0,0.568,1*] and [*1-210*], since there are no simple integer combinations for this orientation and the coefficients are functions of the lattice parameters. This non-integer index of direction is determined from the vector product. Clearly deposition on any surface with different lattice parameters, other than (*0001*), will not distort evenly in the plane, and consequently the layer symmetry will no longer be hexagonal.

These detailed calculations give a range of distortion factors within the bounds of 0.18 – 0.45. A reasonable approximation for lower symmetry materials can be made within these bounds if the stiffness coefficients are unknown, assuming there is only a small directional anisotropy or the symmetry is isotropic in the interface plane. Alternatively if an "engineering" Poisson ratio, ν, is known then this can give a distortion coefficient given by:

$$\varepsilon_{zz} = \frac{-\nu}{1-\nu}(\varepsilon_{xx} + \varepsilon_{yy}) \qquad 1.6.13$$

For rigid materials, ν takes on a low value <0.33 and for flexible materials, e.g. rubber ν can be as high as 0.5, which is also the maximum value it can take. A Poisson ratio of 0.5 indicates that constant volume is maintained during distortion, however the structural form for common semiconductors of this Poisson ratio can vary considerably with structure and orientation, because of the nature of the highly directional covalent bonding. Structures of similar form and orientation however do not vary too excessively, so approximate estimates of a Poisson ratio can be assumed from similar structures if no values are available. Clearly as the molecular form becomes more complicated and the bond directions more random this anisotropy will decrease, making a more general distortion coefficient / Poisson ratio a reasonable approximation. Additional shear actions come into effect along directions of lower symmetry, however these may be of less concern in a homogeneous thin layer of large lateral dimensions. For example, cubic materials with a surface orientation of [*001*], from equation 1.6.13, the Poisson ratio is given by

$$\nu = \frac{c_{12}}{c_{11} + c_{12}} \qquad 1.6.14$$

For patterned wafers, which have small lateral dimensions that are comparable to the layer thickness, the stresses are not simply relieved at the top surface, and the full matrix, equation 1.6.2 should be used, e.g. as in finite element analysis, to estimate the distortions in more than one direction. This is important for analyzing surface quantum dots, free-standing quantum wires, etc.

The concentration here is for planar structures, which are the most common application for these analyses, where the strains in the plane of the interface can be expressed as

$$\varepsilon_{xx} = \frac{{}_L d_x - {}_{L0} d_x}{{}_{L0} d_x}, \varepsilon_{yy} = \frac{{}_L d_y - {}_{L0} d_y}{{}_{L0} d_y} \qquad 1.6.15$$

and in the direction normal to the surface

$$\varepsilon_{\perp} = \varepsilon_{zz} = \frac{{}_L d_z - {}_{L0} d_z}{{}_{L0} d_z} \qquad 1.6.16$$

${}_L d_x$, etc., are the actual atomic plane spacings along x, and ${}_{L0} d_x$ are the unstrained or free standing atomic plane spacings along x, etc.

1.6.2. The epitaxial relationship

The atoms of material deposited on a substrate will try and bond to those in the substrate and the lowest energy configuration will result in the most likely arrangement. The energy is a combination of the direct bond energy (related to the difference in bond length to the ideal), the torsion energy (effective next-nearest neighbour atom distances from the ideal), deviations from the ideal dihedral angle (effectively the next-next-nearest neighbours) and so on, plus the elastic compliance of how the whole structure responds to this interaction and the influence of any structural faults, e.g. grain boundaries. Examples of a simplified energy calculation will be given later, but initially simpler systems will be considered, that have similar atom arrangements and small differences in atom spacings from layer to substrate.

Let us firstly consider a cubic (*001*) GaAs substrate with a thin film of cubic AlAs on top. Both structures have the same space group, the same arrangement of atoms, but slightly different lattice parameters and elastic parameters. If a thin layer (~0.2 μm) is deposited then the atoms will align with those of the substrate and the structure will appear as a continuous lattice with an abrupt change in lattice parameter and composition at the substrate interface. The alignment of the atoms in the interface plane will define the lattice parameter of the layer in the interface plane and through the appropriate elastic stiffness combinations will define the lattice parameter normal to the interface plane, figure.1.6.2.

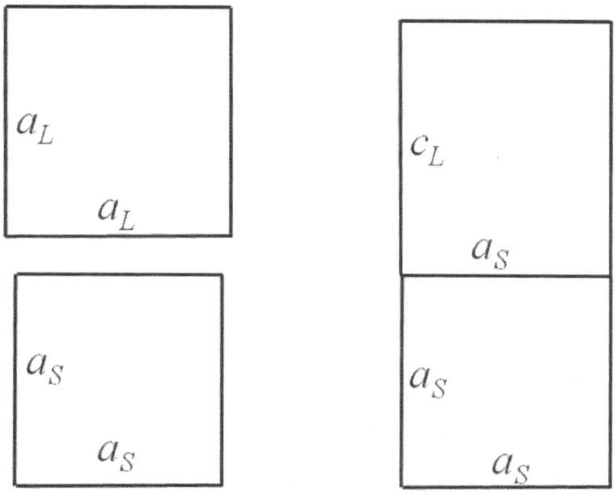

Figure 1.6.2. The undistorted (before deposition) and a distorted (after deposition) unit cell for a simple cubic layer on a cubic substrate; both are orientated along a cubic edge direction.

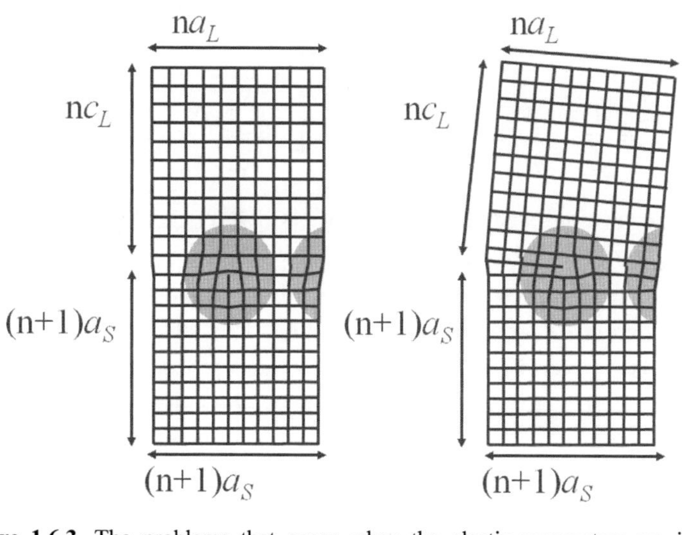

Figure 1.6.3. The problems that occur when the elastic parameters are incapable of accommodating the distortions necessary for perfect epitaxy.

As the thickness of the layer increases the layer will become progressively more reluctant to distort. Eventually the elastic limit will be reached and only partial registry will exist. When this situation arises there will be more rows of atoms in the substrate than in the layer (the lattice parameter of GaAs is less than that of AlAs). The average lattice parameter in the plane of the interface will therefore differ and some distortion will extend into the layer and substrate with possible tilting between the two, figure 1.6.3. Clearly the structure becomes quite complex even for this simple system. For perfect epitaxy we are trying to match the layer interatomic spacings to that of the layer or substrate below, ${}_Sd_x$, etc., however if the layer has partially relaxed back to its strain-free state then we can rewrite equation 1.6.15 as

$$\varepsilon_{xx} = \frac{{}_Sd_x - {}_{L0}d_x}{{}_{L0}d_x}[1\text{-}R_x], \varepsilon_{yy} = \frac{{}_Sd_y - {}_{L0}d_y}{{}_{L0}d_y}[1\text{-}R_y] \qquad 1.6.17$$

where R_x is the relaxation in the misfit along x, etc. The relaxation is then zero for perfect matching and unity when the layer relaxes to its unconstrained shape. Substituting equation 1.6.17 into 1.6.5 will therefore give the perpendicular strain from knowledge of the original lattice parameters of the component layers and the degree of relaxation in two orthogonal directions.

$$\varepsilon_{\perp} = \frac{{}_Sd_z - {}_{L0}d_z}{{}_{L0}d_z} = \frac{F_1}{F_3}\frac{{}_Sd_x - {}_{L0}d_x}{{}_{L0}d_x}[1\text{-}R_x] + \frac{F_2}{F_3}\frac{{}_Sd_y - {}_{L0}d_y}{{}_{L0}d_y}[1\text{-}R_y] \quad 1.6.18$$

When the unit cells of the two materials differ significantly then the registry of the atoms becomes very complex. Consider for example (*0001*) GaN on (*0001*) sapphire, both are hexagonal structures but the lattice parameters differ quite considerably. The atom arrangement of both materials is given in figure 1.6.4. Now the orientation of the GaN on the sapphire (*0001*) surface is determined by the interface free energy, which depends on the thermodynamics of the system, this is covered in considerable detail in Sutton and Balluffi (1995). For the purposes of illustration here a simple approximation is made by assuming the bond energy is at a minimum at some equilibrium distance, but increases as the square of this separation. Similarly a second order effect of the torsion angle can be treated in a similar manner. Further interactions "next-next-nearest neighbours" associated with a dihedral angle are ignored to simplify the process, so it then becomes possible to suggest possible orientation arrangements. The scaling between contributions is a potential source of error and the validity of this simple square-rule relationship at large distances is dubious, although a non-bonding cut-off is imposed. However it is not the purpose here of defining anything with precision, but rather suggest likely configurations of a layer on a substrate. It must be remembered that these poorly matched structures can have a complex microstructure and differing response to elastic distortion, etc., i.e. there are many uncertainties. The energy to be minimized is given by:

$$E \approx E_B(l - l_0)^2 + E_T(\Phi - \Phi_0)^2 \qquad 1.6.19$$

where l and l_0 are the closest approach and ideal bond length, Φ and Φ_0 are the torsion angles for the next-nearest neighbour and that for the ideal configuration and E_B and E_T are scale-factors ($E_B >> E_T$). The calculation is based on determining the ideal bond length and torsion angle for an atom within the layer, and then determining the energy E for various configurations for the atoms at the interface.

The best atomic match appears when the two lattices are aligned along different directions, i.e. the x-direction is rotated through 30^0 with respect to the other. However we can see that this gives approximate alignment of the Al in the sapphire to the Ga in the GaN. The mismatch is so large however that the GaN is very heavily relaxed towards its unstrained state and will be full of defects associated with this poor match.

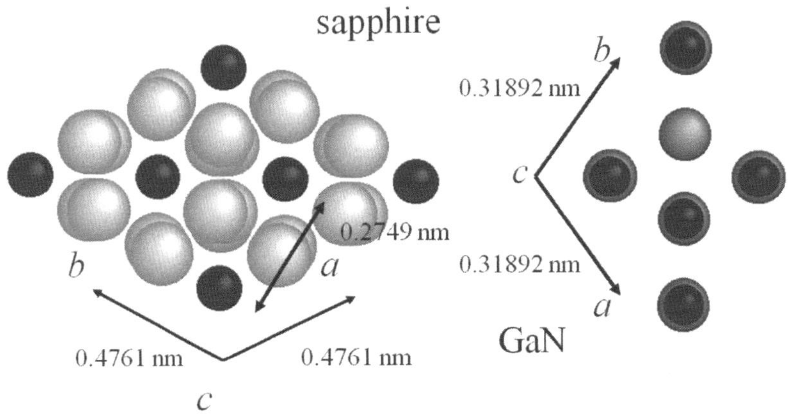

Figure 1.6.4. The basal plane view of GaN and sapphire; this indicates the rotation necessary to accommodate the alignment of atoms for epitaxy. Al and Ga are coloured black.

Pashley (1956) has given a very full account of the possibilities in epitaxy and encompasses the early theories. A full all-encompassing theory explaining the nucleation and orientation dependence is still elusive, but there are some general guidelines that can be given. Generally an orientation dependence occurs when the mismatch (the fractional difference in the lattice plane spacing in the plane of the interface) between the overlayer and underlayer is less than ~14%. Theoretical models and experimental evidence on a wide range of systems support this. The orientation depends on the relative alignment of atoms in the overlayer and underlayer and is not governed by integer relationships of atomic plane spacings of the two lattices. The thickness of the layer influences the extent to which the elastic distortion can be accommodated, the greater the misfit the thinner the layer should be to maintain good epitaxy. Once epitaxial growth is established in the fabrication of a structure then a full understanding of these nucleation processes may seem irrelevant, however some structures make use of some of the nucleation properties of certain materials.

We can consider growth to occur in three principle ways. The first is a two-dimensional mechanism, i.e. the layer is built up atomic layer by atomic layer, and this relies on the atoms migrating across the surface and preferring to locate at atomic layer steps. A complete atomic layer coverage will create a very smooth surface, whereas for intermediate coverage the surface is atomically rough. This oscillation in smooth and rough surfaces explains the oscillating specular reflectivity observed in Reflection High Energy Electron Diffraction (RHEED) during growth by Molecular Beam Epitaxy (MBE), Neave, Joyce, Dobson and Norton (1983). When the surface does not "wet" easily with the deposited atoms of the overlayer, the growth can occur in distinct islands that gradually enlarge and eventually coalesce. This is termed three-dimensional growth and can lead to mosaic or columnar growth with defects concentrated at the boundaries. Another interesting growth mode is a mixture of both three- and two-dimensional growth first described by Stranski and Krastanow (1938). This mechanism is characterised by the initial formation of a wetting layer (two-dimensional growth) that is very thin (no more than a few atomic layers) and the subsequent growth of islands. Examples of these mechanisms can be seen in semiconductor materials. InGaAs deposited on GaAs at low In compositions, <10%, will grow two-dimensionally up to about 70 nm before misfit dislocations are formed at the interface, whereas InAs will grow by the Stranski-Krastanow mechanism. These differing growth modes may appear troublesome but can be used to advantage in creating structures defined in all three dimensions by optimising the growth method. These can have very special properties and offer another challenge to analytical methods.

Determining the growth mode can only be accomplished with precise surface diffusion data and bond strengths, etc. It is then possible to construct a surface by modelling the whole process and extracting a statistical significance. These approaches have been very successful at predicting some of the general observations of surface topography, Itoh, Bell, Avery, Jones, Joyce and Vvedensky (1998).

Predicting the situation when defects form at the interface between two materials, i.e. when the elastic limit has been exceeded, has been the subject of many studies. This of course is a very important parameter because defects in general are detrimental to semiconductor devices. Knowledge of the bounds of lattice parameter misfit and thickness define whether a device is possible to fabricate. Hull and Bean (1992) have reviewed the mechanisms of dislocation generation and propagation and discussed the definitions and derivation of the "critical" thickness defining their onset. Of course there are many experimental studies that have questioned the theoretically derived values. Dunstan, Kidd, Howard and Dixon (1991) have taken a very pragmatic approach to the evaluation of critical thickness and compared the residual strain as a function of thickness. The resulting curve is remarkably predictable for a large range of material systems and offers a very quick procedure for predicting the onset of relaxation.

Understanding and controlling the influence of surface misorientations to reduce defects has been the subject of many studies, e.g. LeGoues, Mooney and Tersoff (1993) who studied the distribution and arrangement of dislocations and related these parameters to the proximity of glide planes. Stringfellow (1982) has reviewed the relationship of small misorientations compared with low-index planes and how these promote step-edge growth, and Schukin and Bimberg (1999) have used this growth control to influence the growth of quantum wires. If the mismatch is large or the substrate is amorphous then the orientation dependence of the layer can be governed by very different criteria and the layer can become essentially textured polycrystalline or even random polycrystalline. Knowledge of the likely form of these materials will define the type of experiment necessary to obtain detailed structural information.

Chapter 2 will describe the theoretical basis of scattering from various structures typically encountered in the field of semiconductor physics as well as more general material forms, e.g. powders. The concentration on semiconductors is of particular interest, because these materials represent the case when a large amount of information can be extracted, although the techniques are applicable to any material and not specific to semiconductors.

Bibliography

Barrington-Leigh, J and Rosenbaum, C (1976) Ann. Rev. Biophysics and Bioengineering **5** 239

Cloetens, P, Ludwig, W, Baruchel, J, Guigay, J-P, Pernot-Rejmankova, Salome-Pateyron, M, Schlenker, M, Buffiere, J-Y, Maire, E and Peix, G (1999) J Phys. D: Appl. Phys. **32** A145

Davis, T J, Gao, D, Gureyev, T E, Stevenson, A W and Wilkins, S W (1995) Nature **373** 595.

Dunstan, D J, Kidd P, Howard, L K and Dixon, R H (1991) Appl. Phys. Lett. **59** 3390.

Fewster, P F (1996) Rep. Prog. Phys. **59** 1339.

Hornstra, J and Bartels, W J (1978) J Cryst. Growth **44** 513.

Hull, R and Bean, J C (1992) Critical Reviews in Solid State and Materials Sciences **17** 507.

Itoh, M, Bell, G R, Avery, A R, Jones, T S, Joyce, B A and Vvedensky, D D (1998) Phys. Rev. Lett. **81** 633.

LeGoues, F K, Mooney, P M and Tersoff, J (1993) Phys. Rev. Lett. **71** 396

Miao J, Charalambous, P, Kirz J and Sayre D, (1999) Nature **400** 342

Neave, J H, Joyce, B A, Dobson, P J and Norton, A (1983) Appl. Phys. Lett. **A31** 1.

Nye, J F (1985) Physical Properties of Crystals – Their representation by Tensors and Matrices Oxford Science Publications: Oxford University Press

Pashley D W (1956) Adv. Phys. **5** 173.

Schukin, A V and Bimberg, D (1999) Rev. Modern. Phys. **71** 1125

Stranski, J N and Krastanow, L (1938) Ber. Akad. Wiss. Wien **146** 797

Stringfellow, G B (1982) Rep. Prog. Phys. **45** 469

Sutton, A P and Balluffi, R W (1995) Interfaces in Crystalline Materials: Clarendon Press, Oxford.

Whatmore, R W, Goddard, P A, Tanner, B K and Clark, G F (1982) Nature **299** 44

Chapter 2

THE THEORY OF X-RAY SCATTERING

This chapter presents the scattering theories applicable to analyzing thin film and polycrystalline materials. The 2-beam dynamical theory is derived, because of its usefulness in simulating scattering from epitaxial semiconductors. The limitations of this model are discussed along with extensions and alternatives that are more exact, including an approach that generates all the appropriate reflections making it a truly multiple-beam dynamical theory. The kinematic model is discussed and its applicability to thin films and imperfect materials, and how the optical theory is adequate for specular reflectivity. An alternative theory for X-ray diffraction is also presented that considers the scattering throughout space, which is particularly relevant for studying polycrystalline materials. It will also become obvious how the whole process is intricately linked to the instrument. The distorted-wave Born Approximation is discussed along with applications in modeling diffuse scattering.

2.1. The interaction of X-ray photons with the sample

The X-ray photon interacts with the sample in many different ways and the form of interaction depends on the photon energy and the nature of the sample. X-ray photons are electromagnetic and it is the electric field vector that interacts most strongly with the sample. The magnetic interaction is small and is only observable under special conditions with very intense X-ray sources. There are several forms of interaction depending on the photon energy and the nature of the electron state. Electrons loosely bound to atoms, for example the valence electrons, may absorb part of the energy of a photon and the emitted photon will have a lower energy and longer wavelength. If it is assumed that the electron is stationary and totally unbound then this wavelength change is given by;

$$\Delta\lambda = \lambda_{SCATTERED} - \lambda_{INCIDENT} = \frac{h}{mc}(1 - \cos 2\theta) \qquad 2.1.1$$

This basically reflects the kinetic energy taken up by the electron. This interaction is termed Compton scattering, Compton (1923). The wavelength change is therefore independent of the wavelength of the incident photon but varies with scattering angle, 2θ, and is small (~0.024Å at most). An electron is not stationary or totally unbound in a solid and this will influence the energy (and wavelength) spread of the scattered photon. This makes the Compton scattering process a very useful tool for studying electron momenta in solids, etc. Because the wavelength change is so small, typical X-ray detectors used in diffraction experiments cannot discriminate this contribution from elastic scattering processes, therefore Compton scattering appears as a background signal. Each photon involved in this process will scatter independently. The scattering probability of coherent and Compton scattered photons for any given atom are of the same magnitude. However waves scattered in phase redistribute this intensity into sharp maxima that give intensities approximately related to N^2 (where N is the number of contributing atoms) compared to N for Compton scattering. Since N is generally very large the proportional contribution of Compton scattering is negligible, unless we are dealing with samples of very poor crystallinity. Equation

2.1 indicates that the scattering along the incident beam direction is zero. Compton scattering therefore increases with increasing scattering angle.

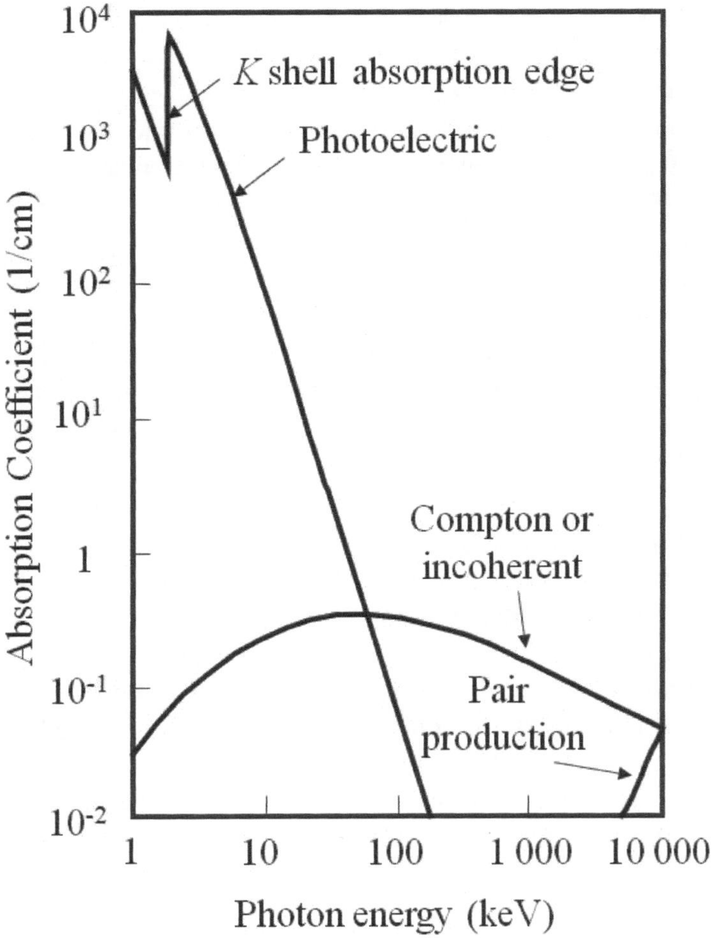

Figure 2.1.1 The various ways in which X-rays interact with matter as a function of their energy.

The tightly bound electrons will appear as a large immovable mass to a photon and therefore the energy transfer on interaction is very small (Rayleigh scattering), that is, the quantised energy state of the electrons is unchanged. If the wavelength of the photon is greater than that of the energy levels in the atoms of the sample, then photoelectron absorption can occur which is particularly strong when the energy exactly matches one of these energy transitions. For photon energies greater than a transition the electron takes up the remaining energy. This is the true absorption process when the photon is lost, although depending on the existence of free energy levels, the recovery of the electrons to lower energy levels can result in X-ray emission (fluorescence) or by involving an extra transition and electron emission (Auger process). Figure 2.1 shows the dominant absorption processes for different energy photons.

Clearly at high energies the generation of electron-hole pairs increases and incoherent scattering rises to a maximum at 100keV in Si, whereas photoelectric absorption dominates at low energies. These low energy photons are primarily scattered by the localised electrons and each interaction will represent an instantaneous snapshot of the atomic positions. The temperature-dependent vibrational frequency of the atoms about their average site is many orders of magnitude lower than the sampling time of the X-ray photon: this has important consequences on how the intensity is averaged. An incoherent scattering process is one in which the emitted photon has a significantly different energy from the incident photon and the coherent process corresponds to the case where the energy of the emitted and the incident photons are the same. One very useful incoherent process is the generation of fluorescent radiation whose energy is element specific and therefore a very useful chemical analysis tool. However the coherent scatter or elastic scattering is the main emphasis of this chapter since this gives us access to the structural information of materials and their molecular configuration. At the low energies given in Figure 2.1, the wavelengths are comparable to the interatomic distances and provide a very useful probe of these lengths.

2.2 The nature of the scattered X-ray photon with no energy loss

We can consider the source of X-rays as a provider of photons that have a distribution of directions and energies. The location of each photon from the source to the sample and on to the detector is indeterminate and therefore a coherent relationship is maintained between all the possible paths to the detector. The possible paths are defined by the collimation between the source and sample and those between sample and the detector and from the interaction of the photon with the sample. These collimators could be crystals or slits depending on the application. However in this chapter we are concerned with the interaction of a photon with the sample under investigation. The instrument used to carry out the experiment will then average the contributions of all these photons and create the observed diffraction pattern. During the diffraction process the photon has the probability of several paths such that it exists along all probable paths until it is detected. What this means is that the probing X-ray photon is phase coherent over all its probable paths. The photon will now average over these probable paths and therefore the scattering will reflect the average of the regions in the sample that are probable paths.

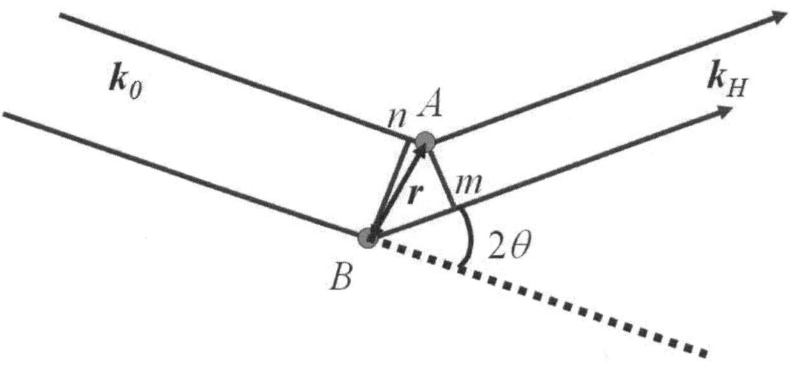

Figure 2.2.1. How two possible beam paths give rise to a phase difference characteristic of the separation of the scattering centres.

To understand coherence it is important to realise that a photon can only interact with itself, which cannot be understood in the classical sense but only with an understanding of quantum theory. If we therefore consider a photon to exist as a spatial distribution of probable paths then if all paths undergo different influences before recombining then interference can occur due to the fact that there is always a phase relationship between the same photon. If the paths cannot be recombined or cannot exist at the same time then there is no phase coherence. The distances over which phase coherence can exist is often termed the coherence length. *This differs from the correlation length, which is a sample dependent parameter and relates to the distances over which a phase relationship can be maintained.* Outside this region the photon paths cannot easily recombine, due to large orientation effects or phase averaging effects when the phases combined with random relationships; this effectively creates a null or noisy signal at the detector.

Another consideration is that the photon has a finite length, so that the difference in path lengths have to be less than this length to maintain coherence. The length of the photon can be deduced from its generation. A photon is generated by an electron falling from a high to low energy level, however no energy level can be a precisely defined value (there will always be some uncertainty) and therefore a photon will have a corresponding energy distribution, and therefore wavelength distribution. From a purely classical derivation the uncertainty in the energy is given by

$$\Delta E \approx -9.5\times10^{-9} E^2 \qquad 2.2.1$$

This derivation is based on high-energy electromagnetic waves where the Auger process is small, i.e. energies >1 keV, Hedin (1974). Also in practice this radiative width is an overestimate and is closer to 15% below this value. From the Uncertainty Principle, Heisenberg (1927), the dimensions of a photon can only be estimated from the level of certainty in its momentum, i.e.

$$\Delta x \approx \frac{h}{\Delta p} \qquad 2.2.2$$

Now from the de Broglie (1925) $p=h/\lambda$ relationship between momentum and wavelength we can deduce

$$\Delta p = h\frac{\Delta\lambda}{\lambda^2} \qquad 2.2.3$$

Therefore the "length of a photon" which is effectively the coherence length is given by combining equations 2.2.2 and 2.2.3

$$\Delta x \approx \frac{\lambda^2}{\Delta\lambda} = -\frac{E}{\Delta E}\lambda \qquad 2.2.4$$

For example Cu $K\alpha_1$ radiation will have an energy distribution of 0.62 eV and a wavelength spread of 1.1×10^{-5} nm giving a coherence length of ~2 μm. Therefore any of the possible beam paths for a photon, will have a diminishing ability to create interference beyond this length.

Consider the scattering of X-rays from two electrons separated by a distance $\boldsymbol{r}$, figure 2.2.1. Suppose an incident wave with wave vector $\boldsymbol{k_O}$ of magnitude $1/\lambda$ impinges on two electrons A and B, where $\boldsymbol{r}$ is given by:

$$r = \mathrm{u}\,a + \mathrm{v}\,b + \mathrm{w}\,c \qquad 2.2.5$$

$\boldsymbol{a}$, $\boldsymbol{b}$ and $\boldsymbol{c}$ are unit cell translations and u, v and w are integers. If the electrons are set into vibration and excited they will emit secondary radiation defined by $\boldsymbol{k_H}$ of magnitude $1/\lambda$, then there will be a path difference between the scattering from A and that from B given by :

$$A\mathrm{n} - B\mathrm{m} = \lambda(r.k_0 - r.k_H) = \lambda r.S \qquad 2.2.6$$

$\boldsymbol{S}$ =($\boldsymbol{k_O}$ - $\boldsymbol{k_H}$) is the scattering vector. For the waves from A and B to be in phase then $\boldsymbol{r.S}$ must be integer and hence:

$$(\mathrm{u}\,a + \mathrm{v}\,b + \mathrm{w}\,c).S = \text{integer} \qquad 2.2.7$$

Since this must be true for all u, v and w then

$$\begin{aligned} a.S &= h \\ b.S &= k \\ c.S &= l \end{aligned} \qquad 2.2.8$$

These are Laue's equations, where h, k and l are integers. Suppose we take the first two equations, rearrange them and subtract them from each other

$$\left(\frac{\boldsymbol{a}}{h} - \frac{\boldsymbol{b}}{k}\right).\boldsymbol{S} = 0 \qquad 2.2.9$$

Therefore (***a**/h* - ***b**/k*) must be perpendicular to ***S***, and similarly for all the combinations. We can now consider a plane that intercepts the *'a'* axis at 1/*h*, the *'b'* axis at 1/*k* and *'c'* axis at 1/*l*, which we can denote with Miller indices (*hkl*). Equation 2.2.9 can only be satisfied if ***S*** is orthogonal to the plane (*hkl*), figure 2.2.2. The spacing between these sets of crystal planes containing the electrons is the projection of ***a**/h* on ***S*** (i.e. the distance between the plane and the origin), therefore

$$d_{\mathrm{hkl}} = \frac{a.S}{h|S|} = \frac{1}{|S|} \qquad 2.2.10$$

by combining with Laue's equation.

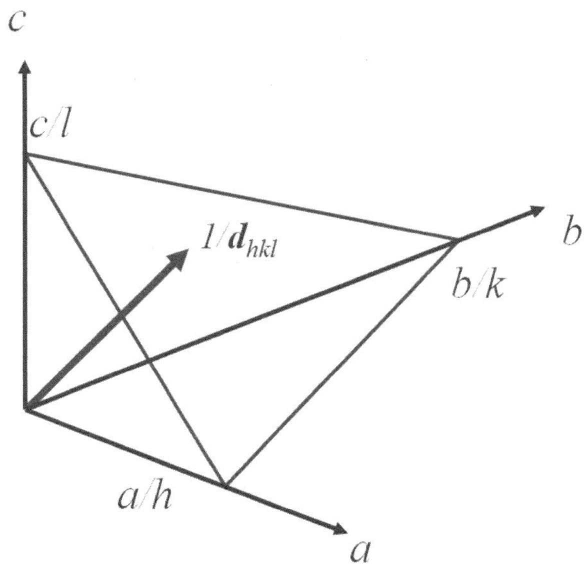

Figure 2.2.2. The plane (designated *hkl*) has a characteristic vector that has a length given by the inverse of its distance from the origin and is normal to this plane.

If the scattering angle 2θ, for the phase coherent scattering condition, is defined as the angle between $\boldsymbol{k_O}$ and $\boldsymbol{k_H}$ from figure 2.2.3.

$$|\boldsymbol{S}| = \frac{1}{\boldsymbol{d}_{hkl}} = \frac{2\sin\theta}{\lambda} \qquad 2.2.11$$

or

$$2\theta = 2\sin^{-1}\left\{\frac{\lambda}{2d_{hkl}}\right\} \qquad 2.2.12$$

This is Bragg's equation, Bragg (1913). From this approach we can therefore consider the scattering process as a reflection.

Suppose that we have a crystal, then a vector of length $1/\boldsymbol{d}_{hkl}$ can represent each plane *hkl* with a direction along the plane normal. The end points of all these vectors will form a periodic array that relates to the periodicity of the crystal. This is a very convenient representation of the crystal and from equation 2.2.12 we can see that this lattice in "reciprocal space" will only be a small distortion of its image in "diffraction space" over small regions. This concept will be developed through the course of this chapter and discussed further in Chapter 4.

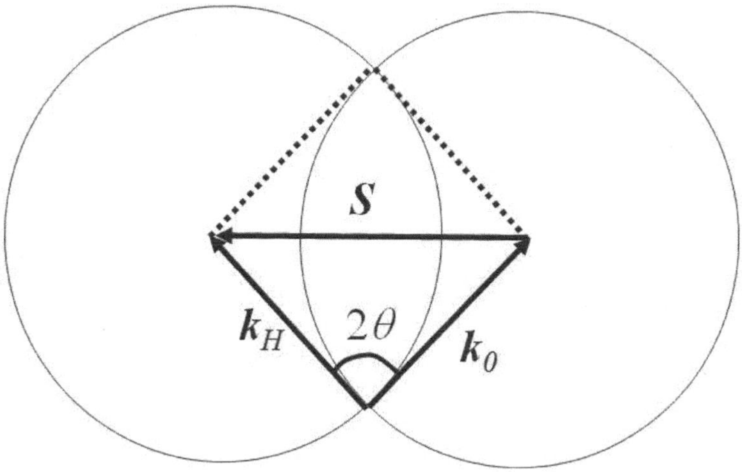

Figure 2.2.3. The relationship between the incident and scattered wave-vectors with the scattering vector.

So far we have considered the scattering from individual electrons, whereas in reality the electrons are distributed about the atomic positions. Also the atoms will vibrate about their average position and these factors must be included. Consider first that the electron cloud associated with the atom occupies a simple sphere then scattering from two different regions will interfere, figure 2.2.4. Debye (1915) showed how this can give rise to diffraction effects in gases and Compton (1917) indicated how atomic sizes could be estimated based on these ideas. The phase difference will therefore reduce the resultant scattering as the scattering angle 2θ increases. The strength of the scattering from an atom is therefore proportional to the number of electrons, which is further modified by this phase effect that depends on the size of the electron cloud. The ratio of the strength of the scattering from an atom to that of an individual electron is the scattering factor, which is given by the expression:

$$f(S) = \int \psi_2^*(r)\exp(iS.r)\psi_1(r)dV \qquad 2.2.13$$

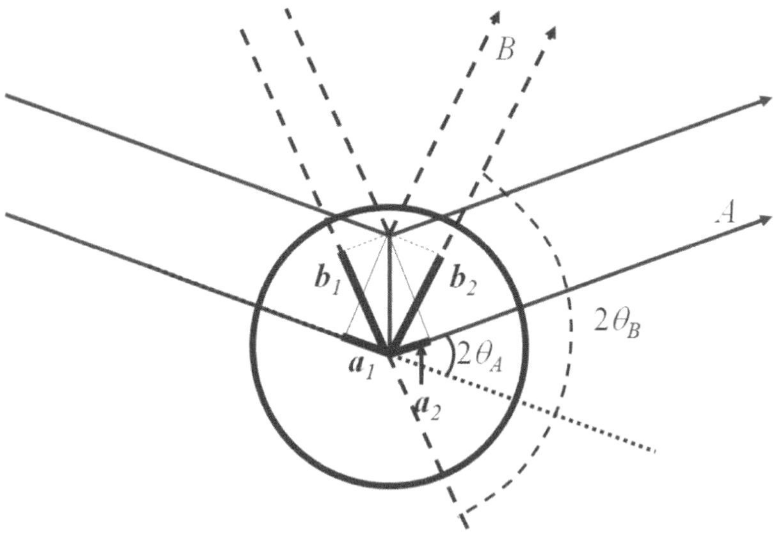

Figure 2.2.4. The finite atomic scattering volume (represented by circle) gives rise to a phase difference of possible beam paths that varies with scattering angle (compare beam paths *A* and *B*).

This equation does not include relativistic effects and is therefore not valid for very high energy (high frequency) X-rays and becomes less reliable in accounting for the scattering from the inner electrons of heavy atoms. ψ_1 and ψ_2 represent the initial and final states of the wave-function, but since we are concerned with coherent scattering these are the same. We can therefore consider the scattering factor (form factor) to be the Fourier transform of the charge distribution ($\psi^*\psi$) around an atom. The effective charge distribution is modified in the relativistic case, (Hartree, 1935; Fock, 1930). Most calculations of the scattering factor are based on a spherical approximation by averaging the various orbital contributions to effectively create that for a free atom or ion. Many authors have now calculated these scattering factors and to an excellent approximation they can be represented by an expansion of exponential terms

$$f_i = \sum_{j=1}^{4} a_j \exp\left(-\left\{\frac{b_j \sin^2\theta}{\lambda^2}\right\}\right) + c_i \qquad 2.2.14$$

The scattering factor for each atom in various states of ionisation can therefore be determined from a tabulated set of values (International Tables for X-Ray Crystallography **IV** p71 (1968)). The scattering factors for asymmetric distributions have also been calculated but for ease of calculation we will assume the spherical distribution is a good approximation.

So far we have considered the atoms to be static, whereas in reality the finite temperature of any sample will mean that they vibrate, section 2.8.1. Since X-rays are very high frequency electromagnetic waves the sample will appear static to each photon, although the atoms in general will be displaced from their average positions. These displacements will create a phase difference between scattering atoms and can therefore be treated in a similar manner to the scattering factor. We are only interested in the vibrations parallel to the scattering vector, however if we assume that the vibrations are essentially isotropic then they will influence all intensity maxima in a similar way. We can then define a single "temperature factor" influence to the scattering:

$$f = f_i \exp\left(-\left\{\frac{8\pi^2 < u^2 > \sin^2\theta}{\lambda^2}\right\}\right) \qquad 2.2.15$$

Where $<\boldsymbol{u}^2>$ is the root mean square displacement of the atoms, based on the geometrical model of Debye (1914). Again some of these values for $(8\pi<\boldsymbol{u}^2>)$ have been calculated and some determined experimentally for various structures, since these values are sensitive to the molecular environment, Reid (1983). Many semiconductor materials have small amplitudes of vibration because of the strong sp^3 hybridised bonding that exists within many extended lattices. For typical semiconductors a common isotropic value applicable to all atoms in the structure is generally sufficient for modeling the intensities. However for flexible molecules in a lattice dependent on van der Waals or hydrogen bonding then this temperature factor can be highly anisotropic and very different for each atom site.

Let us now consider the validity of the geometrical description of diffraction. Bragg's equation is only approximate because the measured scattering angle is outside the crystal, whereas the actual scattering angle is between incident and scattered waves inside the crystal that are modified by the refractive index. The influence of the refractive index will vary with scattering angle and the angle between the entrance surface and the exit surface of the photon. This can be understood most simply on the understanding that the wavelength is modified in the sample and the Bragg angle must change to compensate, since all other parameters in the Bragg equation are invariant. We will see later that the refractive index is less than unity, i.e. the wavelength of the photon is larger than in vacuo.

From Bragg's equation we can see that the minimum interplanar spacing that we can measure is determined by the wavelength, i.e. $\boldsymbol{d}_{hkl} > \lambda/2$ hence for a typical X-ray wavelength of 0.15 nm the smallest interplanar spacing measurable is 0.075 nm, which is less than all interatomic bond lengths. Shorter wavelength X-ray sources closer to half these values are commonly used for molecular structure determination to obtain atomic scale resolution with ease. Measurements of thicknesses in multi-layer composites structures are also possible with these wavelengths but require very high-angular resolution and in the limit a good understanding of the shape of the scattering profile. The limit to the maximum dimension that can be determined depends on the width of the diffraction profile, which is composed of the smearing effects of the diffractometer (the instrument probe dimensions) and the intrinsic scattering width that is a function of the sample and scattering conditions. As we will see a typical intrinsic scattering width for a perfectly crystalline sample is of the order of seconds of arc ($\sim 0.001^0$). This permits the measurement of lengths up to many microns.

At this stage we can see that the interplanar spacing within the sample can be determined from the positions of the scattering maxima through Bragg's equation, however the strength of these maxima depends on the internal structure. Clearly scattering planes that have high electron density will scatter more strongly than those with low electron density. Also when the scattering is stronger the probability of a photon penetrating any great depth is diminished. As we shall see, the width of the scattering maxima depends on the depth of penetration in perfect materials. We shall first consider the scattering from an ideally perfect crystal to illustrate the scattering process.

The scattering vector given in equation 2.2.6 and 2.2.11 can be expressed as:

$$S = k_H - k_0 \qquad 2.2.16$$

and be represented graphically as two spheres of radius $1/\lambda$ with their centres separated by $\boldsymbol{S}$, since the magnitude of each wave vector is $1/\lambda$, figure 2.2.3. The magnitude of the scattering vector is $|\boldsymbol{S}| = 1/\boldsymbol{d}_{hkl}$. The points of intersection of the circles containing these vectors represent conditions when Bragg's equation is satisfied. Clearly if $\boldsymbol{d}_{hkl}$ is too small the spheres do not overlap. If they do overlap then any intersection results in a singularity that suggests that the physics at this point will require a more exacting description. The second point to recognise is that the average refractive index is less than unity, and hence the wave-vectors within the crystal have different magnitudes and their wavelength is longer. We therefore have two concentric spheres associated with each end of the vector ***S*** representing the internal and external wave-vectors.

We will consider five conditions:

(1) The dynamical diffraction model that best describes the physics of the waves within a nearly perfect sample.
(2) The kinematical diffraction model that is applicable to weak scattering from small crystals.
(3) Scattering from polycrystalline aggregates and show the scattering is not necessarily from the Bragg condition.
(4) Specular scattering where the X-rays are reflected from changes in electron density and is not necessarily sensitive to atomic plane spacings.
(5) The combination of specular scattering to obtain the strength of the X-ray wave at positions from which crystal distortions can scatter kinematically.

2.3. The near exact theoretical description of scattering

Consider a sample with a series of atomic planes that are parallel to each other, figure 2.3.1. If we considered the beam path AB of a photon to be at an incident angle, θ_{hkl}, to scatter it in the direction BC then it will also be at the correct incident angle to be scattered from the underside of these atomic planes. Immediately we can understand that we have a complex wave-field building up in the sample where energy is swapping back and forth between the incident and scattered directions. We therefore have a dynamic situation where the energy of the incident beam is diminished with depth due to losses to the scattered beam and interference between the multiple scattered beams interfering with each other along the direction parallel to AB. This is the basis of the model proposed by Darwin (1914a and b) and can be understood on purely geometrical terms. The scattering model that accounts for this is the dynamical scattering theory. Additional energy losses from the X-ray wave-fields arise from photoelectric absorption, Prins (1930).

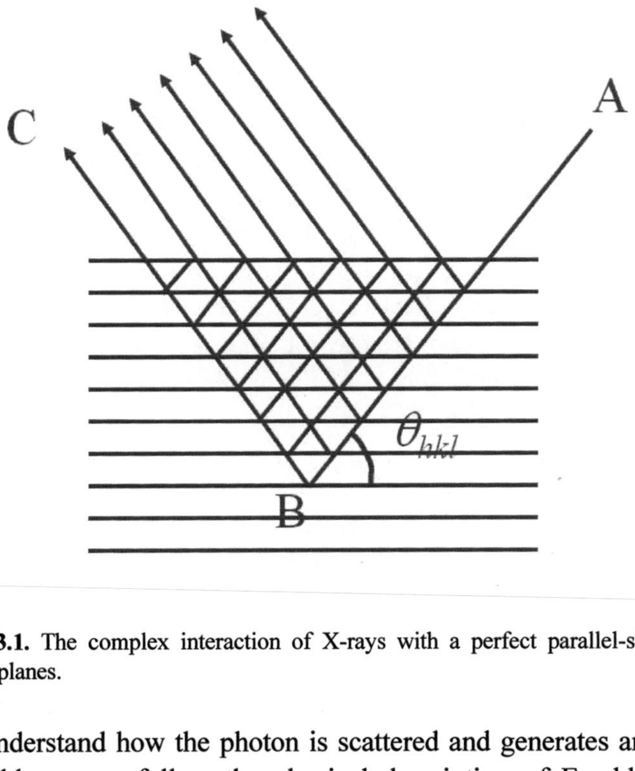

Figure 2.3.1. The complex interaction of X-rays with a perfect parallel-sided set of scattering planes.

To understand how the photon is scattered and generates an internal wave-field we can follow the physical description of Ewald or Laue (Ewald (1916a, 1916b, 1917), Laue (1931)). Ewald considered each atomic site to be occupied by a dipole that is set into oscillation by a passing photon. Each oscillating dipole emits radiation that adds to the total radiation field. We therefore have an array of dipoles oscillating which Ewald called "dipole-waves" all emitting electromagnetic radiation that interacts with other dipoles. The whole problem is considered by relating the resultant "dipole-wave" field to the electromagnetic field giving rise to it. Each dipole on a plane is assumed to emit with a phase relationship that will create two plane wave-fronts. We therefore have two wave types: an electromagnetic wave that is created by the dipole wave and the dipole wave itself.

The "dynamical" aspect of this physical model is that the wave-field created by the dipole oscillations should be just sufficient to maintain them.

However we can see from Ewald's model that the dipoles are located at the atom sites, where in reality the sample has a distributed electron density and should be considered as a dielectric. Laue took this latter approach and the results are essentially the same as those determined by Ewald. We consider the crystal to exist of a continuous negative charge with shielded positive charges (the atomic nuclei) in a periodic array. When no incident photon exists, any position in the crystal can be considered as neutrally charged. But when an electric-field is applied there will be a relative displacement of the charges resulting in an electric polarisation and therefore the induced electric field (electric displacement) $\boldsymbol{D}$ is the resultant of the applied electric field $\boldsymbol{E}$ and the polarising field $\boldsymbol{P}$ and ε_0 is the permittivity

$$D = \varepsilon_0 E + P \qquad 2.3.1$$

If the electric field strength is not too strong then the induced electric field will be proportional to the local electric field strength. A strong electrostatic field can be defined as a significant proportion of the ionisation potential of an atom on the atomic scale. This is the condition relevant to X-ray frequencies because of the short wavelength oscillating high energy electric fields. Hence we cannot consider a simple dielectric constant of proportionality, where the electron density is isotropic and a variable electron density $\rho(\boldsymbol{r})$ is introduced. Firstly we assume that the restoring force on the electron is given by

$$m\frac{d^2x}{dt^2} = -eE = -eE_0\exp(2\pi i\nu t) \qquad 2.3.2$$

The electron displacement from its average undisturbed site through integrating is

$$x = -\frac{e}{4\pi^2 m(\nu_0^2 - \nu^2)}E \qquad 2.3.3$$

where and ν is the oscillation frequency of the X-rays and ν_0 represents the natural frequency of vibration for the electron. This equation represents the resonance condition when strong absorption of the X-rays can take place. Clearly an infinite displacement is unrealistic so we can add in an imaginary term ($i\delta\nu$) to the bracketed term of the denominator, this can be derived from assuming the damping is proportional to the displacement. This imaginary term can be included by assuming that the damping term can be related to the individual scattering factors from each atom type and charge state. This is clearly wavelength dependent.

The consideration given above is for a single electron atom and this has a strong bearing on the values associated with ν_0. For a many electron atom the natural or resonant frequency can be derived from the energy of excitation of a *K* shell electron, for example, to an unoccupied outer shell. These outer shells of any reasonable size atom are very close in energy and therefore form a series of unoccupied states giving a band of energies. Consequently ν_0 is not single valued, but rather complicated. However equation 2.3.3 does illustrate the general form of the resonance effect. It is also worth noting that atoms within a sample will create an almost continuous band of possible frequencies, however the dominant scattering is from the innermost localised electrons that are less sensitive to the immediate atomic environment.

Let us now rewrite the scattering factor as

$$f = f_0 + f' + if'' \qquad 2.3.4$$

f_0 is the term expressed in equation 2.2.14. Since we are mainly considering modifications to the scattering from inner shell resonance conditions, the normal reduction of scattering with angle is very small for these two additional terms. The term f' is a direct modification to the scattering strength due to the proximity of the resonance condition and, f'' introduces a phase lag in response to the incident X-rays. As the incident X-ray frequency moves further from this condition of resonance, the electrons appear less tightly bound and behave rather like free electrons. Therefore the resonant frequency appears weak.

The polarising field will therefore relate to the product of the displacement, the charge and density of the charge carriers, i.e. electrons

$$P = xe\rho(r) = -\frac{e^2}{4\pi^2 mv^2}\rho_e(r)E \qquad 2.3.5$$

Therefore from equation 2.3.1, the induced electric field is given by

$$D = E(1+\chi)\varepsilon_0 \qquad 2.3.6$$

where

$$\chi = -\frac{e^2\lambda^2}{4\pi^2\varepsilon_0 mc^2}\rho_e(r) \text{ or } F = V\rho_e(r) \qquad 2.3.7$$

from equations 2.3.1, 2.3.5 and 2.3.6 and $\nu = c/\lambda$. F is the structure factor. This equation is equally applicable to the atomic nucleus but with its high mass, *m*, we can see that its influence on the electric field is close to a factor of 2000 times weaker than that from electrons. χ is the effective polarisability or the electric susceptibility of the crystal and will therefore vary throughout the crystal with the same periodicity as the electron density

$$\rho_e(x,y,z) = \frac{1}{V}\sum_{h=-\infty}^{\infty}\sum_{k=-\infty}^{\infty}\sum_{l=-\infty}^{\infty} F_{hkl}\exp(-2\pi i(hx+ky+lz)) \qquad 2.3.8$$

where *x,y,z* are the fractional position co-ordinates within the unit cell of volume *V*. The crystal can therefore be considered as a structure with an anisotropic periodic complex susceptibility. The assumption in using a summation and not an integral here is that the electron density is strongly associated with the atomic sites, i.e. the inner electrons dominate. Clearly from equation 2.3.8 the molecular structure in terms of the electron density distribution can be determined. However the summation is over all *hkl* i.e. all crystallographic planes. In practice the scattered intensity from only a finite number of "reflections" can be measured and this limits the resolution in determining the electron density. Since this assumption now strongly associates the scattering centres close to the atomic sites the physical model of the sample is equivalent to that proposed by Ewald (1916a, 1916b, 1917).

The quantity F_{hkl} is the structure factor for the unit cell or periodic repeat unit of volume V. The structure factor is the sum of all the scattering contributions from all the electrons in the unit cell

$$F_S = \int_r f_r \exp\{-2\pi i S.r\}dr \sim \sum_{j=1}^{N} f_j \exp\{-2\pi i S.r_j\} \qquad 2.3.9$$

N is the number of atoms in the unit cell. The structure factor $\boldsymbol{F_S}$ in this equation is given in terms of the variable $\boldsymbol{S}$ and this is more general. The integral represents the full description assuming a distributed electron density. If however we assume the electrons are very localised the summation is adequate. The atomic size effect is accommodated by modifications to the scattering factor as well as thermal vibrations. The location of an atom j in terms of the fractional co-ordinates x, y and z, for a unit cell defined by the vectors $\boldsymbol{a}$, $\boldsymbol{b}$ and $\boldsymbol{c}$, can be written as

$$r = xa + yb + zc \qquad 2.3.10$$

then from Laue's equations 2.2.8

$$r.S = xa.S + yb.S + zc.S = hx + ky + lz \qquad 2.3.11$$

$\boldsymbol{F_{hkl}}$ will therefore be complex because it contains phase information associated with the atom positions within the unit cell. In describing the condition of resonance that dramatically changes the polarising field we established that a complex term could be included associated with the electron density. Since this resonance damping term is element specific the most appropriate way of including this is through the scattering factor, equation 2.3.4.

The electric field built up inside the crystal must obey Maxwell's equations

$$\mathrm{curl}\, B = \frac{1}{c^2}\left(\frac{\partial D}{\partial \mathrm{t}} + 4\pi J\right) \sim \frac{1}{c^2}\left(\frac{\partial D}{\partial \mathrm{t}}\right)$$

$$\mathrm{curl}\, E = -\frac{\partial B}{\partial \mathrm{t}} \approx \mathrm{curl}\left\{\frac{D}{1+\chi}\right\} \qquad 2.3.12$$

$$\mathrm{div}\, D = \frac{\rho}{\varepsilon_0} \approx 0$$

$$\mathrm{div}\, B = 0$$

B is the magnetic flux (=μH) and c the velocity of the wave. The conductivity, ρ, is assumed to be zero at X-ray frequencies therefore the current density J and the charge density are assumed to be zero. This also suggests that there will be no resistive heat loss. The approximations given here are those given by Laue (1931), where it is assumed that vectors ***E*** and ***D*** have the same direction. The displacement field, ***D***, the magnetic field, ***H***, and the wave vector $\boldsymbol{K}_{\mathrm{m}}$ form an orthogonal set, whereas to be precise ***E*** lies within the plane of ***D*** and ***K*** and therefore can have a component along the propagation direction. Hence ***D*** forms a transverse wave and ***E*** does not. This introduces a small deviation at the 10^{-5} level, this is discussed later.

The concept of the theory is that the resulting field is the sum of plane waves, which can be defined as:

$$D = \exp(2\pi \mathrm{i}\,\nu t)\sum_{\mathrm{m}} D_{\mathrm{m}} \exp\{-2\pi\mathrm{i}[K_{\mathrm{m}}.r]\}$$

$$H = \exp(2\pi \mathrm{i}\,\nu t)\sum_{\mathrm{m}} H_{\mathrm{m}} \exp\{-2\pi\mathrm{i}[K_{\mathrm{m}}.r]\} \qquad 2.3.13$$

These represent the total electric displacement and magnetic fields at time t and position ***r*** for a total of m waves propagating in the crystal. The frequency of the electromagnetic wave is ν and the scattered wave vector, $\boldsymbol{K}_{\mathrm{m}}$ satisfies Bragg's equation within the crystal, i.e.

$$K_{\mathrm{m}} = K_0 + S_{\mathrm{m}} \qquad 2.3.14$$

where $\boldsymbol{S}_{\mathrm{m}}$ is a scattering vector inside the crystal and $\boldsymbol{K}_0$ is the forward-refracted beam. If we combine equations 2.3.6 and 2.3.12 we obtain

$$\text{curlcurl}\,D = \text{curlcurl}\,E + \text{curlcurl}\,\chi E = -\frac{1}{c^2}\frac{\partial^2 D}{\partial t^2} + \text{curlcurl}\frac{\chi D}{1+\chi}$$

2.3.15

and from the standard vector relationship we can state that curlcurl***D*** = graddiv***D*** - ∇^2 ***D***, and since div***D*** = *0* we can now write

$$\nabla^2 D - \frac{1}{c^2}\frac{\partial^2 D}{\partial t^2} = -\text{curlcurl}\left(\frac{\chi D}{1+\chi}\right) \approx -\text{curlcurl}(\chi D) \quad 2.3.16$$

The approximation is justified since $\chi \sim 10^{-5}$ for most materials. Now we can substitute ***D*** (equation 2.3.13) into equation 2.3.16, however we should consider the expansion of χ***D*** first i.e.

$$\chi D = \exp(2\pi i\,\nu t)\sum_{n}\sum_{m}\chi_{n-m}D_m \exp(-2\pi i K_m.r) \qquad 2.3.17$$

where $\nu = c/\lambda = ck$ and $k^2 = \boldsymbol{k}^2$. This substitution gives the following equation:

$$\left(k^2 - K_m^2\right)D_m - \sum_{n}\chi_{n-m}\left[K_m \times \left(K_m \times D_m\right)\right] = 0 \qquad 2.3.18$$

The expression in square brackets can be rewritten using the vector rule:

$$\left[K_m \times \left(K_m \times D_n\right)\right] = K_m(K_m.D_n) - D_n K_m^2 \qquad 2.3.19$$

For a transverse wave ***D*** is orthogonal to ***K*** and therefore $\boldsymbol{K}_m.\boldsymbol{D}_n = 0$ and therefore this equation can be simplified to:

$$\frac{K_m^2 - k^2}{K_m^2}D_m = \sum_{m}\chi_{m-n}D_{n(\perp K_m)} \sim \frac{K_m^2 - k^2}{k^2}D_m \qquad 2.3.20$$

k is the magnitude of the incident wave vector (= |***k***| = $1/\lambda = \nu/c$) and is very similar in magnitude to the wave vector ***K*** inside the crystal. Referring back to the assumption of Laue concerning the use of the displacement field we can derive a similar expression for the electric field:

$$\frac{K_m^2 E_{m(\perp K_m)} - k^2 E_m}{K_m^2} = \sum_{m}\chi_{m-n}E_n \sim \frac{K_m^2 E_{m(\perp K_m)} - k^2 E_m}{k^2} \qquad 2.3.21$$

This equation differs from equation 2.3.20 in the fifth order due to the small longitudinal component in the electric field $\boldsymbol{E}_{\mathrm{m}}$ in the scattering plane. However the boundary conditions can be simpler and are often preferred in multiple beam diffraction calculations.

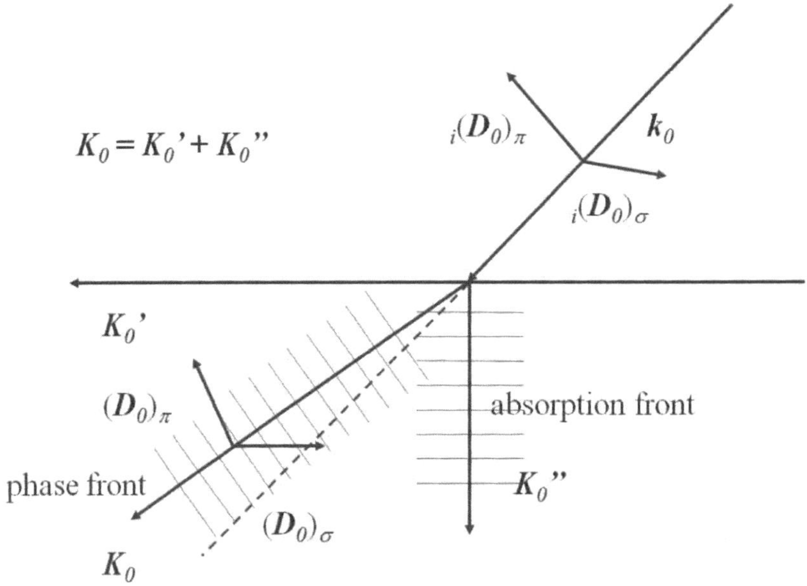

Figure 2.3.2. An incoming electromagnetic wave from a non-absorbing medium to an absorbing medium will create an absorption front traversing normal to the surface and a phase front along the propagation direction.

We can understand this latter point by considering a wave incident on a crystal, figure 2.3.2. If we assume that the external incident wave is in vacuum and suffers no absorption then it is totally real and on entering the crystal it undergoes absorption. Since the component of the wave parallel to the vacuum / crystal boundary must be continuous the imaginary component of the wave vector must be perpendicular to the surface. The electric field component of the incident wave will be parallel to the displacement, since the susceptability is zero in an unpolarisable medium. The forward-refracted wave, because of its small deviation from the incident beam direction will now include a component of the electric field along the wave vector and is therefore partially longitudinal. However the transverse component is much more significant. The real part of the wave vector essentially defines the direction and the phase of the wave varies along this direction. We can therefore consider that the regions of constant absorption are parallel to the surface and those of constant phase normal to the real component of the wave vector.

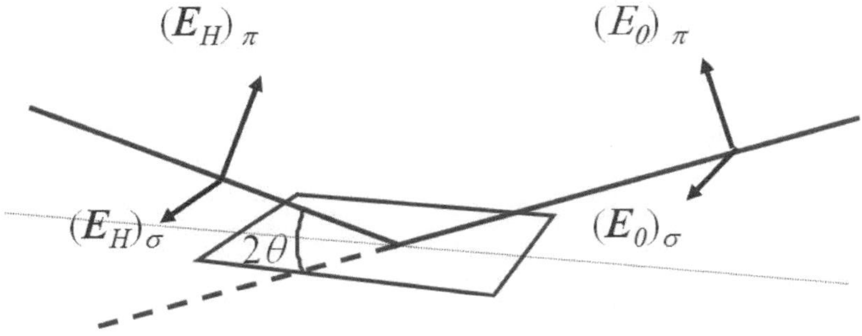

Figure 2.3.3. The change in the wave direction on scattering will influence the two electric field components to different extents depending on the scattering angle.

So far we have not considered the alignment of the transverse electric field direction and how this interacts with the sample. If the X-rays are generated by a laboratory source then the electric field will be circularly polarised, i.e. this direction is random but normal to the wave-vector. X-rays generated in a synchrotron have a very strong polarisation direction in the plane of the storage ring. We shall resolve the electric field polarisation into electric field vectors in the plane of scattering and normal to the direction, figure 2.3.3. The electric field component normal to the plane of scattering is unchanged apart from the reduction related to the reflecting power, the σ component. The electric field direction in the scattering plane, the π component is altered on scattering and effectively the electric displacement field is modified whenever it interacts with the scattering planes from above or below. The strength of the scattering from atomic planes is directly related to the susceptibility χ_{hkl} and χ_{-h-k-l}, and we can determine this reduction factor geometrically, figure 2.3.3.

$$\{\chi_{hkl}\}_{\pi} = \chi_{hkl}\cos 2\theta$$
$$\{\chi_{hkl}\}_{\sigma} = \chi_{hkl} \qquad 2.3.22$$

Clearly as the scattering angle moves close to 90° the contribution of this π polarisation component becomes negligible compared with the σ component, which is unchanged on scattering. Similarly as more reflections are involved as in multiple crystal diffractometers the π component is reduced. We can now consider the equations above to be composed of two polarisation states and the scattered wave-fields are modified by a factor $C = \cos 2\theta$ for the π component and $C = 1$ for the σ component.

We will now consider several wave-field conditions.

2.3.1. The condition of a single wave generated in a crystal

Suppose a wave is incident on a crystal and no diffracted wave is generated, then we have the displacement field at a position ***r*** to be given by

$$D = D_0 \exp(-2\pi i K_0 . r) \exp(-2\pi i \nu t) \qquad 2.3.23$$

Substituting this into our general equation 2.3.20 we obtain:

$$\frac{(K_0^2 - k^2)}{K_0^2} D_0 = \chi_0 D_0$$

i.e. with $K_0 \sim K$

$$\{(1-\chi_0)K_0^2 - k^2\}D_0 = 0 \qquad 2.3.24$$

Since $\boldsymbol{D_0}$ exists we have a simple expression:

$$|K_0| = \{1-\chi_0\}^{-1/2}|k| \sim \left(1 + \frac{\chi_0}{2}\right)|k| \qquad 2.3.25$$

This solution just defines an average refractive index $(1+\chi_0/2)$ and the resultant dispersion surface is simply a sphere of radius $|\boldsymbol{K}_0|$. This sphere is complex because the polarisability (susceptibility) is complex, however a good representation and understanding can be obtained from considering just the real component. The excitation point on the dispersion surface is found by constructing a surface normal that intersects this sphere, i.e. at a point that creates a transmitted wave with a wave vector of length $|\boldsymbol{K}_0|$, figure 2.3.4. The angle of the incident beam $\boldsymbol{k}_0$ with respect to the surface normal is $\pi/2 - \omega$, we can therefore see that a second intersection point occurs on the larger diameter sphere. This second intersection point represents the specular reflected wave, with a wave vector $\boldsymbol{k}_R$, and it is clear that this makes an angle of $\pi/2 - \omega$ with respect to the surface normal. This therefore represents a pure reflection with respect to the surface. The difference in the directions of the wave vectors $\boldsymbol{K}_0$ and $\boldsymbol{k}_0$ arise from the refractive index. This construction of the surface normal on the dispersion surface must and does satisfy the condition that the components of the internal and external wave vectors parallel to the surface are the same. This is a necessary and an obvious boundary condition.

From figure 2.3.4, we can see that the surface normal intersects each of these two concentric spheres in two positions. The incident wave, $\boldsymbol{k}_0$, creates a directly reflected wave, $\boldsymbol{k}_R$, and a transmitted wave, $\boldsymbol{K}_0$. The other wave, dash line in figure 2.3.4 represents a wave that would be created by the transmitted wave being reflected; this becomes important for thin crystals when the propagating wave in the crystal reaches an interface. The amplitude of this wave increases with depth. We can now see that we have two competing processes once the incident wave reaches the sample surface; that is the balance between the transmitted wave and the reflected wave.

If we consider the case when the surface normal only intersects the external sphere, line *CD*, then there is no internal wave vector and no wave field exists within the sample. This is the condition of total external reflection and for the geometric representation of the dispersion surface in figure 2.3.4 all the incident beam is specularly reflected and the real part of $\boldsymbol{K_0}$ does not exist. However the true dispersion surface is complex and, it is this imaginary component that is associated with a finite absorption that results in a small exponential decaying wave-field to enter the sample.

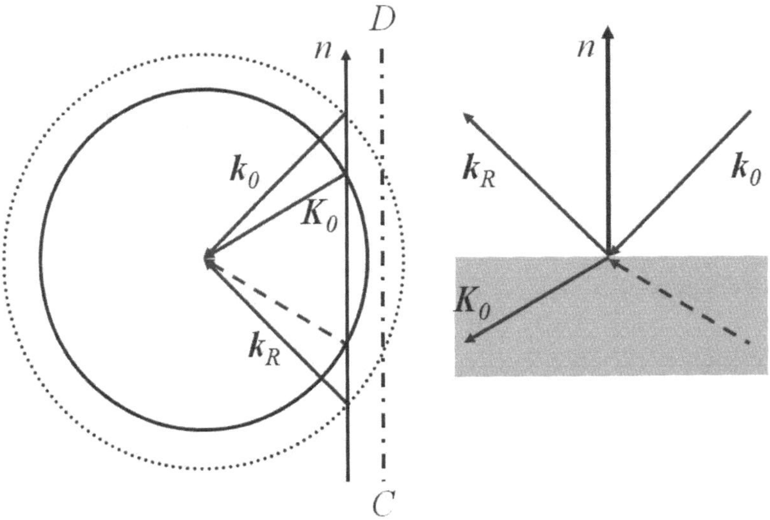

Figure 2.3.4. The dispersion surface construction for the one-wave case that just has a specular component.

It is clear from these arguments that specular reflection occurs throughout all ranges of angles but is only really significant at very low angles when the incident wave can be totally externally reflected. At this stage we can see that we are assuming some average refractive index. The method of calculating the intensity as a function of angle can be considered very simply, section 2.10, using the more convenient optical theory. The specular profile also emerges naturally from the dynamical theory by including all interactions with the dispersion surface, this will be covered in section 2.4.3.1

2.3.2. The condition of two waves generated in a crystal

In this case we consider that the incident wave generates a scattered wave and therefore we have two waves of appreciable amplitude in the crystal:

$$D = \exp(2\pi i \nu t)\{D_0 \exp(-2\pi i K_0.r + D_m \exp(-2\pi i K_m.r\} \quad 2.3.26$$

Hence substituting this into equation 2.3.20, for m = *0* and m = *H*, we have

$$\frac{K_0^2 - k^2}{K_0^2} D_0 = \chi_{-H} D_H + \chi_0 D_0$$
$$\frac{K_H^2 - k^2}{K_H^2} D_H = \chi_H D_0 + \chi_0 D_H \quad 2.3.27$$

To include the influence of polarisation into these equations the susceptibility that relates to the strength of the scattering from a set of crystallographic planes, χ_H and χ_{-H}, are multiplied by the factor *C*. If these are rearranged and the effects of polarisation are included then

$$C\chi_{-H} D_H + \left(\chi_0 - \frac{K_0^2 - k^2}{K_0^2}\right) D_0 = 0 \quad 2.3.28$$

and

$$C\chi_H D_0 + \left(\chi_0 - \frac{K_H^2 - k^2}{K_H^2}\right) D_H = 0 \quad 2.3.29$$

For these two equations to have a common solution the determinant of these must be zero

$$\left(\chi_0 - \frac{K_0^2 - k^2}{K_0^2}\right)\left(\chi_0 - \frac{K_H^2 - k^2}{K_H^2}\right) = C^2 \chi_H \chi_{-H} \quad 2.3.30$$

This equation can be represented as two spheres whose surfaces are the loci of all the excitation points that can exist in the crystal at the chosen X-ray wavelength and polarisation. This surface is the dispersion surface and is analogous to the Fermi or iso-energy surface of electron states in solids. This surface is given in figure 2.3.5 and is a shape of revolution about the scattering vector *OH*. The intersection of these spheres is no longer a singularity (simple cross-section) as in the simple geometrical understanding given above and this modification gives rise to a very good explanation of the profile shape and extinction effects that will be explained later.

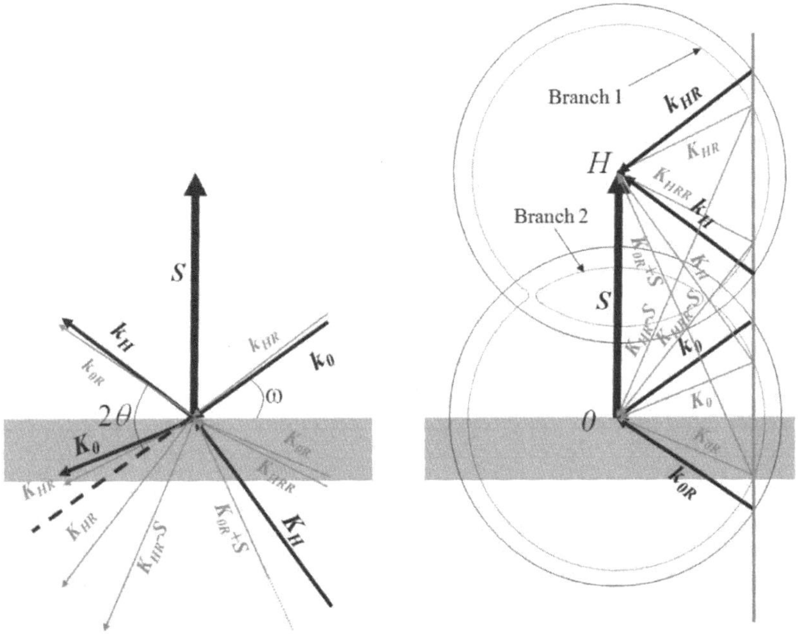

Figure 2.3.5. The dispersion surface removes the singularity and splits into two branches. The single external wave results in four external waves k_0, k_{OR}, k_H and k_{HR}, i.e. the incident, reflected, diffracted and diffracted reflected waves, solid black lines. There are also eight internal waves excited (grey lines in the right-hand diagram) including K_0, K_{OR}, K_H, K_{HR} that are the refracted equivalents of the external waves, and a further four reflected and scattered combinations, K_{HRR}, K_{HR}-S, K_{HRR}-S, K_{OR}-S. The only significant waves in the present derivation are given in black solid lines in the left-hand diagram.

The singularity is now a gap and represents the condition of total reflection; i.e. no wave-field exists in a non-absorbing crystal. The dispersion surface model greatly aids the understanding of the energy flow in a crystal. In general the flow is normal to the dispersion surface at the tie excitation points. As we increase the incident beam angle from a value below the Bragg condition the energy is gradually redirected from the forward-refracted direction towards the midpoint of the $\boldsymbol{K_0}$ and $\boldsymbol{K_H}$ vectors. All the energy is then reflected and no waves are excited in the crystal. A further increase in the angle excites tie points on the lower branch (branch 2) of the dispersion surface and the energy flow moves back towards the forward-refracted direction. As the excitation points cross from one branch to another the phase of the scattered amplitude changes dramatically. The main concentration here is on the reflection condition, when only one branch is excited at any time. However for the transmission condition both branches are excited and the phase difference can lead to very rapid changes in the amplitude depending on the scattering conditions (for example sample thickness). It is important to remember that the dispersion surface is complex and this takes account of absorption, i.e. perfect reflectivity is not possible and some sample penetration does exist.

Equation 2.3.30 is the satisfying condition for X-ray scattering and now we will derive the magnitude of the scattering as a function of incident angle. An abbreviation is taken, see equation 2.3.20,

$$\beta_H = \frac{K_H^2 - k^2}{2K_H^2} \qquad 2.3.31$$

that is called the deviation parameter. This following derivation is largely historical but does give a clear link to the basis of conventional dynamical theory. The position of the surface normal as before can be used to describe the excitation points. The wave vector $\boldsymbol{K}_H = \boldsymbol{k}(1+\varepsilon_H)$, where ε_H is a very small value representing the relation between the external and internal wave vectors, figure 2.3.5. We can simplify equation 2.3.31, etc., with the approximation

$$2\beta_H = \frac{K_H^2 - k^2}{K_H^2} \sim \frac{K_H^2 - k^2}{k^2} \sim 2\varepsilon_H \qquad 2.3.32$$

The assumption in this step is simply that $\varepsilon_H^2 \sim 0$ and as we shall see later $\varepsilon_H \sim 10^{-5}$ for X-ray wavelengths and typical materials under analysis, especially close to the scattering condition. This assumption makes the equation linear and more transparent.

The boundary condition defines the surface normal with respect to the dispersion surface, as before. That is the component of the $\boldsymbol{k_0}$ and the $\boldsymbol{K_0}$ wave vectors parallel to the surface must be equal. We can therefore construct $\boldsymbol{K_H}$ as being composed of two components

$$K_H = k(1+\varepsilon_H) = k(1+\alpha_H) - kg\sin(\omega - 2\theta) \qquad 2.3.33$$

The parameter α_H is a geometrical factor that can be related to the angular deviation of the incident beam from the Bragg condition. The parameter g is the component of the vector ($\boldsymbol{k_0}$ - $\boldsymbol{K_0}$) in terms of $|\boldsymbol{k}|$. This is normal to the surface direction. Similarly the refracted wave vector can be written as

$$K_0 = k(1+\varepsilon_0) = k - kg\sin\omega \qquad 2.3.34$$

The parameter g is small and by using Snell's law for refraction and equation 2.3.25 we can derive an approximate magnitude.

$$g = \sin\omega - \left(\sin^2\omega + \chi_0\right)^{1/2} \approx \frac{\chi_0}{2\sin^2\omega} \qquad 2.3.35$$

Since the susceptibility is small the second terms in both equation 2.3.33 and 2.3.34 are very small compared with the first terms. The geometrical factor α_H can be determined precisely using the construction in figure 2.3.6. The length $\boldsymbol{k}(1+\alpha_H)$ is the distance from the tie-point intersection on the external wave vector sphere to the end of the scattering vector, i.e. PH. Using the cosine rule in triangle PHL_0 we have:

$$\alpha_H = \left\{1 + 4\sin\frac{\omega-\omega_0}{2}\left[\sin\frac{\omega-\omega_0}{2} + \sin\left(2\theta - \frac{\omega-\omega_0}{2}\right)\right]\right\}^{\frac{1}{2}} - 1 \qquad 2.3.36$$

If we assume that the angular deviation is small then we can make the approximation

$$\alpha_H \sim 2\sin\frac{\omega-\omega_0}{2}\sin 2\theta \sim (\omega-\omega_0)\sin 2\theta \qquad 2.3.37$$

ω_0 is the incident angle that satisfies the Bragg condition. This approximation given in the original theory is adequate for small strains and small angular ranges ~ 0.5^0, and will be considered in more detail in section 2.4.1.

We can eliminate the g parameter by introducing an amplitude ratio ($X = \boldsymbol{D}_H/\boldsymbol{D}_0$) combining equations 2.3.33 and 2.3.34 with 2.3.28 and 2.3.29. The resultant equation is given by:

$$\chi_{-H}\frac{\gamma_H}{\gamma_0}X^2 + \left\{\chi_0\frac{\gamma_H}{\gamma_0} - \chi_0 + 2\alpha_H\right\}X - \chi_H = 0 \qquad 2.3.38$$

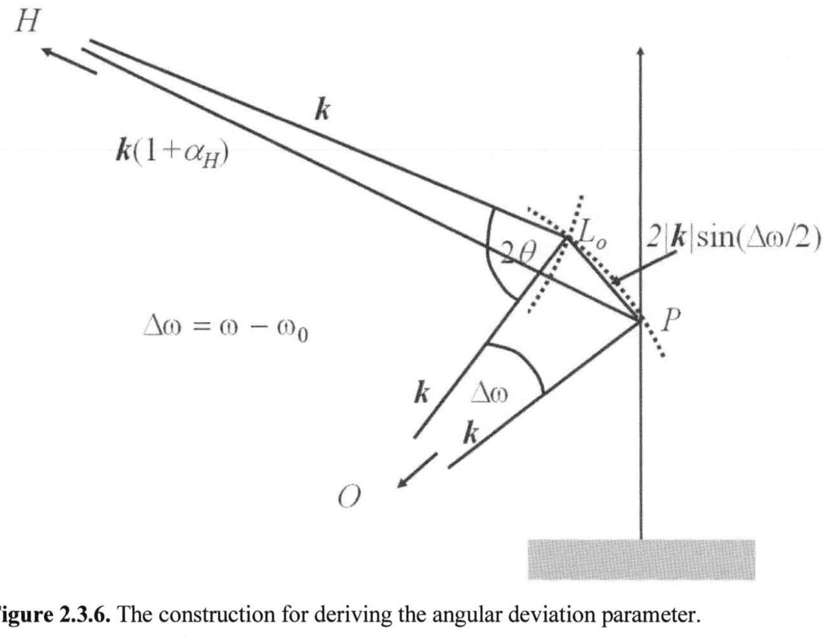

Figure 2.3.6. The construction for deriving the angular deviation parameter.

Assuming that β_0 is negligible, since the component of $(\boldsymbol{K_0} - k_0)_{//} \approx 0$, equation 2.3.31 and 2.3.34, because the refractive index is close to unity; then this is solved simply to give the reflectivity as a function of the deviation from the Bragg condition

$$X = \frac{-\eta \pm \sqrt{\eta^2 + 4\chi_H \chi_{-H} \dfrac{\gamma_H}{\gamma_0}}}{2\chi_{-H} \dfrac{\gamma_H}{\gamma_0}} \qquad 2.3.39$$

The $\sin(\omega - 2\theta)$ and $\sin(\omega)$ have been substituted by the direction cosines γ_H and γ_0, and

$$\eta = 2\alpha_H - \left\{1 - \frac{\gamma_H}{\gamma_0}\right\}\chi_0 \qquad 2.3.40$$

From these equations we can therefore calculate the scattering profile and the measured intensity at each position ω is now given by:

$$I(\omega) = X(\omega)X^*(\omega)\left|\frac{\gamma_H}{\gamma_0}\right| \qquad 2.3.41$$

The inclusion of the direction cosines accounts for the effects of beam compression or expansion, effectively the ratio of the projected area of the incident beam and the scattered beam on the sample. The original wave-field energy relates to a unit area and therefore if there is beam expansion and all of the scattered beam is captured then we have to account for this.

The simulation of a perfect GaAs crystal with an $Al_{0.4}Ga_{0.8}As$ layer on top, *004* reflection and Cu $K\alpha_1$ radiation, produces a profile as in figure 2.3.7. In the figure the position of the Bragg angles for the layer and substrate are included. There is a small shift in peak position with respect to the Bragg position due to the refractive index, which is similar for both materials and has little influence on the separation of the peaks. However there is a larger effect that becomes more pronounced as the layer becomes thinner and that is the apparent movement of the layer peak towards the substrate, Fewster and Curling (1987). This effect is due to the coupling of the wave-field within the layer not being fully established, due to the influence of the substrate or other adjacent layers. This suggests that dynamical simulation of these structures is necessary for good evaluations of peak separation, for obtaining composition and strain measurements.

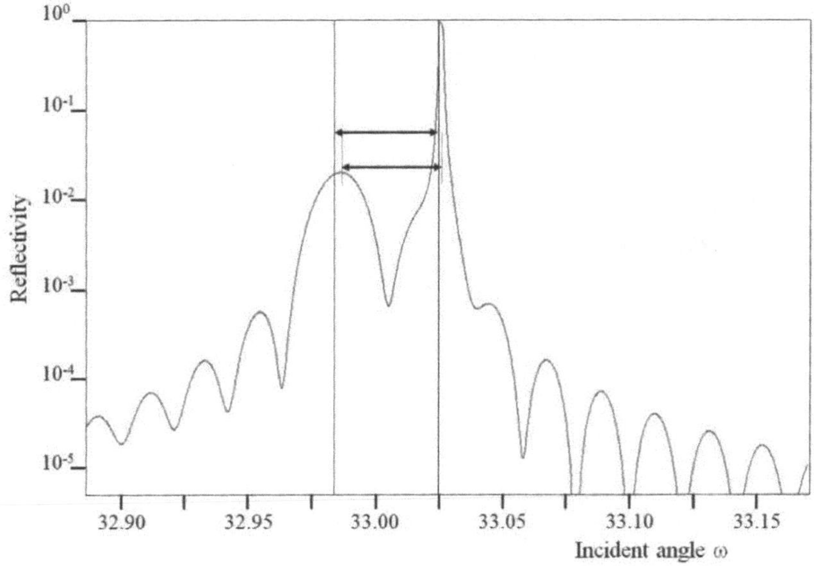

Figure 2.3.7. The simulation of the *004* reflection from the *(001)* surface of a 0.25 μm $Al_{0.4}Ga_{0.6}As$ layer on a GaAs substrate, based on the 2-beam dynamical theory. The actual Bragg angles are given by the vertical lines; each peak is displaced due to the refractive index and the peak separation is reduced by the effect first found by Fewster (1987).

2.4. The assumptions in the above model

The above derivation includes many assumptions that bring uncertainties and inaccuracies to the simulation of a diffraction profile. Some of the most important assumptions are given below:

(1) The evaluation of the deviation parameter is only valid over small angular ranges and an improved derivation will extend the applicability of this profile simulation. This is important for combining epitaxial layers that have large differences in their lattice parameters.

(2) Suppose that the material is inhomogeneous, e.g. strained with depth, then a more general model is needed that will allow the structure factor to fluctuate with depth. The work of Takagi initiated this, and is the dynamical formulation that is widely used.

(3) Equation 2.3.16 introduces an approximation that is valid for small χ, but not larger values, e.g. in the case of long wavelength (low energy) X-rays. The approximation is convenient in that it creates a transverse wave based on $\boldsymbol{D}_{\mathrm{m}}$ that makes it valid for both polarization states σ and π, figure 2.3.2.

(4) Another assumption in the above derivation is that only two significant waves are excited, but from figure 2.3.5 there are several interactions with the dispersion surface. Consequently the assumption that just an incident wave and a diffracted wave dominate the process is sometimes insufficient.

(5) Extending the point raised in (4), the above approach assumes that only a single reflection is excited at a time, however several reflections can contribute to the intensity in a scattering profile. This can lead to a significantly more complicated dispersion surface.

(6) Equation 2.3.9 indicates an assumption, which is especially inappropriate for simulating the scattering from thin layers and that is the simplification of the structure factor / polarizability (or the alternative name of susceptibility) χ. In this equation it is assumed that the scattering is dominated by very closely bound electrons and the integral can be approximated by a summation. This approximation has several consequences:

   a. This assumes that all unit cells contribute equally, which is only strictly valid when end effects (i.e. interfaces) are ignored as in an extended uniform crystal.

b. For periodic thin layers, the effects of (4) and (5), as well as (6a) are most pronounced, especially remote from the Bragg condition when satellite reflections and fringes spread out over large angles. Consequently the dynamical theory outlined above does not produce the correct profile, if the scattering from several reflections interfere. This occurs when the surface is accurately parallel to the scattering planes.
c. Another aspect of changing the structure factor from an integral to a summation, is that it does not accommodate non-isotropic distributions of electronic density, e.g. sp^3 hybridized bonding that are common in semiconductors. This can result in some of the assumed "systematically absent" reflections being observed (any intensity at these positions may also be explained as in section 2.9.3 and the Epilogue).

These points will now be discussed in more detail and now these have been resolved, in the following sections.

2.4.1. Improving the angular range assumption

By working directly with equation 2.3.39 and applying the properties of the construction of the dispersion surface; i.e. the surface normal that intersects the dispersion surface purely represents the condition that the components of the wave-vectors parallel to the surface outside and inside the crystal are equal. Hence we can rewrite the equation 2.3.32 as:

$$\beta_H \approx \frac{K_H^2 - k^2}{2k^2} = \frac{\{({}_{\parallel}K_H + {}_{\perp}K_H) - ({}_{\parallel}k_H + {}_{\perp}k_H)\}\{({}_{\parallel}K_H + {}_{\perp}K_H) + ({}_{\parallel}k_H + {}_{\perp}k_H)\}}{2k^2}$$

2.4.1

remembering of course that $\boldsymbol{k}_H = \boldsymbol{k}$, the external wave vector. Since the parallel components are equal this simplifies to :

$$\beta_H \approx \frac{{}_{\perp}K_H - {}_{\perp}k_H}{2k^2}(K_H + k) \sim \frac{{}_{\perp}K_H - {}_{\perp}k_H}{k} \qquad 2.4.2$$

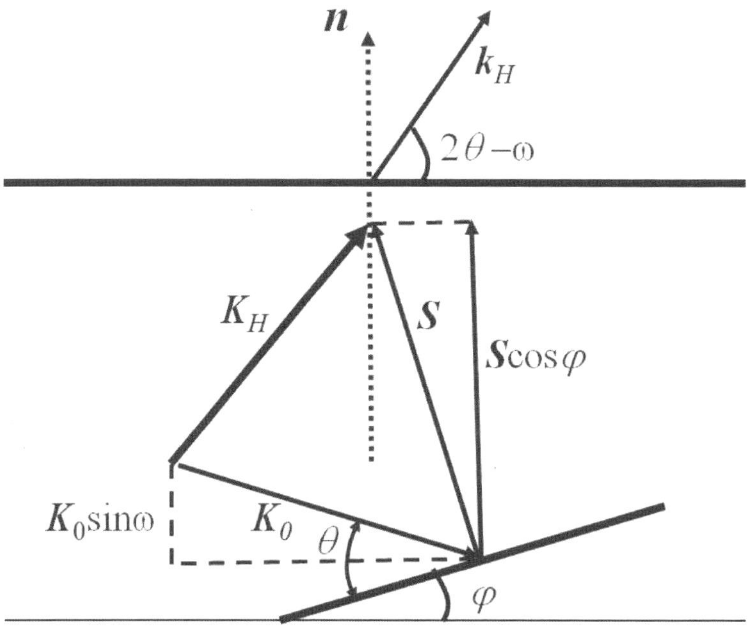

Figure 2.4.1. An alternative method of deriving the deviation parameter.

Now consider figure 2.4.1, which represents the scattering vector and the incident and scattered wave vectors for the Bragg condition and a position deviated from this. From this figure we can see that

$$_{\perp}K_H = |S|\cos\varphi - |K_0|\sin\omega \qquad 2.4.3$$

$$_{\perp}k_H = k\sin(2\theta - \omega) \qquad 2.4.4$$

Hence

$$\beta_H \approx \frac{1}{|k|}\left[|S|\cos\varphi - |K_0|\sin\omega - |k|\sin(2\theta-\omega)\right] \qquad 2.4.5$$

Now the scattering vector is given by Bragg's equation 2.2.11, $|\boldsymbol{S}|=1/d=\{2\sin\theta_B/\lambda\}(1+\chi_0)^{1/2}$, where the wavelength within the crystal has been modified with a refractive index term. The internal incident wave vector is refracted and modified similarly by this term. The magnitude of the external wave vector $k = 1/\lambda$. Substituting into equation 2.4.5, and taking the component normal to the surface we have

$$\beta_H \approx \left\{2\sin\theta_B \cos\varphi(1+\chi_0)^{1/2} - (1+\chi_0)^{1/2}\gamma_0 - \gamma_H\right\}\gamma_H \qquad 2.4.6$$

If we neglect the refractive index contribution, since this is very small we have the expression derived by Zaus (1993). The scattered beam direction must be known and it will be derived below in section 2.5.

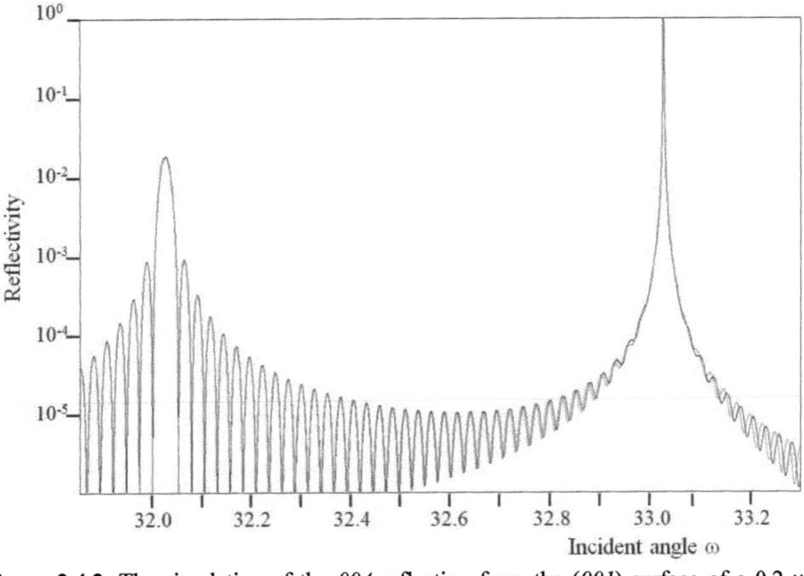

Figure 2.4.2. The simulation of the *004* reflection from the (*001*) surface of a 0.2 μm $In_{0.2}Ga_{0.8}As$ layer on a GaAs substrate, based on the 2-beam dynamical theory assuming the deviation parameter given in equation 2.3.36, grey, and that in equation 2.4.6, black. The fringes begin to go out of phase at an angle of -1.2^0 from the Bragg angle of the layer and also close to the substrate Bragg peak.

The deviation parameter given in equation 2.4.6, is compared with the approximation given in equation 2.3.36. It should be recognized that the differences are only observed at large angular deviations, larger strains and increasing surface to scattering plane angles, figure 2.4.2. For example in figure 2.3.7, both equations produce coincident profiles over the angular range plotted. Equation 2.4.6 has been compared with simulations based on the solution of the full dispersion surface, Holý and

Fewster (2003). The agreement is precise except close to grazing incidence when the specular component is significant or when the scattering from another reflection interferes coherently. These conditions are important for some of the analyses and therefore will be considered in more detail later, section 2.4.3.

2.4.2. A scattering theory to accommodate real crystals

So far we have been assuming that these derivations are based on an extended perfectly periodic crystal without distortions. There have been many attempts to extend the model to accommodate distortions and perhaps the most successful is that based on the concepts put forward by Takagi (1962). This will now be discussed.

The model above gives good agreement with experimental results but is rather restricted when applied to crystals that are distorted or consisting of multi-layers of different materials, etc. The conceptual approach first proposed by Takagi (1962) and expanded by Taupin (1964) and Takagi (1969) and others will be presented here.

From equations 2.3.7 and 2.3.8, the susceptibility at position $\boldsymbol{r}$ in the sample is given by:

$$\chi(r) = \sum_{S_m} \chi_{S_m} \exp(-2\pi i S_m . r) \qquad 2.4.7$$

$\boldsymbol{S}_m$ being a reciprocal lattice vector and

$$\chi_{S_m} = -\frac{r_e \lambda^2}{\pi V} F_{S_m} \qquad 2.4.8$$

and

$$r_e = \frac{e^2}{4\pi\varepsilon_0 mc^2} \qquad 2.4.9$$

i.e. the electron radius. A structure that is distorted can be considered to have an average and a locally perturbed susceptibility, χ' due to a scattering centre displaced to the position co-ordinate $(\boldsymbol{r} - \boldsymbol{u}(\boldsymbol{r}))$, hence

$$\chi'(r) = \sum_{S_m} \chi_{S_m} \exp\{-2\pi i(S_m.r - S_m.u(r))\} \qquad 2.4.10$$

$\boldsymbol{S_m.r}$ has the periodicity of the average lattice and is therefore rapidly varying on the scale of the interatomic spacing, whereas grad($\boldsymbol{S_m}.\boldsymbol{u}(\boldsymbol{r})$) is assumed to be a variable on a larger macroscopic scale. The displacement fields can therefore be given as:

$$D(r) = \exp(2\pi i v t)\sum_{m} D'_m(r)\exp\{-2\pi i(K_m.r - S_m.u(r))\} \quad 2.4.11$$

This sum of plane waves must satisfy Maxwell's equations as before and by deriving expressions for curlcurl($\boldsymbol{D}/(1+\chi)$), $\nabla^2\boldsymbol{D}$ and substituting into equation 2.3.16 we obtain:

$$\sum_{m}\left\{(k^2 - K_m^2)D'_m - iK_m.\text{grad}(D'_m) + \sum_{n}\chi_{m-n}[D'_n]_m\right\}\exp(-2\pi i(K_m.r - S_m.u(r)) = 0$$

2.4.12

The term $[\boldsymbol{D'}_n]_m$ represents the component of $\boldsymbol{D}_n$ perpendicular to $\boldsymbol{K}_m$. To make the manipulation possible there are a few assumptions that are included to achieve this result; i.e. the electric displacement $\boldsymbol{D}$ and the polarisability $\boldsymbol{P}$ and their first derivatives are macroscopic in their variation. Hence the second order derivatives of $\boldsymbol{D}$, first order derivatives of $\boldsymbol{P}$ ($\boldsymbol{P}$ being very much smaller than $\boldsymbol{D}$) and the divergence of the gradient of the deformation strain field ($\nabla^2\boldsymbol{S}_m.\boldsymbol{u}$) are assumed to be negligible.

Equation 2.4.12 can now be integrated with respect to $\boldsymbol{r}$ over a unit cell repeat since $\boldsymbol{u}(\boldsymbol{r})$ is assumed to be almost constant within these dimensions. This yields a more general result than that given in equation 2.3.20

$$s_m.\text{grad}(D'_m) = \pi i\frac{K_m^2 - k^2}{K_m^2}K_m D'_m - \pi i K_m \sum_{n}\chi_{m-n}[D'_n]_m \qquad 2.4.13$$

($\boldsymbol{s}_m$.grad) $= \partial/\partial s_m$ is simply a differential operator since $\boldsymbol{s}_m$ is a unit vector along the wave vector $\boldsymbol{K}_m$. As with the former approach we can now solve this for the case of two significant waves in the sample, i.e. for m = *0* and *H* the forward refracted wave and the scattered wave

$$\frac{\mathrm{i}\lambda}{\pi}\frac{\partial D_H'}{\partial s_H} = \chi_0 D_H' + C\chi_H D_0' - \frac{K_H^2 - k^2}{K_H^2} D_H' \qquad 2.4.14$$

and

$$\frac{\mathrm{i}\lambda}{\pi}\frac{\partial D_0'}{\partial s_0} = C\chi_{-H} D_H' + \chi_0 D_0' - \frac{K_H^2 - k^2}{K_H^2} D_0 \sim C\chi_{-H} D_H' + \chi_0 D_0' \qquad 2.4.15$$

where $|\boldsymbol{K}_{\mathrm{m}}| \sim 1/\lambda$ and the polarisation modification C has been included, equation 2.3.22. We can simplify these equations by introducing an amplitude ratio ($X = \boldsymbol{D}_H'/\boldsymbol{D}_0'$) similar to that in equation 2.3.38. Note the similarity of these equations to 2.3.28 and 2.3.29, which are simply the limiting case of these equations 2.4.14 and 2.4.15. The parameters $\boldsymbol{s}_0$ and $\boldsymbol{s}_H$ are the length vectors along the incident and scattered beam directions and can be substituted by t/γ_0 and t/γ_H respectively, where t is the depth into the crystal normal to the surface plane. We therefore obtain

$$\frac{dX}{dt} = \frac{\pi}{i\lambda}\left\{\frac{1}{\gamma_H}(\chi_0 X + C\chi_H - 2X\beta_H) - \frac{1}{\gamma_0}\left(\chi_0 X + C\chi_{-H} X^2\right)\right\} \qquad 2.4.16$$

This equation can be solved by integration after separation of the variables and use of partial fractions. The resulting equation gives:

$$X(Z,\omega) = \frac{SX(z,\omega) + i\{E + BX(z,\omega)\}\tan(GS[z-Z])}{S - i\{AX(z,\omega) + B\}\tan(GS[z-Z])} \qquad 2.4.17$$

Where

$$A = C\chi_{-H}\frac{|\gamma_H|}{\gamma_0} \qquad 2.4.18$$

$$B = 0.5\left\{\chi_0\left(1 + \frac{|\gamma_H|}{\gamma_0}\right) - 2\beta_H\right\} \qquad 2.4.19$$

$$E = -C\chi_H \qquad 2.4.20$$

$$S = (B^2 - AE)^{0.5} \qquad 2.4.21$$

and

$$G = \frac{\pi}{\lambda \gamma_H} \qquad 2.4.22$$

The deviation parameter in this case is a little more complex than above since $\boldsymbol{K}'$ refers to a distorted wave vector. We should add an additional term into the deviation parameter β_H since $\boldsymbol{r} = \boldsymbol{r}_0 + \boldsymbol{u}(\boldsymbol{r})$, the local incident wave vector will be changed to $\boldsymbol{K}' = \boldsymbol{K} - \text{grad}(\boldsymbol{S}_H \,.\boldsymbol{u})$. The deviation parameter becomes

$$\beta_H' = \beta_H - \frac{1}{k}\frac{\partial}{\partial t}(S_H.u(r)) \qquad 2.4.23$$

This additional term can be neglected if we can consider our sample in terms of layers of isotropic material. The whole system is greatly simplified in this way by reducing the problem to layers sufficiently thin to be considered isotropic. In this way very complex multilayer structures can be modeled.

This theory leads to a fundamental difference with that of the original dynamical theory given in section 2.3.2. Equation 2.4.11 for the wave has an extra exponential term which allows the gradually variation in the displacement field with position. This essential difference predicts subtle details observed in topographic images that the earlier plane wave theories could not explain and were left to phenomenological approaches. This makes this theory the basis for most studies from statistical dynamical theory, to model defect scattering behaviour, through to simulating section topographs. Of course it will also work well in modelling perfect crystalline materials.

The amplitude ratio is now a function of a depth co-ordinate and the parameters refer to the local susceptability, geometrical factors and of course distortion. The co-ordinate z refers to some depth in the structure for which the amplitude ratio $X(z,\omega)$ is known and Z (equation 2.4.17) refers to a depth for which the amplitude ratio is required. The distortion is contained in the parameter β_H that is given in equation 2.4.6. The resulting scattered intensity is therefore given by

$$I(\omega) = X(\omega)^* X(\omega)\left|\frac{\gamma_H}{\gamma_0}\right| \qquad 2.4.24$$

This derivation is a useful step because now we can analyse the scattering from multilayer structures very easily. This is the theory used in the majority of examples in this book, since in practice it is sufficient for nearly all experiments. Of course this Tagaki model is only sufficient for relatively small deviations from perfection, however when large distortions occur the "local dispersion surface" can become quite complex leading to inter-branch scattering, i.e. waves can undergo multiple scattering for a single Bragg reflection, Shevchenko (2010).

2.4.3. Multiple-beam dynamical theory:

Consider the situation for a simulated profile from a periodic multilayer that has an extensive array of satellites and fringes that extend along the diffraction tails. Suppose the sample has a (*00l*) surface, then the tails of the *000*, *002*, *004* and *006* reflections will begin to interact. The profile simulation based on 2-beam dynamical theory, discussed above, can only include one of these reflections. The extension to more beams, with multiple beam dynamical theory, can accommodate more reflections. From inspection of figure 2.3.5, it is clear that even with a single diffracted beam, there are 4 intersections of the dispersion surface (tie-points) for each of the two polarization states. Two are associated with the diffracted beam and two associated with the specularly reflected beam. This then gives rise to a multitude of internal wave-vectors.

2.4.3.1 Combination of specular scattering and one diffracted wave:

From a purely phenomenological viewpoint a beam set to scatter from a beam at grazing incidence, will have a significant specular component, section 2.3.1. Below some critical angle the incident wave does not exist inside the crystal, and cannot therefore diffract, resulting in the energy mostly going into the specular beam. This is easily explained by reference to figure 2.3.5, where the "original" 2-beam dynamical theory assumed that only the incident and diffracted beams are significant, i.e. only the points close to the dispersion surface intersections are considered. However this second-order description of the dispersion surface is an assumption and the actual shape in figure 2.3.5 is of fourth-order, Härtwig (1981), Holý and Fewster (2003);

$$k^2 E_m^2 + K_m \times (K_m \times E_m) = -k^2 \sum_n \chi_n E_{m-n} \qquad 2.4.25$$

Hence there are 4 intersections with the dispersion surface as in figure 2.3.5, whereas up until now it has been assumed that there were only two. The above equation is derived using $\boldsymbol{E}_m$ waves to also resolve the error associated with the magnitude of χ raised in assumption (3), section 2.4, and therefore this equation should be compared with the approximation in equation 2.3.21. The solution to this will produce a better representative profile, illustrating the specular and diffracted profile.

In the first instance we will consider the case of a perfectly aligned sample where the tails of the diffracted beam, e.g. *002* and the specularly reflected beam, *000* overlap and interfere. The simulation of the profile for a perfect superlattice structure is given in figure 2.4.3. However if the surface is not parallel to the (*001*) plane then the tails do not overlap and the profile will change since the interference is broken and the specular and diffracted beams are more independent. The latter is probably the most likely situation for many samples, since the chance of a perfectly orientated surface is rare, although there could well be certain azimuthal orientations that can introduce this interference. This occurs when the misorientation is set to be normal to the scattering plane and the photon can either take the specular path or the diffraction path, within the spread of the axial divergence. This configuration will be described in more detail in section 5.4.5.5.5.

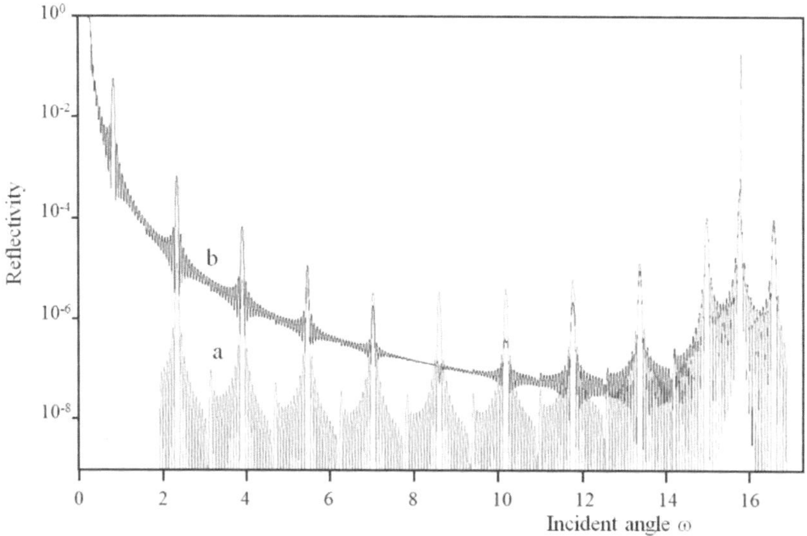

Figure 2.4.3. The simulation of an {AlAs (0.00283 μm) + GaAs (0.002831 μm)}×11 superlattice structure, i.e. 11 periods of 5 unit cells of GaAs and AlAs, on GaAs, for the *002* reflection using just 2-tie points (a) and using all 4-tie points (b) in the 2-beam dynamical theory. This assumes the surface is exactly parallel to the *(001)* plane, such that the specular and *002* reflection tails interfere.

It is worth noting that the experimental profile of the *002* reflection, when the tails do not overlap, agrees well with the simulation based on the theories discussed in section 2.3.2 and 2.4.2, two strong beams (2-tie points), i.e. ignoring the specular component. However if these tails do overlap as in figure 2.4.3, these two profiles deviate significantly in agreement beyond a few degrees from the Bragg condition, when the interference of the specularly reflected component has influence. Now the influence can only occur if the surface specular rod of intensity aligns with the tails of the *002* reflection, i.e. when the orientation of the scattering planes is aligned perfectly with the surface. This is a very stringent condition, so in most cases the theoretical description given in sections 2.3.2 and 2.4.2 appears to be sufficient. It is also worth mentioning that in general the period thickness does not correspond to an integer number of the average lattice parameter, as in figure 2.4.3, so the satellites from the specular profile will not be coincident with those from the diffraction profile. This leads to some doubling of the peaks, and because of the incommensurate nature of the periodicity: this is in general indicative of interface steps and roughening. This will diminish the satellite intensities with distance from the average peaks, i.e. *000* and *hkl*. This incommensurate condition is discussed in more detail in section 5.4.5.2.

This fall in diffracted intensity and rise in specular intensity is quite clear to understand, however what happens in grazing exit is a little more subtle.

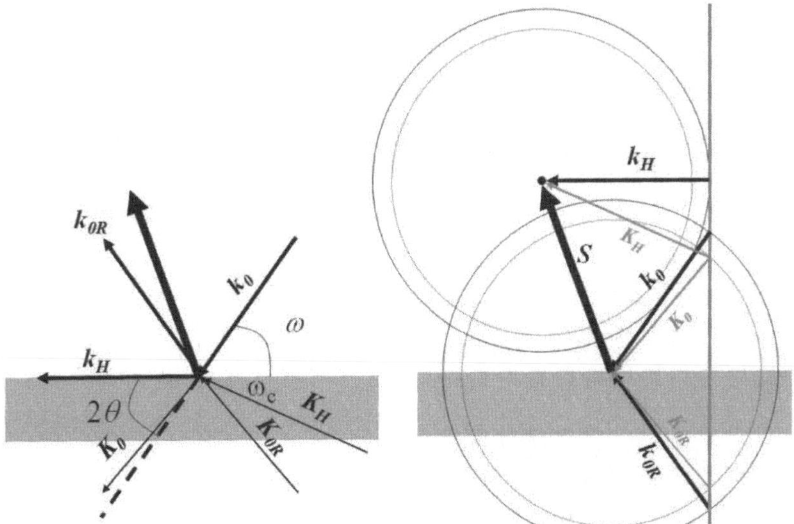

Figure 2.4.4. The geometry (a) and the dispersion surface (b) for the condition of very small diffracted beam exit angles. The grey lines are the internal wave-vectors (b), and the thin black lines in (a). The diffracted beam cannot escape from the surface unless 2θ-$\omega > \omega_c$, where ω_c is the critical angle in this construction, although when the absorption (imaginary component) is included there will be a decaying residual intensity escaping at angles below the critical angle.

The scattering in grazing exit is particularly interesting for some of the diffraction geometries that will be described later in this book. If we consider the geometry as shown in figure 2.4.4a, where the diffracted beam emerges at very low angles, then the dispersion surface appears as in figure 2.4.4b. Now clearly if the external diffracted wave is not directed out of the surface above the internal angle then it will not appear, so will not be detected, despite its internal wave direction giving the appearance of a detectable beam. There is a region above the surface up to the "critical angle" where there is a small leak of intensity. This is one case when the solution relying on 4 intersections with the dispersion surface is most important, since the refractive index of the internal wave is crucial to modeling the profile precisely.

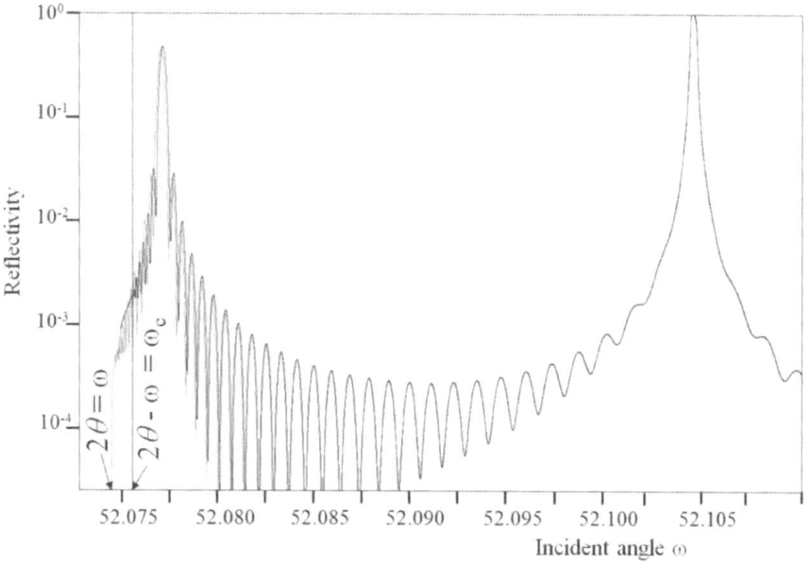

Figure 2.4.5. The scattering from the (*113*) of GaAs with a 0.2 μm layer of $In_{0.2}Ga_{0.8}As$ close to grazing exit, calculated with the 2-tie point 2-beam dynamical theory (grey) and the 4-tie point 2-beam dynamical theory (black). There are some subtle changes, including a small peak shift that introduces a small error in the derived In concentration (0.198 rather than 0.2, i.e. $\varepsilon_\perp$ = -0.00022) and a smooth fall in the intensity below the critical angle rather than the oscillations observed in the 2-tie point 2-beam model. The critical angle is given by the vertical line.

The profile of a simple structure close to this condition is given in figure 2.4.5. It is worth noting that the peaks close to the critical angle are shifted to a small extent, see figure caption. The layer peak position in figure 2.4.5 relates to a composition of x = 0.2 in $In_xGa_{1-x}As$, where a shift is noticeable, however for x = 0.15 the models based on 2 tie-points and 4 tie-points are virtually coincident.

If the investigating incident beam is brought close to the edge of the sample then the internal diffracted wave will appear through the side of the sample. This has been measured experimentally and could potentially yield some interfacial information, but this is outside the scope of this chapter.

2.4.3.2 Analysing the scattering between two diffraction peaks

To analyse the intensity between two diffraction peaks requires the addition of two reflections contributing and the 3-beam dynamical diffraction solution is required, i.e. an incident beam and two strong diffracted beams. So now the relevant waves are:

$$E(r) = E_0 e^{ik_0 . r} + E_H e^{ik_H . r} + E_G e^{ik_G . r} \qquad 2.4.26$$

This equation 2.4.26 is then substituted into equation 2.4.25 and the solution then defines the dispersion surface for this 3-beam case and is given by:

$$\begin{aligned}&\left(K_0^2 - k^2\right)\left(K_H^2 - k^2\right)\left(K_G^2 - k^2\right) - \left(K_0^2 - k^2\right)k^4 \chi_{G-H}\chi_{H-G}\\&- \left(K_H^2 - k^2\right)k^4 \chi_G \chi_{-G} - \left(K_G^2 - k^2\right)k^4 \chi_H \chi_{-H}\\&- k^6 \chi_H \chi_{-G} \chi_{-G-H} - k^6 \chi_G \chi_{-H} \chi_{H-G} = 0\end{aligned} \qquad 2.4.27$$

This is obviously an equation of 6th order. This can be compared to equation 2.3.30 that can be described geometrically as two intersecting spheres to accommodate the reflectivity and one diffraction condition, and now the inclusion of a further diffracted wave requires three intersecting spheres.

The dispersion surface for the 3-beam case or any multiple-beam condition becomes increasingly complex: it effectively becomes a series of spheres with centres at the end of the diffraction vector, e.g. *OH* in figure 2.3.5. The intersections of these spheres then take a similar form to that in figure 2.3.5, i.e. all the singularities of these multiple spheres are removed. Because of the increasingly complexity the equations are best visualized in matrix form by combining the wave-vectors and wave-fields, and making use of the continuity of these components parallel to the interfaces and surface. Also because the waves are not strictly transverse, representation of the π polarization is more complicated, so what is given above is only exact for the σ polarization. For the 2-beam case including the specular component, the matrix is 4×4, as given above, and in the 3-beam dynamical theory this becomes a 6×6 matrix, see equation 2.4.27. The complexity starts to increase considerably, and the profile is getting closer to the true profile of a perfect crystal.

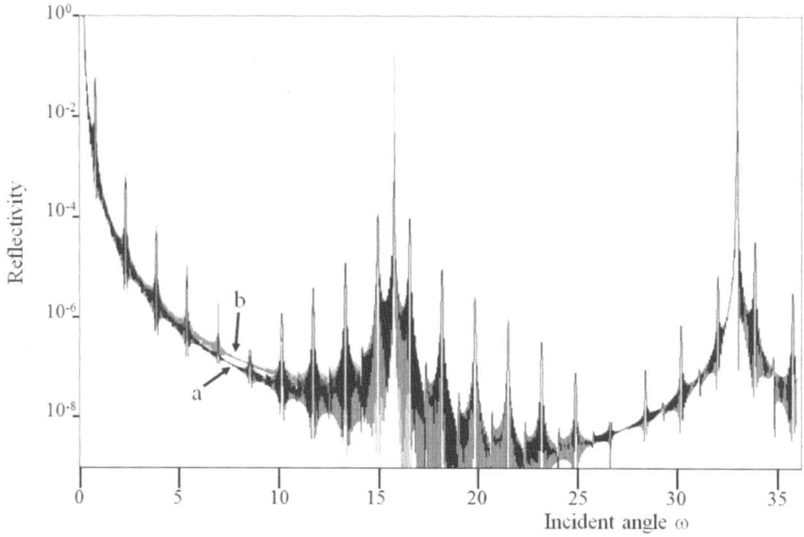

Figure 2.4.6. The simulation of the scattering from planes parallel to the surface for an {AlAs (0.002831 μm) + GaAs (0.00283 μm)}×11 superlattice structure, i.e. a period of 5 unit cells of GaAs and AlAs, on GaAs, including the *000*, *002* and *004* reflections using 3-beam dynamical theory (a) from 0^0 to 36^0, compared with the 2-beam dynamical theory just including the *000* and *002* reflections (b) from 0^0 to 17^0.

This becomes an interesting challenge mathematically as the number of beams is increased, and the profile does take better account for the continuity of the intensity from Bragg peak to Bragg peak, figure 2.4.6. It is possible to see subtle changes, even between the *000* and *002* reflections due to the inclusion of the *004* reflection, which is especially obvious at an incident angle of ~ 8^0 in this structure. So how many reflections should be included to model a structure that is on perfect orientation? This is where we require a different approach. It is also important to overcome the fundamental problem raised in the approximation given in (6) in section 2.4, i.e. the simplification of the structure factor / polarizability. This can have a profound effect on the intensity profile.

2.4.4. A more complete dynamical theory:

The assumption discussed in (6) of section 2.4 in equation 2.3.9 is that the structure factor or polarisability (susceptibility) exists as a continuous value, which is truncated for each layer with abrupt edges. This is a reasonable assumption when the layers are thick and the edge effects at the unit cell dimension are not significant, and the region in reciprocal space being investigated is insensitive to this scale, e.g. close to diffraction Bragg peaks. However for the case given above, between Bragg peaks, these effects can be significant. A kinematical model can be used to determine the scattering between Bragg peaks, since this can be calculated on an atom by atom basis, but this cannot model the scattering close to the Bragg peaks with sufficient accuracy to rely on the profile.

The approach here is based on the work of Holý and Fewster (2008). The evaluation of the structure factor / polarisability in standard dynamical theory is based on the unit cell as the smallest unit, however if the scattering is considered on a far finer scale, say 0.01 of a unit cell, then this assumption can be removed. The standard formulism for dynamical theory is replaced by the Fresnel formulae, by taking each lamella as the average electron density parallel to the surface. For reflections from planes parallel to the surface this is fairly straightforward and will be defined for compactness in matrix form as in Holý and Fewster. This approach can be compared with section 2.10 where reflectometry is discussed as a separate methodology. So the electric field wave $\boldsymbol{E}$ in one layer is related to the layer below by its reflectivity and its phase (since this varies as a function of depth), and hence:

$$E_{j-1} = \Phi_{j-1} R_j E_j \qquad 2.4.28$$

where Φ contains the phase shifts for the transmitted and reflected waves between the boundaries of the lamellae:

$$\Phi_j = \begin{bmatrix} e^{iK_z^j \Delta z} & 0 \\ 0 & -e^{iK_z^j \Delta z} \end{bmatrix} \qquad 2.4.29$$

where Δz is the thickness of the lamella chosen. The reflection matrix of the interface *j* to *j-1* is given by:

$$R_j = \frac{1}{t_r} \begin{bmatrix} 1 & r_j \\ r_j & 1 \end{bmatrix} \qquad 2.4.30$$

where *t* and *r* are the Fresnel coefficients for transmission and reflection and are given by:

$$r_j = \left(\frac{K_{j-1} - K_j}{K_{j-1} + K_j} \right)_\perp$$
$$t_j = \left(\frac{2K_{j-1}}{K_{j-1} + K_j} \right)_\perp \qquad 2.4.31$$

These are valid for the σ polarization; for the π polarization these include refractive index terms, which are also given in section 2.10:

$$r_j = \left(\frac{n_{j-1}^2 K_{j-1} - n_j^2 K_j}{n_{j-1}^2 K_{j-1} + n_j^2 K_j} \right)_\perp$$
$$t_j = \left(\frac{2n_{j-1}^2 K_{j-1}}{n_{j-1}^2 K_{j-1} + n_j^2 K_j} \right)_\perp \qquad 2.4.32$$

The electric field vector is composed of two components, the incident field and specularly reflected field, which can be represented by:

$$E_j = \begin{bmatrix} E_{T,j} \\ E_{R,j} \end{bmatrix} \qquad 2.4.33$$

Deep down inside the sample, or at the undersurface, there will be a transmitted wave and no reflected wave, and this is the initial starting condition for the calculation, i.e.

$$E_N = \begin{bmatrix} E_{T,N} \\ 0 \end{bmatrix} \qquad 2.4.34$$

So the electric field vector, is composed of an incident wave, $\boldsymbol{E_0}$, and a reflected wave, $\boldsymbol{E_R}$, at the surface and is given by:

$$\begin{bmatrix} E_0 \\ E_{0R} \end{bmatrix} = R_1\Phi_1 R_2\Phi_2 R_3\Phi_3 \ldots\ldots\ldots R_N\Phi_N R_{N+1}\begin{bmatrix} E_T \\ 0 \end{bmatrix} \qquad 2.4.35$$

or very simply:

$$\begin{bmatrix} E_0 \\ E_{0R} \end{bmatrix} = M\begin{bmatrix} E_T \\ E_0 \end{bmatrix} \qquad 2.4.36$$

The phase term Φ_0 at the surface "air" layer is clearly a unitary matrix, since the lamella thickness can be set to zero. This equation 2.4.35 is relatively simple to evaluate, however for a typical structure including the substrate, N could be very large $\sim 10^9$. Therefore the perfect extended regions that can be represented by unit cells, and can be reduced to:

$$M_{cell} = R_1\Phi_1 R_2\Phi_2 R_3\Phi_3 \ldots\ldots\ldots R_n\Phi_n \qquad 2.4.37$$

Where n is the number of lamellae in one unit cell, such that a thick layer (e.g. the substrate) can be represented as:

$$M_{perfect} = \left(M_{cell}\right)^m \qquad 2.4.38$$

where m is the number of unit cells in a layer or substrate, so that equation 2.4.35 could for example become:

$$\begin{bmatrix} E_0 \\ E_{OR} \end{bmatrix} = R_1\Phi_1 R_2\Phi_2 R_3\Phi_3 \ldots\ldots R_{M-1}\Phi_{M-1} R_M \Phi_M M_{perfect} \begin{bmatrix} E_T \\ 0 \end{bmatrix} \qquad 2.4.39$$

for M lamellae representing the structure that could be sub-unit cell dimensions, etc. By trial and error it can be shown that for GaAs, for example, the smallest number of lamellae in a unit cell necessary for producing a non-changing profile over the whole scattering range is ~30. The reflection from the bottom surface is assumed to be zero, thus creating the starting condition to evaluate the coefficients throughout the structure.

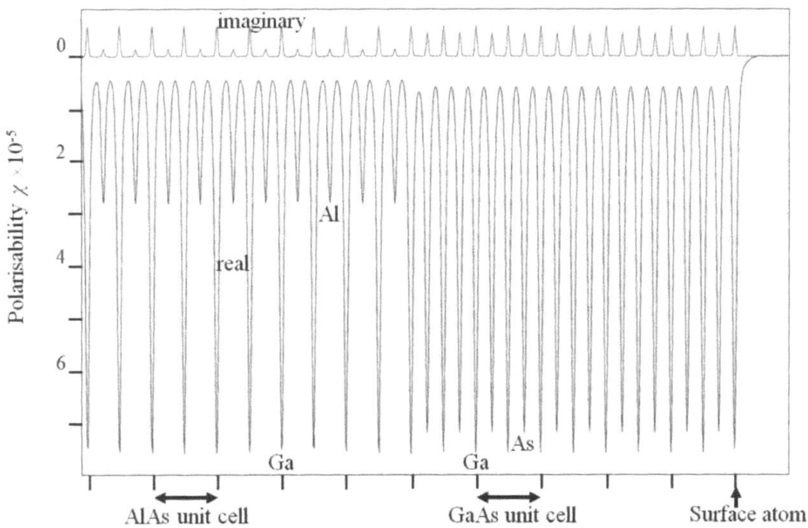

Figure 2.4.7. The variation in the polarisability through a period of 5 unit cells of GaAs and 5 unit cells of AlAs used in the multi-beam dynamical theory to produce the profile of figure 2.4.8. The smoothly varying polarisability compares with the delta-functions (single values) at each atom position used to obtain an average unit cell polarisability for each layer, which is used in conventional theories.

Now it is interesting to note that no mention is made of reflections, and that is because the calculation will produce all the reflections by virtue of the appropriate phase and reflectivity relationships. The polarisability profile for performing the calculation appears as in figure 2.4.7. Thus the whole problem of deciding how many beams to include is removed. The profiles are correct and exact between Bragg peaks and the profiles of the Bragg peaks match those using the more conventional dynamical theories, figure 2.4.8. However remote from the Bragg peaks the agreement is not good, suggesting in this case that the more conventional theories underestimate the satellite intensities in this example.

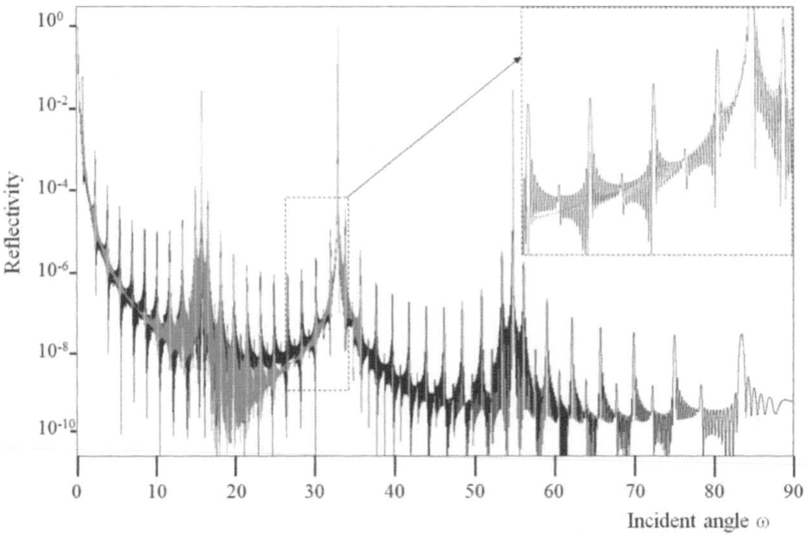

Figure 2.4.8. The simulation of the scattering from planes parallel to the surface for an {AlAs (0.002831 μm) + GaAs (0.00283 μm)}×11 superlattice structure on GaAs, including the *000*, *002*, *004*, and *006* reflections using this new dynamical theory. The diffraction peaks are not specifically entered into the model, but occur naturally from the fluctuation of the polarisability described on the 1/100th scale of the unit cell, and by choosing to evaluate the scattering from a defined angular range, i.e. 0^0 to 90^0. The 3-beam "conventional" simulation of the structure is superimposed in grey. It is interesting to note that the intensities away from the Bragg peaks do not agree.

For scattering from planes inclined to the surface the situation is more complicated since these planes intersect the lamellae and will have an inhomogeneous variation and not a simple average. The approach proposed in Holý and Fewster, is to represent the electron density in each lamella as a Fourier sum that is periodic, $\rho(x)$, which can be represented as a polarisability, see equation 2.3.7, i.e.

$$\chi(x) = -\frac{e^2\lambda^2}{4\pi^2\varepsilon_0 mc^2}\rho(x) = \sum_G \chi_G e^{iG.x} \qquad 2.4.40$$

where $\boldsymbol{G}$ is the vector representing a two-dimensional reciprocal lattice and $\boldsymbol{x}$ is the lateral position vector. This lateral periodicity is different for every lamella.

Now if we consider the transmitted and reflected waves in this lamella, then the electric field at any position $\boldsymbol{r}$ (i.e. terms of x, y and z) can be represented as a Bloch wave

$$E(r) = \sum_G \left(E_{GT}e^{iK_z z} + E_{GR}e^{-iK_z z}\right)e^{i(k_{//}+G).x} \qquad 2.4.41$$

where $k_{//}$ is the in-plane component of the incident wave-vector. The solutions for the reflected and transmitted waves are

$$\begin{aligned} K_0^2 E_{GR} - K_{GR} \times (E_{GR} \times K_{GR}) + k^2 \sum_{G \neq J} \chi_{G-J} E_{JR} = 0 \\ K_0^2 E_{GT} - K_{GT} \times (E_{GT} \times K_{GT}) + k^2 \sum_{G \neq J} \chi_{G-J} E_{JT} = 0 \end{aligned} \qquad 2.4.42$$

where $\boldsymbol{K_0}$ is given in equation 2.3.25, and $\boldsymbol{K_{GT}}$ and $\boldsymbol{K_{GR}} = (\boldsymbol{k_{//}} + \boldsymbol{G}, \pm \boldsymbol{K_{0z}})$, where $\boldsymbol{K_{0z}}$ is the vertical component of the transmitted wave-vector. Clearly $\boldsymbol{G}$ can take on an infinite number of values, i.e. there is no limit to the number of waves. However if the problem is simplified to simulating a single column of reflections, as well as those for planes parallel to the surface, e.g. *021*, *022*, *023* for a *001* surface, then the solution becomes a lateral two-beam approximation with two linear equations for each of the reflected and transmitted waves

$$\begin{aligned} (K_0^2 - K_{0R}^2)E_{0R} + k^2\chi_{-G}E_{GR} = 0 \\ (K_0^2 - K_{GR}^2)E_{GR} + k^2\chi_G E_{GR} = 0 \end{aligned} \qquad 2.4.43$$

and

$$\begin{aligned}(K_0^2 - K_{0T}^2)E_{0T} + k^2\chi_{-G}E_{GR} &= 0\\ (K_0^2 - K_{GT}^2)E_{GT} + k^2\chi_G E_{GR} &= 0\end{aligned} \qquad 2.4.44$$

The non-trivial solution will yield the dispersion surface, similar to the derivation of equation 2.3.30. $\boldsymbol{K_{0z}}$ for this condition when probing these two columns of reflections will intersect the dispersion surface in 4 positions (4 tie points), simply explained as two vertical lines passing through a sphere.

$$K_{0zn} = \pm\left\{K_0^2 - \frac{K_{0x}^2 + K_{Gx}^2}{2} \pm \left[\left(\frac{K_{0x}^2 + K_{Gx}^2}{2}\right)^2 + k^4\chi_G\chi_{-g}\right]^{1/2}\right\}^{1/2} \qquad 2.4.45$$

And for each intersection (n) there will be an amplitude ratio given by

$$\frac{E_{GR,n}}{E_{0R,n}} = c_n = \frac{1}{k^2\chi_{0G}}\left\{\frac{K_{0x}^2 + K_{Gx}^2}{2} \pm \left[\left(\frac{K_{0x}^2 + K_{Gx}^2}{2}\right)^2 + k^4\chi_G\chi_{-G}\right]\right\} \qquad 2.4.46$$

This ratio is the same for the transmitted and reflected waves, and $K_{0x}=|\boldsymbol{k}_{//}|$ and $K_{Gx}=|\boldsymbol{k}_{//}+\boldsymbol{G}|$.

Therefore now we have to describe our electric field vector in each lamella as

$$E = \begin{bmatrix} E_{0T1} \\ E_{0T,2} \\ E_{0R,1} \\ E_{0R,2} \end{bmatrix} \qquad 2.4.47$$

The phase matrix, equation 2.4.29, now becomes extended to

$$\Phi_j = \begin{bmatrix} e^{iK_{0z1}\Delta z} & 0 & 0 & 0 \\ 0 & e^{iK_{0z2}\Delta z} & 0 & 0 \\ 0 & 0 & e^{iK_{0z1}\Delta z} & 0 \\ 0 & 0 & 0 & e^{iK_{0z1}\Delta z} \end{bmatrix} \qquad 2.4.48$$

and the reflection matrix at each interface becomes

$$R_j = \left(C_{j-1}\right)^{-1} C_j, \qquad R_1 = C_1, \qquad R_{N+1} = \left[C_N\right]^{-1} \qquad 2.4.49$$

where

$$C_j = \begin{bmatrix} 1 & 1 & 1 & 1 \\ K_{0z1} & K_{0z2} & -K_{0z1} & -K_{0z2} \\ c_1 & c_2 & c_1 & c_2 \\ K_{0z1}c_1 & K_{0z2}c_1 & -K_{0z1}c_1 & -K_{0z2}c_2 \end{bmatrix} \qquad 2.4.50$$

The equivalent equation for the scattering from inclined planes to that of the surface parallel planes, i.e. equation 2.4.36, is

$$\begin{bmatrix} E_0 + E_{0R} \\ k_{0z}\left(E_0 - E_{0R}\right) \\ E_{GR} \\ k_{GRz}E_{GR} \end{bmatrix} = M \begin{bmatrix} E_{0T} \\ k_{0z}\left(E_{0T}\right) \\ E_{GT} \\ k_{GTz}E_{GT} \end{bmatrix} \qquad 2.4.51$$

where k_{0z} is the vertical component of the incident wave-vector and k_{GTz} is the transmitted diffracted wave-vector that emerges from the underside of the sample. The amplitudes $\boldsymbol{E}_0$, $\boldsymbol{E}_{0R}$, $\boldsymbol{E}_{0T}$, $\boldsymbol{E}_{GR}$, $\boldsymbol{E}_{GT}$, are those relating to the incident wave, the specularly reflected wave, the transmitted wave emerging from the underside of the sample, the diffracted wave and the associated transmitted wave emerging from the underside of the sample, figure 2.4.9. Hence the amplitude components, and therefore the intensities, can be evaluated for a given k_{0z} component that relates directly to the incident angle, $\omega = \sin^{-1}(\lambda k_{0z})$, and the in-plane vector $\boldsymbol{G}$.

The electron density is then evaluated for each lamella. This density follows the distribution of electrons through the structure and will not abruptly stop at the top surface, or have null regions at an interface, but will extend and fall in value at a distance from the atoms. This approach has the advantage that any atom displacements can be modeled, although the present model assumes a laterally periodic structure. A displaced surface atom for example, will alter the scattering with the most marked effect occurring between the Bragg peaks. Similarly because it is a true multi-beam method, a reflection on another column of reciprocal lattice points will take intensity out of the profile of the one investigated when the incident beam satisfies the diffracting condition for that reflection. This effect is similar to the Aufhellung effect (when the intensity is reduced) or Umweganregung (when the intensity is enhanced). These multiple beam effects are usually associated with combining diffraction peaks, and can be used to probe the asymmetric distribution of electron density giving rise to nominally "forbidden reflections," Renninger (1937), and obtaining extra structural information, see for example Weckert and Hümmer (1997). These effects in general manifest themselves as sharp reductions or gains in intensity that can be found with careful settings of the sample in the azimuthal direction.

The small intensity modulation between Bragg peaks, discussed above, obviously requires correct modeling between the Bragg peaks and at the Bragg peaks. This intensity modulation has been observed experimentally with a laboratory sealed source and a high-resolution "beam-selection" diffractometer (described in section 4.3.4), figure 2.4.10.

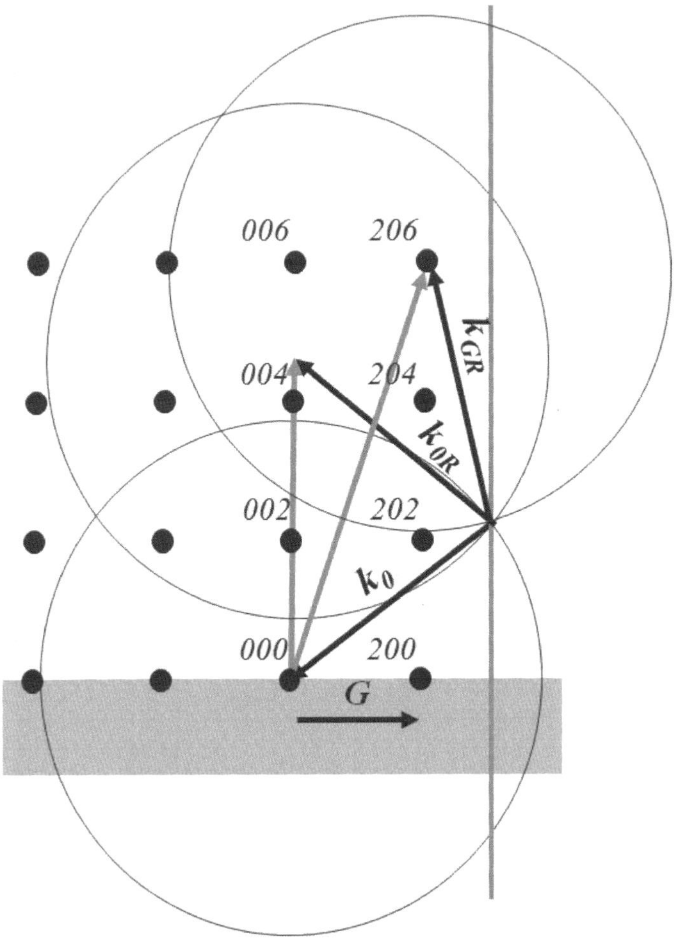

Figure 2.4.9. The construction of the external wave-vectors for evaluating the scattering along the column associated with planes parallel to the surface (*00l*) and from a parallel column containing planes inclined to the surface, (*20l*) separated by the vector *G*. Each scattering vector *S* (e.g. grey arrows) is associated with a spherical dispersion surface of radius $1/\lambda$, in this example the *206* reflection and a position between the *004* and *006* reflections. The incident wave k_0, in this construct has the normal reflected wave k_{0R} as in Figure 2.4.8, and when the dispersion surfaces intersect some of the energy is scattered towards the *206* reflection via k_{GR} causing an influence on the reflected wave amplitude associated with k_{0R}.

The precise agreement with conventional theory close to Bragg reflections and the fact that this approach can generate features found in practice all add to the support that this is a more complete dynamical theory. This approach takes dynamical theory closer to being able to reliably model the scattering from materials on the scale of the atom and could lead to a method of including all forms of distortions and defects.

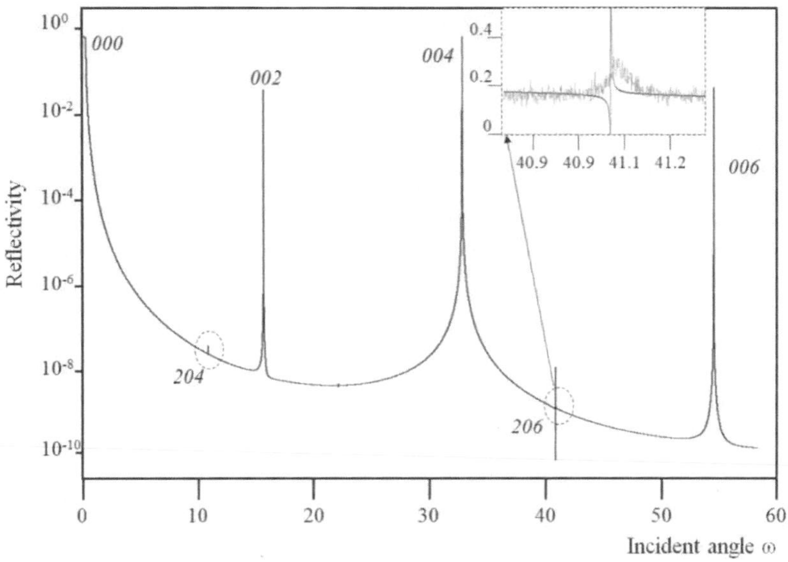

Figure 2.4.10. When two columns are calculated the influence of multiple-diffraction, such that some of the intensity is shared between the intersection points. In this case the *204* and *206* reflections create a modulation of the intensity along the column of intensity at the corresponding incident angles. The insert shows the calculated profile and the actual measured intensity modification at the incident beam angle for the *206* reflection. The measured intensity is the sum of 21 profiles of 500 data points at 20 s per point and displayed in counts s^{-1}, so the dip is lost through noise and the upward inflection dominates because it is 7× more intense than the downward inflection.

Until now we have considered the scattering associated with the incident beam direction, now we shall derive the direction of the scattered beam, i.e. where the detector needs to be placed to capture the information. This leads to the modeling of imperfect structures, since one single incident beam direction can result in a distribution of scattering angles and we move from single profiles to 2-dimensional scattering; diffraction space maps and reciprocal space maps.

2.5. Diffraction and reciprocal space maps

So far we have examined the scattering resulting from an incident wave creating a scattered wave but have not been concerned about the direction of the scattered wave, except in the derivation of the more exact deviation parameter in section 2.4.1. Suppose that we now only collect photons traveling along certain directions, i.e. our detector window is very narrow, then knowledge of the scattered wave direction is vital. We differentiate diffraction space maps from reciprocal space maps, in that the former refers to a distribution of scattering in angular space, these relate directly to the coordinates of the sample and detector angles, whereas a reciprocal space map is a transformation of these angles to relate to the inverse of the length scales. The latter makes the interpretation simpler (provided the incident beam is monochromatic), whereas the former is more general. Both will be used throughout this book depending on the analysis.

Consider figure 2.5.1 where we can once again use our knowledge of the boundary conditions. Since the parallel component of the scattering vector is constant across boundaries then it is constant with depth for any angle. Again we can write the scattering condition:

$$K_H = K_0 + S \qquad 2.5.1$$

Since the parallel components are invariant

$$K_H \cos(2\theta_i - \omega_i) = K_0 \cos\omega_i + \frac{2\sin\theta_i}{\lambda}(1+\chi_0)^{1/2}\sin\varphi = K_0\cos\omega_i + \frac{1}{d}\sin\varphi \qquad 2.5.2$$

The subscript i refers to the equivalent directions due to the refractive index effects. Similarly for the external parallel components that equate to those above we have:

$$k\cos(2\theta - \omega) = k\cos\omega + \frac{2\sin\theta_B}{\lambda}\sin\varphi \qquad 2.5.3$$

The refractive index effects cancel, which is not surprising because now the external scattering angle is associated with a medium with a refractive index that cannot have imaginary components (i.e. assumed to be non-absorbing). Also the change in wavelength and Bragg angle within the crystal compared with the external values must compensate for each other since the interplanar spacing *d* is independent of the measurement method. The scattered wave direction is therefore

$$2\theta = \cos^{-1}\{\cos\omega \mp 2\sin\theta_B \sin\varphi\} + \omega = \cos^{-1}\left\{\cos\omega \mp \frac{\lambda}{d}\sin\varphi\right\} + \omega \qquad 2.5.4$$

where $k = 1/\lambda$.

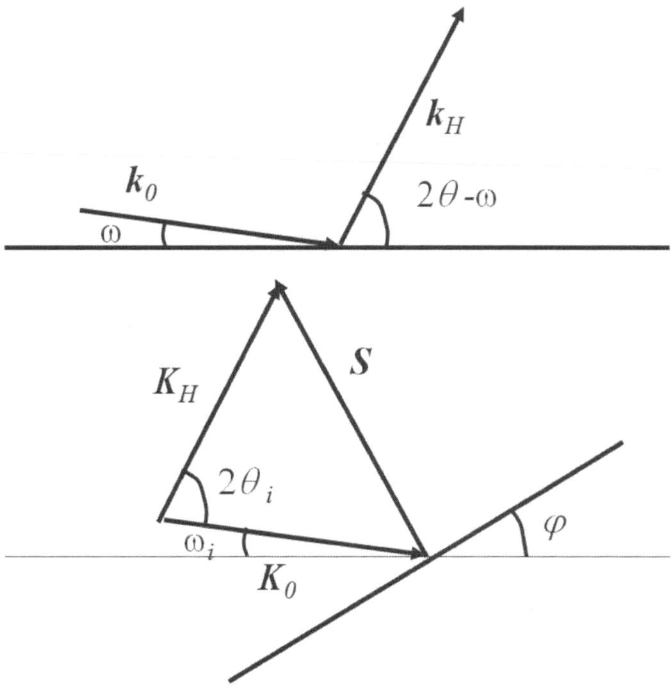

Figure 2.5.1. The internal and external wave-vectors and the angles for deriving the various relationships to satisfy the boundary conditions.

Hence now we have a more complete picture of the scattering process. As far as our experiment is concerned we have to ensure that the incident wave reaches the surface at the required angle and that the detector is placed at the correct angle to receive the scattered photon.

2.6. Transmission geometry

This will be covered very briefly here for completeness although much has been discussed previously on this configuration because of its importance in transmission topography, Authier (1996), Tanner (1996). The theoretical basis is no different from that described previously, section 2.3, except that the boundary conditions are different. In transmission the surface normal in figure 2.3.5 will be, be for example, rotated 90^0 and it will intersect both branches of the dispersion surface. The excitation of additional wave-fields creates fascinating interference effects that have been used to vindicate the assumptions in dynamical theory. The most extensive use of transmission geometry (or often termed Laue geometry) has been for the interpretation of high-resolution topography, e.g. Si wafers, for imaging defects created during processing or growth. In general simulation of the contrast is of less concern to us, whereas direct interpretation of the contrast is of interest, therefore we will take a very pragmatic approach in this section. Initially though, the intensity profile for dynamical diffraction in transmission-mode will be presented so that it can be compared with the reflection-mode and kinematical theory.

The amplitude, X, for diffraction in transmission-mode, to be compared with equation 2.3.39, is based on the derivation given by Zachariasen (1945), and is given by:

$$X^2 = E^2 e^{-z\mu_0} \frac{\{\sin(av) + \sinh(aw)\}^2}{\left|AE + B^2\right|} \qquad 2.6.1$$

The parameters above are given by:

$$a = \frac{\pi z}{\lambda \sin \omega}$$

$$v = \mathrm{Re}(AE + B^2)^{0.5}$$

$$w = \mathrm{Im}(AE + B^2)^{0.5}$$

$$\mu_0 = \frac{2\pi}{\lambda}\mathrm{Im}\left(\frac{r_e \lambda^2 F_{000}}{\pi V}\right)$$

$$z = 0.5\left(\frac{1}{\sin \omega} + \frac{1}{\sin(2\theta - \omega)}\right)t$$

$$A = C\chi_{-H}\frac{|\gamma_H|}{\gamma_0}$$

$$E = -C\chi_H$$

$$B = 0.5\left\{\chi_0\left(1 + \frac{|\gamma_H|}{\gamma_0}\right) - 2\beta_H\right\}$$

where the deviation parameter, β_H, is as described in equation 2.4.6. 'Re' and 'Im' refer to the real and imaginary components of a complex function. The assumption in these equations is that the absorption coefficient has one value; the situation is much more complicated than this and is discussed later in this section.

If we consider the case of a flat parallel-sided crystal, then the change in the profile shape can be seen as the crystal thickness t is varied. Clearly it is not a simple increasing or decreasing peak height or peak width, but is considerably more complicated, figure 2.6.1. It is clear that interpreting intensity contrast, as in topography, is rather straightforward if the crystal is very thin or very thick (or the defect feature is a large or a small distance from the exit surface), but if it is of intermediate thickness (or a defect is an intermediate distance from the exit surface) it is more difficult to predict.

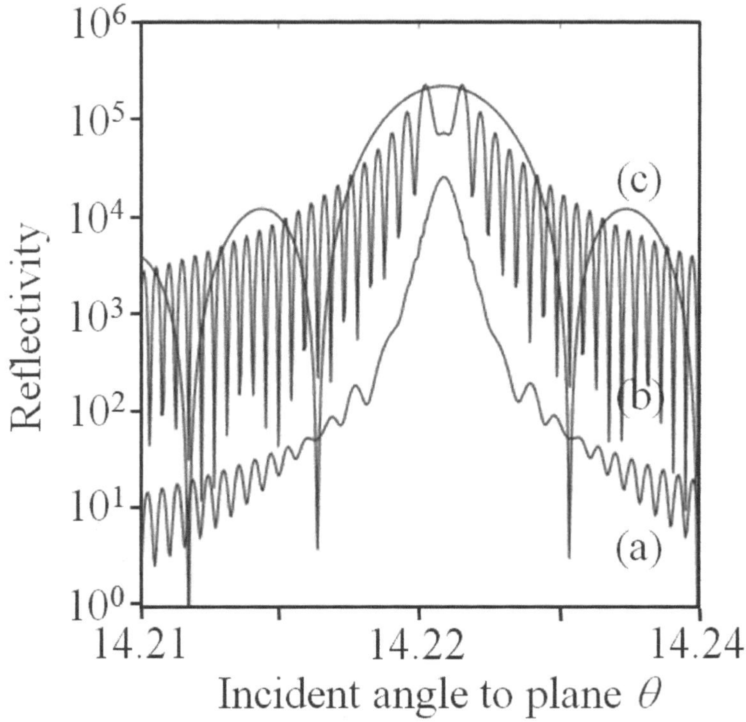

Figure 2.6.1. The simulation of dynamical theory in transmission-mode for various thickness of a parallel-sided plate, from crystal planes parallel to the surface normal. (a) 100 μm, (b) 10 μm and (c) 1 μm, Cu $K\alpha_1$ *111* reflection from Si. The absorption coefficient is based on equation 2.7.12, where in reality there are absorption coefficients associated with each branch of the dispersion surface: equation 2.6.3.

From the Takagi theory, section 2.4.2, it was clear that the displacement field will be influenced but is still maintained close to small distortions. Therefore the path of the X-rays will be deviated, changing the distribution of intensity emerging from the sample. If the distortion is too large as in the case close to the core of a dislocation then the wave is uncoupled and will be scattered out as described in the later section 2.8.2. Of course the strain close to a defect and the associated crystal plane curvature diminishes with distance from the defect centre. Because of the high sensitivity of X-rays scattering to small strains and small curvatures the image is large (~ microns), and therefore is visible on a direct image with no magnification.

We shall consider a few cases to give an outline of how the contrast is created in samples of various thicknesses. The important definition that determines the likely contrast is the product of the thickness, t and the linear absorption coefficient, μ, equation 2.7.13, since this influences the X-ray penetration depth. The basis of X-ray topography, which is our concern in this section, is that the scattered image has a finite size and we wish to examine the contrast within this scattered diffraction spot. The contrast in this spot image approximately relates to the region of the sample that is probed by the incident X-rays.

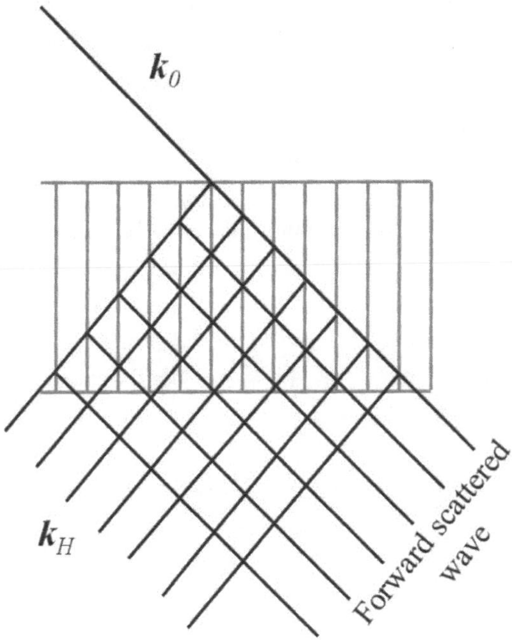

Figure 2.6.2. The complex wave-field created in transmission geometry (cf. Figure 2.3.1). When the absorption increases this Borrmann triangle reduces and transmission is confined to the direction parallel to the scattering planes.

Suppose the sample is very thin or weakly absorbing, $\mu t \sim 1$. From the dynamical scattering theories discussed above we can see that the intrinsic scattering width for a perfect crystal is very small. If we now probe our sample with an incident beam that has a divergence larger than this, we have to visualise two regions; those that are perfect crystal and those that are not. The perfect region will scatter at the correct scattering angle, within a very narrow angular range, i.e. it will extract a very small proportion of the incident X-rays for scattering. However close to defects there is a range of angles that is presented to the divergent forward-refracted wave and thus there are many more possibilities for scattering outside the intrinsic scattering width of the perfect regions. The intensity from the defects is therefore enhanced with respect to the perfect regions, provided that the absorption is not too high. X-ray topographic images of defects closer to the exit surface of the sample are better defined than those close to the entrance surface. This occurs because the proportion of the Borrman triangle, figure 2.6.2 that creates the relative contrast is diminished and the scattered waves from the defects suffer less absorption.

As the sample thickness is increased such that $\mu t \sim 10$, the absorption appears to be so large that no X-rays will pass through the sample. However this average linear absorption coefficient is only a factor used in the kinematical theory, section 2.7.2 and applied in the optical theory, section 2.10. The dynamical model on the other hand makes no such assumption and in fact for the transmission case four tie points can be excited (both branches of the dispersion surface for both polarisation states) and each will have their own characteristic absorption. From section 2.3, we established that the imaginary component of the wave-vector susceptibility relates directly to the absorption and we have derived an expression relating the linear absorption coefficient, equation 2.10.4. If we compare this with equation 2.10.2 then

$$\mu = -4\pi \mathrm{Im}(\chi_0 K_{\mathrm{m}}) \qquad 2.6.2$$

where $\mathrm{Im}(\chi_0 K_{\mathrm{m}})$ represents the imaginary component of $\chi_0 \boldsymbol{K}_{\mathrm{m}}$, $|\boldsymbol{K}_{\mathrm{m}}|=1/\lambda$. The attenuation of interest is in the direction of energy flow into the crystal, whereas equation 2.6.2 represents the influence normal to the surface, figure 2.3.2. We can see from figure 2.3.5, if several points are excited on the dispersion surface (i.e. on branch 1 and branch 2) then the internal wave-vectors will have different components along the surface normal and consequently different linear absorption coefficients. Similarly the polarisation factor creates a further offset of the dispersion surfaces and therefore the four existing waves have their own linear absorption coefficients. Bonse (1964) gives the full expression:

$$\mu_w = \mu \frac{\left\{1 + |X|^2 + 2\left[\frac{C\,\mathrm{Re}({}_w\chi_H X)}{\mathrm{Im}({}_w\chi_0)} + \mathrm{Im}({}_w\chi_H X)\right]\right\}}{\left\{1 + |X|^4 + 2|X|^2 \cos 2\theta\right\}^{1/2}} \qquad 2.6.3$$

where the subscript *w* is the index for the appropriate wave-field and the other symbols have been defined previously. What is found from using this formula is that the linear absorption coefficient for different wave-fields varies dramatically and thick crystals are still able to allow energy to flow. The physical explanation is that the wave amplitude oscillates with the periodicity of the lattice such that the nodes correspond to the atomic sites and the absorption is low. There is of course a similar wave that has its anti-nodes at the atomic sites and is soon absorbed in a thick crystal. The energy flow for this low absorbing wave through the crystal is parallel to the atomic planes.

Although this anomalously transmitted wave is rather weak it is a very powerful probe for imaging defects. Clearly any defect close to the entrance surface will scatter this wave and the intensity will be lost, giving rise to missing intensity in the image. If the degree of perfection is good until close to the exit surface then the curvature of the planes around a defect will "steer" the energy flow. This will give rise to enhanced intensity when it is directed more to the scattered wave direction and reduced intensity when directed more towards the incident beam intensity, giving rise to black-white contrast. If a defect distortion occurs deep inside a crystal that is at an intermediate distance from the exit surface, then the anomalously transmitted wave on scattering will not be simply "steered" or absorbed but be of an intermediate type. This intermediate type corresponds to scattering of intermediate thicknesses, figure 2.6.1b, making the contrast very sensitive to it position. This is wonderful for testing the validity of dynamical theory, but can add complexity to the interpretation. In general these principles can be used in the interpretation of most X-ray topographs taken in transmission geometry. Although a deep theoretical understanding is necessary for modeling images and greater depth is required for interpreting subtle dynamical scattering features, this is not our main concern here.

Until now the theories have been based on perfect materials, however the influence of defects, e.g. misfit dislocations, threading dislocations, mosaic blocks and other distortions, do influence the scattering to varying degrees. The dynamical theories given above assume that the material is nearly perfect, however if the X-ray wave has only limited coherence due to inhomogeneities in the material (i.e. having limited correlation lengths) then the scattering theory can be simplified through assumptions.

2.7. Approximate theory: the kinematical approach

Dynamical theory is based on a wave field approach: that is the incoming wave is attenuated due to scattering and also undergoes interference with the scattered wave. In the kinematical theory the incident wave is only assumed to be attenuated by normal photoelectric absorption, and there is no loss through scattering. Therefore this theory is only strictly valid when the scattering is weak. It is generally perfectly adequate to model the scattering from defects, ideally mosaic samples or highly distorted material. Another aspect that is ignored in the kinematical model is the refractive index. Of course care must be applied with the kinematical theory in that the total scattered intensity can exceed the incident wave intensity if the accumulative scattering is too great.

2.7.1. Comparison between dynamical and kinematical models of diffraction

We will firstly consider the differences between the theories by removing some of the conditions associated with the analysis above. Consider for example the Takagi equations 2.4.14 and 2.4.15. The interference of the scattered wave with the forward refracted wave is a consequence of the magnitude of χ_{-H}, hence if we assume that the scattering strength is zero from the underside of the scattering planes then the real part of χ_{-H} is zero. This is sometimes referred to as primary extinction. The imaginary component is assumed to exist to account for normal photoelectric absorption. The refractive index arises from the finite magnitude of χ_0, therefore the real part of χ_0 is assumed to be zero. We can of course also remove the absorption effects, but in general this can easily be accommodated in the kinematical approximation from tabulated absorption coefficients. Therefore the equation 2.4.17, reproduce here below, can be used in the kinematical approximation,

$$X(Z,\omega)=\frac{SX(z,\omega)+i\{E+BX(z,\omega)\}\tan(GS[z-Z])}{S-i\{AX(z,\omega)+B\}\tan(GS[z-Z])} \qquad 2.7.1$$

although A is purely imaginary and B becomes [the imaginary part of $B + 2\beta_H$]. Also the coupling of the wave fields between layers should be removed, just as in the case of a disrupted interface that is described in the section 2.8.2. Some authors have used a hybrid theory by including the scattering from the substrate as dynamical and from the layers as kinematical. The influence of the refractive index will shift the peak positions of the dynamical and kinematical models. The differences between the dynamical and kinematical theories can be quite crucial and misleading in the analysis of multilayer structures. However for the analysis of weak scattering from small crystallites the kinematical theory can sometimes be perfectly adequate.

If we include the assumptions of the kinematical approximation in equation 2.7.1 and take the assumption of an infinite crystal such that $\tan GS[z-Z] = i$, then

$$X(\omega) \sim \frac{-E}{2B} = \frac{C\chi_H}{\mathrm{Im}\{\chi_0\}\left(1+\frac{|\gamma_H|}{\gamma_0}\right) - 2\beta_H} \qquad 2.7.2$$

We can see from this that it is only the absorption of the incident wave that prevents the amplitude becoming infinite at the Bragg angle (i.e. when $\beta_H = 0$). The intensity is then simply related to the square of the structure factor for the scattering position concerned. We can consider the kinematical model for an infinite crystal lattice to create a scattering pattern that has delta functions at the Bragg angles with a magnitude related in some way to the square of the structure factor for the reflection concerned.

Similarly the transmission case, equation 2.6.1, can be considered in the limit. It is not sensible to take this to the infinite crystal thickness, since no beam will be transmitted, however if χ_{-H} and the imaginary part of χ_0 tend to zero and the absorption is small, then sinh(a.Im(B)) tends to zero and we have:

$$X^2 \sim \frac{E^2 \sin^2(av)}{B^2} \sim \chi_H^2 \left(\frac{\sin(aB)}{B^2}\right)^2 \qquad 2.7.3$$

If we were to take the intermediate thickness case in the reflection mode, that would similarly result in a sin(B)/B function: this becomes clear in the following simulations.

The information concerning the peak shape and true intensity is usually lost when applying the kinematical model. We can though include some form of shape function and impose this on the magnitude that we derive from the structure factors. As described above the dynamical and kinematical theories produce similar results for weak scattering, figure 2.7.1. However the influence of the substrate peak influences the fringing significantly. If a layer is considered as scattering in isolation then a more direct comparison can be made. The dynamical simulation is broader by 3.7% for a 0.5 μm layer compared with kinematic theory, and this reduces to 2.3% for a 0.4 μm layer and 0.96% for a 0.3 μm layer. In general the comparison of intensities is relative and the reference is lost in kinematic theory (the reflectivity can be greater than unity, Figure 2.7.1), but interpretation of the peak shape to obtain a correlation length, crystallite size or layer thickness can start to introduce significant errors, significantly above these example values.

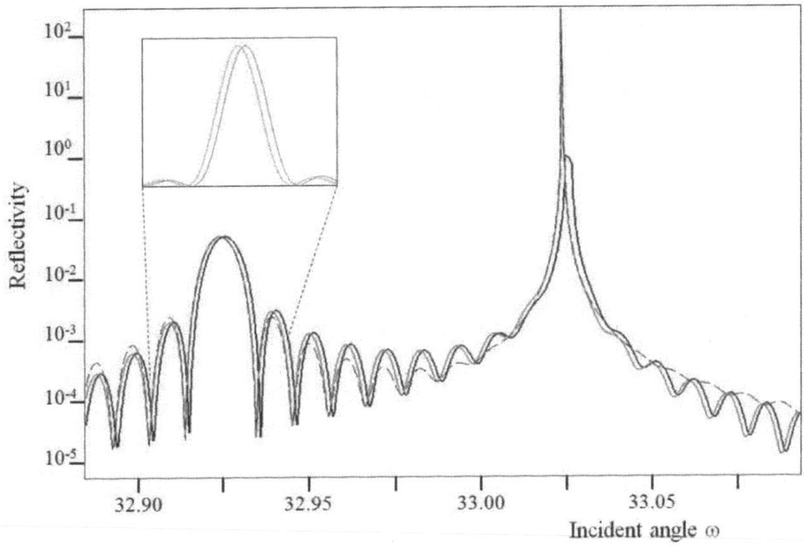

Figure 2.7.1. The calculated scattering for a AlAs layer (0.5 μm) on GaAs *(001)* substrate structure for the *004* reflection, using the dynamical model (black), the kinematic model assuming the interfaces are coupled (grey) and with the interfaces uncoupled (black-dash). The offset is due to the lack of refractive index correction in the kinematical case. If the substrate amplitude is uncoupled from that in the layer the fringes differ considerably and there is a further small offset in the layer peak position (inset plotted on a linear intensity scale).

2.7.2. The important derivations of the kinematical theory

Since we have established that this theory is based on weak scattering and crystals of limited extent in order to obtain realistic intensities, we shall lay out the appropriate assumptions. The intensity in a diffraction pattern from an infinite crystal is concentrated at the Bragg points and has a magnitude that is proportional to the square of the structure factors

$$I(S) = KF_S^* F_S \delta(S - r) \qquad 2.7.4$$

K is a constant of proportionality. Clearly this is significantly simpler than the dynamical model and can be a very powerful approximation for interpreting weak scattering.

Of course weak scattering is usually associated with small crystals and therefore we have to modify our model. Consider a sample with N scattering planes separated by a distance d, figure 2.7.2. We can now write down the phase difference between scattering planes at scattering angles ε from the Bragg angle for deriving the path difference in this figure

$$\Phi = \frac{2\pi d}{\lambda}\{\sin(\theta+\varepsilon)+\sin(\theta+\varepsilon)\} \sim \frac{2\pi}{\lambda}\{\lambda+2d\varepsilon\cos\theta\} \equiv \frac{4\pi d}{\lambda}\{\varepsilon\cos\theta\} \qquad 2.7.5$$

for small deviations from the Bragg angle.

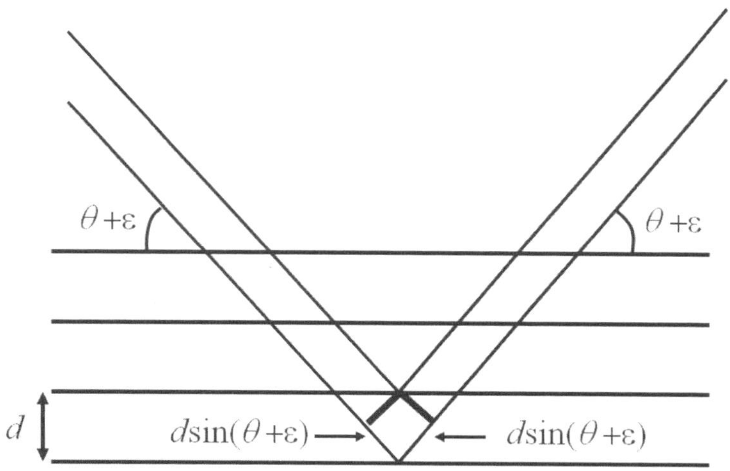

Figure 2.7.2. The influence on the path length and hence phase relationship for small deviations from the optimum scattering condition.

Suppose now we are interested in the shape of the profile, then this can be obtained by using the equations from optical theory for a multiple-slit. The profile is composed of a combination of the scattering from a single slit (interplanar spacing) and the interference from the number of contributing slits (N planes). For the single slit, there will be many identical contributing amplitudes across the slit with the same phase difference between each. In an amplitude plot this will appear as vectors of length 'a' at an angle to each other defined by the phase angle; in the limit this will result in an arc. The resultant is given by the start and end points and can be represented as a chord. The scalar sum of all these amplitudes a around the arc will equate to the incident amplitude A_0, whereas the chord joining the start and end points will be the resultant vector, A. Thus the ratio of A/A_0 will be the ratio of the chord to the arc of the circle, i.e.

$$\frac{A}{A_0} = \frac{2R\sin\beta}{R2\beta} \qquad 2.7.6$$

Where R is the radius of the circle and 2β is the angle subtended at the circle centre. β is therefore half the resultant phase difference. This phase difference is small because it relates to the interplanar spacing, d, so that $2\beta = 2\pi d/\lambda$, and hence $\sin\beta/\beta \sim 1$. However the combination of multiple slits will be the product of the single slit (separation between planes) and the combination of all the contributing planes. The resultant amplitude from many equally spaced parallel planes is the sum of all the individual amplitudes with a phase term, which is $\exp(-in\Phi)$ for the nth plane, where Φ is the phase change from plane to plane. This is equivalent to the multiple slit or grating in optics and gives the second term in the amplitude here:

$$A_N = A_0 \frac{\sin\beta}{\beta} \frac{\sin\frac{N\Phi}{2}}{\sin\frac{\Phi}{2}} \sim A_0 \frac{\sin\frac{N\Phi}{2}}{\sin\frac{\Phi}{2}} \qquad 2.7.7$$

The form of the intensity, assuming the kinematical approximation $I=|A_N|^2$ is represented in figure 2.7.3. It is now possible to relate the depth over which scattering takes place and the thickness of the layer to the width of the scattered profile. The interatomic spacing gives the phase angle Φ, from equation 2.7.5, and the layer thickness is represented by N

interatomic spacings. Now since Φ is small, $\sin\Phi \sim \Phi$, and the result is still very close to a sine cardinal, i.e. sinA/A.

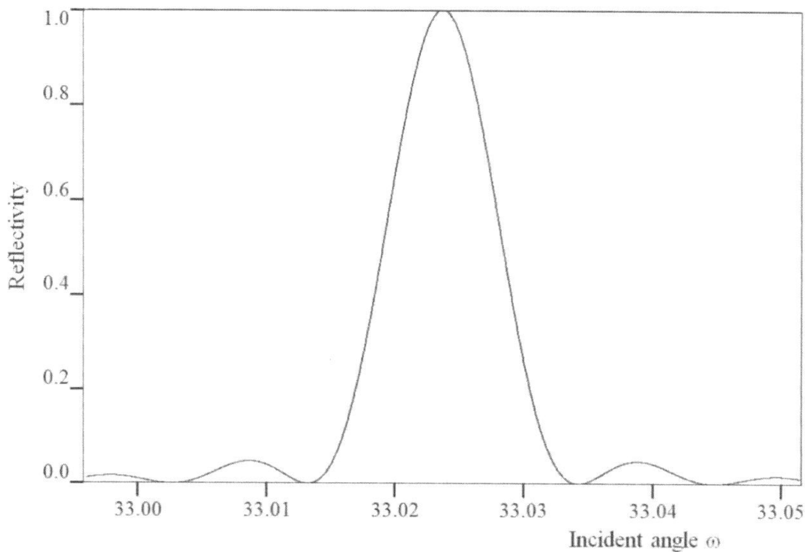

Figure 2.7.3. The shape of the interference function results from the finite size of the scattering region, in this case a 0.5 μm layer of GaAs.

We can now make a further approximation by inspection of figure 2.7.3. We can deduce from equation 2.7.7 that when $N\Phi/2 = \pi$ the amplitude becomes zero, and this refers to the first minimum above the peak angle position in figure 2.7.3. At deviations further from this position the intensity of the profile becomes small, therefore we can write from equation 2.7.5:

$$\left(N\frac{\Phi}{2}\right)_{A=0} = \frac{2\pi}{\lambda}\{Nd\varepsilon_{A=0}\cos\theta\} = \pi \qquad 2.7.8$$

The crystal dimension, L, or the region of coherent scattering is given by Nd, consequently we have

$$\varepsilon_{A=0} \sim \frac{\lambda}{2L\cos\theta} \qquad 2.7.9$$

The full width at half maximum intensity is related by a factor of 0.434 to the separation of the first upper and lower zero points, i.e. $2\varepsilon_{A=0}$ defined by the shape of this sine cardinal (sinc function) profile. If we express this relationship on the scattering angle (2θ) scale, then the depth of coherent scattering can be determined directly from this profile width from

$$L = \frac{0.868\lambda}{\varepsilon_{FWHM(2\theta)} \cos\theta} \qquad 2.7.10$$

This is the Scherrer equation, Scherrer (1918), for a parallel sided layer. For different crystallite shape this dimension can be scaled accordingly. If the crystallites are spherical, then the average dimension being probed is 0.8 of the diameter, etc.

So far we have ignored the aspects of absorption due to photoelectric effects and other non-scattering processes. We can consider the absorption to be related to the total path length of the X-rays in the sample. We will not consider the detailed processes here but rather take a pragmatic approach and consider the incident beam to lose energy (statistical loss of photon numbers) as it passes through the sample. If we have a small depth of material dt, with a density ρ that changes the incident beam intensity from I to $I–dI$, we can assume that the loss of intensity is a proportional reduction and that the loss is proportional to dt. Hence we have:

$$dI = -\mu I dt \qquad 2.7.11$$

where μ is the total linear absorption coefficient. The ratio (μ/ρ), which we will term the mass absorption coefficient, is found to be roughly independent of the form of the material. Also to a good approximation the total losses are equal to the sum of the individual losses within the sample, i.e.

$$\frac{\mu}{\rho} = \sum_i \left(\frac{\mu}{\rho}\right)_i g_i \qquad 2.7.12$$

where g_i is the mass fraction of atom i. From the mass absorption coefficient and knowledge of the density of the sample we can derive the total linear absorption coefficient. Integrating equation 2.7.11 and defining the incident and transmitted intensity as I_0 and I, we have

$$I = I_0 \exp(-\mu t) \qquad 2.7.13$$

The mass absorption coefficients have a complicated form but have been determined and tabulated in detail in International Tables for Crystallography (1962) for example, for each atom type for a large spread in wavelengths. These mass absorption coefficients are proportional to the cube of the wavelength, Bragg and Pierce (1914), except close to resonant absorption, section 2.3.

The calculation of the total attenuation in multi-layers must include these effects by including this attenuation through the whole structure above the scattering material. In periodic superlattices we can approximate this to

$$I = I_0 \prod_i \exp\left\{-\mu_i t_i \left(\frac{1}{\sin(2\theta-\omega)} + \frac{1}{\sin\omega}\right)\right\} \qquad 2.7.14$$

Fewster (1986), where i is the layer index.

2.7.3. Adding complexity to the kinematical approach

This relatively simple theory can be very useful for giving an indication of the scattering from complex structures. This following example indicates how complexity can be built up using kinematical theory.

Suppose we have a series of periodically placed features on the surface that scatter and their separation is less than the coherence length of the probing X-rays, figure 2.7.4, then we can make use of the description for the scattering profile similar to that given in equation 2.7.7. In this case we are interested in two lateral length scales. The first is associated with the shape of the hillocks, and we shall for the sake of simplicity assume they are rectangular and there are N_B scattering centres within each. The second length scale is the period of the 'grating', which is composed of N_G hillocks. For the simple case of scattering from planes parallel to the surface the phase is given by:

$$\Phi = \frac{2\pi\delta}{\lambda}\{\cos(2\theta - \omega) - \cos\omega\} \qquad 2.7.15$$

The parameter δ is some small dimension and if l is the lateral dimension of the hillock, then $l = N_B\delta$. The periodicity of the grating, Λ, is given by $N_G\delta$ and therefore making use of equation 2.7.15 for the two phase terms we then have the amplitude of the scattering given by:

$$A = a\frac{\sin N_G\Phi_G}{\sin\Phi_G}\frac{\sin N_B\Phi_B}{N_B\Phi_B} \qquad 2.7.16$$

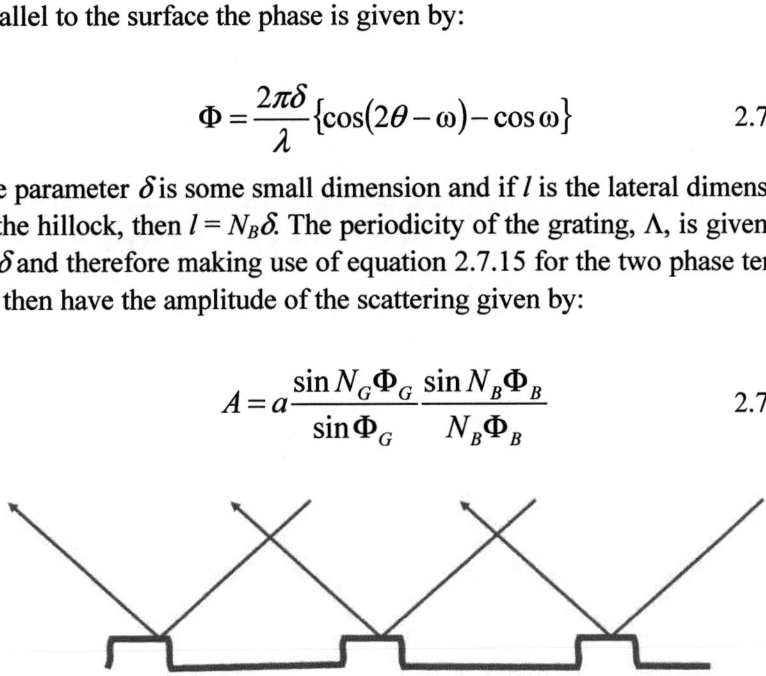

Figure 2.7.4. A laterally periodic structure will create a further modulation in the scattering pattern.

The periodic contribution from the grating G is modulated by the smoother interference function of the hillocks B. The dramatic effects are presented in figure 2.7.5, where this periodicity is superimposed on a simulation of a profile based on dynamical theory. Of course here we are working with something which is perfectly periodic, yet we can include dispersion of the grating periodicity and hillock size by including several contributions to the amplitude term with different lengths:

$$\overline{A} = A_1 + A_2 + A_3 + A_4 + \ldots\ldots \qquad 2.7.17$$

It is obvious that we are now able to build up a model of our sample and with the combination of the dynamical and kinematical models to predict the scattering that we might expect.

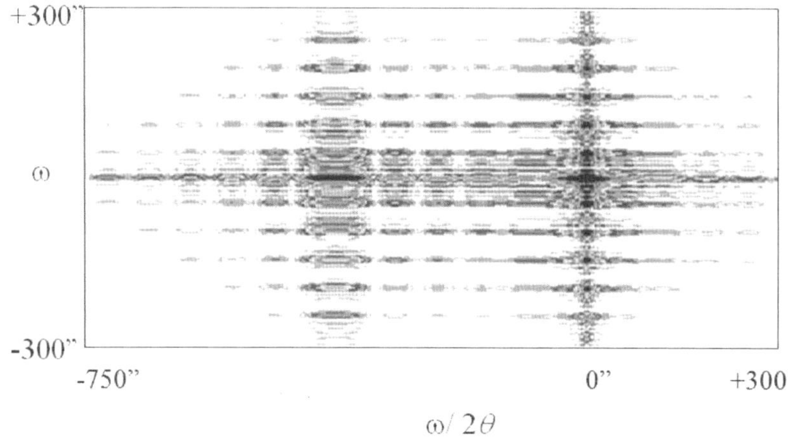

Figure 2.7.5. The complex pattern from a laterally periodic structure. The simulation of 0.3μm thick and 500nm wide AlAs hillocks repeated laterally 5 times with a period of 600nm on a GaAs substrate creates a complex pattern. This calculation is based on equation 2.7.16 and dynamical theory.

It is important to realize that this approach of generating a pseudo-dynamical profile and using the local amplitudes to create a kinematical modification is in many ways just an approximation, and strictly all possible waves that can exist should be included, making the whole process and pattern generation very complex, Galihanou et al (1993).

For example the incident wave could pass through a hillock and scatter from the substrate, also the strain distribution will exist due to the free standing hillocks, etc. This simplistic approach presented in this section will though yield some of the length scales, but is unlikely to explain all the observed features.

The next section considers various forms of defects and how they can influence the scattering pattern. This is where the dynamical and kinematical theories are sometimes combined, although often the scattering is weak and dispersed and the kinematical theory alone is preferred, and of course simpler.

2.8. Scattering theory for structures with defects

In this section we shall consider dilute distributions of defects, point defects, and then consider the influence of relaxation with arrays of dislocations and finally mosaic structures. This will then lead on naturally to the scattering from polycrystalline aggregates and powders.

2.8.1. Scattering by small defects

Materials are likely to contain point defects (substitutional impurities, interstitial atoms and vacancies) and small dislocation loops, threading dislocations, and small precipitates, etc. These features will influence the scattering pattern and will be considered in this section. The scattering is generally very weak and can be treated kinematically, although some authors have extended this to dynamical theory, (Kato, 1980; Olekhnovich and Olekhnovich, 1981; Holý, 1982; Khrupa, 1992; Pavlov and Punegov, 1998), but since we wish to give a general understanding we will concentrate on the kinematical approximation. The differences in the two theories are in the very fine detail that is usually obscured by the Bragg scattering, although of course it is very important to establish the validity of the kinematical theory approach and these articles largely confirm this.

Suppose we have a defect, this could be an interstitial or a vacancy for example; then the associated scattering will be related to the structure factor for the defect multiplied by the number per unit volume. If the defects scatter independently (i.e. there is no phase relationship between them, because they are randomly distributed) then the structure factor of the ith defect is given by:

$$F_i = \sum_{d=1}^{D} f_d \qquad 2.8.1$$

D is the number of atoms within the defect, where it is most likely formed of independent scatterers, i.e. there is no recognizable crystallinity because of its size. This will give rise to scattering that will be distributed throughout space with a total integrated intensity of

$$I = N\left\{\sum_{d=1}^{D} f_d\right\}^2 \qquad 2.8.2$$

For a low concentration of N defects with intensity distributed throughout the scattering pattern, this will be a rather insignificant background. Of more significance is that the presence of a defect will displace the atoms of the matrix and this can be represented by a modification to the structure factor of the matrix, such that equation 2.3.9 becomes

$$F_S = \sum_{j=1}^{N} f_j e^{-2\pi i S.\{r_j + u_j\}} \qquad 2.8.3$$

This represents a perturbation on the average lattice and in the kinematical approximation the intensity is given by

$$I = F_S^* F_S = \sum_n \sum_m f_n f_m \exp\{-2\pi i S.(r_n - r_m)\}\exp(-2\pi i S.(u_n - u_m))$$

2.8.4

Now the atom movements can never be considered independent because of the bonding in a solid will result in a restoring force to any movement. This analysis is equally valid for thermal agitation or distributed defects, although the distribution of stationary waves may differ, but they can be represented by a Fourier sum of phonon waves.

$$u_n = \sum_k A_k \cos\{\omega_k t - r_n.k - \delta_k\} \qquad 2.8.5$$

where $\boldsymbol{k}$ in this context is the wave-vector for a phonon wave with a maximum amplitude of vibration A_k along the direction of the scattering vector $\boldsymbol{S}$. The phase term δ_k is to account for the differences observed by each photon during a measurement for a vibrating (thermally agitated) crystal lattice. Therefore we can now write

$$(u_n - u_m)^2 = \left[\sum_k A_k \cos\{\omega_k t - r_n.k - \delta_k\} - \sum_k A_k \cos\{\omega_k t - r_m.k - \delta_k\}\right]^2 \qquad 2.8.6$$

Any static distortion due to defects can change the average separation in the atoms and this will appear as a change in the average lattice parameters and add to the Bragg scattering. We have less interest in this at the moment although it does offer a way of determining the defect density or alloy composition in semiconductors. Our purpose here is to examine the influence of the variations from the average. If we now determine the average of equation 2.8.6 we obtain

$$\overline{(u_n - u_m)^2} = \sum_k A_k^2 \{1 - \cos([r_n - r_m].k)\} \qquad 2.8.7$$

The time dependent and phase averaging are assumed to cancel on averaging and the average of $\sum\cos(x)=0.5$. Now using the approximation for an exponential when the displacements are small:

$$\overline{\exp(-2\pi i S.(u_n - u_m))} \approx 1 - 2\pi^2 S^2 \overline{(u_n - u_m)^2} \qquad 2.8.8$$

Again we have removed those terms that just move the average, or average to zero, since these as described above just result in an expansion or contraction of the overall lattice structure.

$$\overline{\exp(-2\pi i S.(u_n - u_m))} \approx 1 - 2\pi^2 S^2 \sum_k A_k^2 \{1 - \cos([r_n - r_m].k)\}$$

$$\approx \exp\left\{\frac{-16\pi^2 \sin^2\theta}{\lambda^2}\overline{u_S^2}\right\} \exp\left\{2\pi^2 S^2 \sum_k A_k^2 \cos([r_n - r_m].k)\right\} \qquad 2.8.9$$

We have substituted $S = 2\sin\theta/\lambda$, equation 2.2.11. If we compare this with equation 2.2.15 the first exponent in the left-hand side of equation 2.8.9 is the general temperature factor or Debye-Waller factor that reduces the intensity of the Bragg scattering. The factor of two arises from this equation relating to intensity whereas equation 2.2.15 relates to the scattering amplitude, i.e. exp(-M)exp(-M) = exp(-2M). The displacement $u_S^2=0.5\langle\sum A_k^2\rangle$, that is the average of half the maximum amplitude.

We now want the time average of equation 2.8.4, since the X-ray sampling is time averaged apart from when femto-second pulses are analysed independently with an X-ray laser, so substituting equation 2.8.9 into 2.8.4 we obtain

$$\langle I \rangle = \sum_n \sum_m f_n f_m \exp\{-2\pi i S.(r_n - r_m)\}\exp(-2M)\left\{1 + \left\langle 2\pi^2 S^2 \sum_k A_k^2 \cos([r_n - r_m].k)\right\rangle\right\}$$

2.8.10

The intensity is therefore composed of Bragg scattering with a reduction from a mixture of thermal effects and some deviation from perfection in the lattice and an oscillatory *cosine* term. Let us consider this *cosine* term in more detail.

These phonon waves create a modulation in the intensity rather similar to those of a synthetic modulated structure or superlattice described in chapter 5, section 5.4.5. This modulation will produce satellites centred on each Bragg peak at distances related to the phonon wavelengths that are relatively long. This becomes clear when we expand the *cosine* term

$$\exp\{-2\pi i S\Delta r_{n,m}\}\cos\left\{\frac{\Delta r_{n,m}}{\lambda_p}\right\} = \exp\{-2\pi i S\Delta r_{n,m}\}\frac{\exp\left\{\frac{i\Delta r_{n,m}}{\lambda_p}\right\} + \exp\left\{-\frac{i\Delta r_{n,m}}{\lambda_p}\right\}}{2}$$

$$= 0.5\left[\exp\left\{-2\pi i S\Delta r_{n,m} + \frac{i\Delta r_{n,m}}{\lambda_p}\right\} + \exp\left\{-2\pi i S\Delta r_{n,m} - \frac{i\Delta r_{n,m}}{\lambda_p}\right\}\right]$$

$$= 0.5\left[\exp\left\{-2\pi i\Delta r_{n,m}\left(S + \frac{1}{\lambda_p}\right)\right\} + \exp\left\{-2\pi i\Delta r_{n,m}\left(S - \frac{1}{\lambda_p}\right)\right\}\right]$$

2.8.11

This term is composed of a fixed term associated with the Bragg condition, ***S***, and satellites appearing at a position inversely related to the phonon wavelength λ_p, figure 2.8.1.

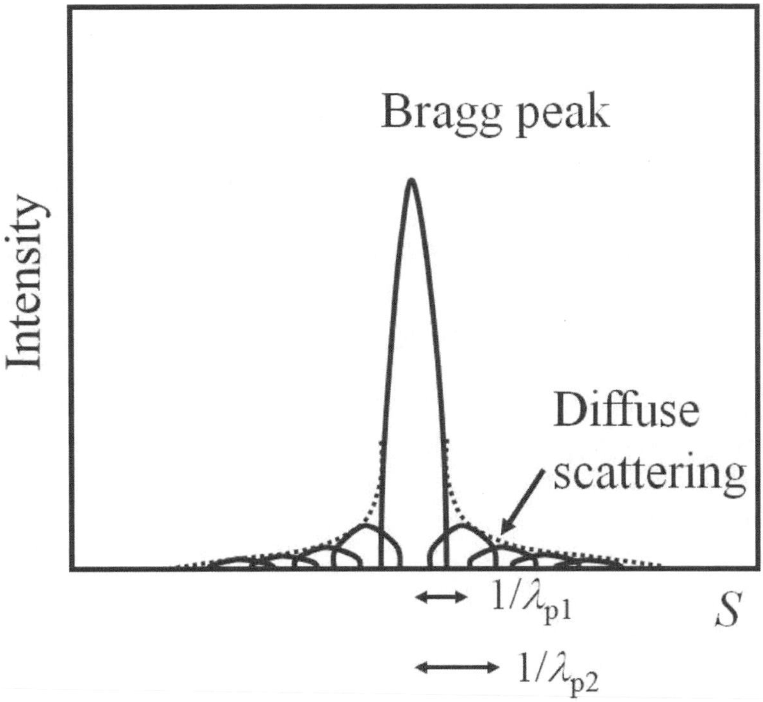

Figure 2.8.1. The origin of diffuse scatter comes from the sum of phonon waves.

The phonon wavelengths are large on the scale of the X-ray wavelength and will therefore produce broadening at the base of each Bragg peak. Also the distances over which these lengths are correlated and hence maintain an X-ray phase relationship will also add to the broadening of the scattering, equation 2.7.9. The correlation length for a short wavelength phonon would be expected to be small and therefore produce a weak broad profile, whereas the longer wavelength phonons have a longer correlation length and hence a sharper profile. The distribution of phonon wavelengths and their amplitudes will be a function of the defect or thermal vibration. In general therefore the intensity falls rapidly with decreasing phonon wavelengths and gives rise to the characteristic decline in diffuse scattering from the Bragg peak maximum, characteristic of many experimental profiles, figure 5.3.15.

For weak deformations (e.g. small isolated elastic distortions) the intensity falls as $1/(\Delta S)^2$ and is characteristic of the scattering a long way from the core of a defect, Huang (1947). Closer to the defect core or for a low concentration of cluster defects the deformations will be strong and the intensity falls as $1/(\Delta S)^4$, etc. These fall-off rates are the theoretically ideal and can be reduced with poor instrumental resolution, e.g. significant axial divergence. The intersection of the change in slope from these different rates of decline can give an indication of the defect radius, R. This is simply the inverse of the distance in reciprocal space, i.e. $R \sim 1/(\Delta S)$, where the radius is defined as the transition of the strong to weak deformation. Thermal vibration falls into the category of weak modulation and will fall as $1/(\Delta S)^2$; this is commonly termed thermal diffuse scattering, TDS. We have only considered the diffuse scattering projected onto the scattering vector and clearly for a full analysis of the distribution of the diffuse scattering we should look in many directions (most conveniently with reciprocal space mapping, section 5.4.4.1) to evaluate the strain fields created by defects. Dederichs (1971) has considered some possible defects and how they influence the diffuse scattering. Since this diffuse scattering relates to the correlated movement of atoms detailed analysis have been possible for deriving the elastic properties and Debye temperatures of solids, etc.

2.8.2. Scattering from relaxing interfaces

From equation 2.4.17 it is simple to see that we can determine the scattering from layer structures. At some depth into the sample, for example the bottom of the substrate or wafer, the scattered wave-field will be insignificant and this gives a starting boundary condition for the recursive calculation. Determining the amplitude ratio at each interface will create the initial ratio for the layer above; when the surface is reached the amplitude ratio will be related to the intensity that can be measured by the detector.

Consider now that at one of these interfaces the distortions are so great that the wave-field no longer sees an underlying periodic susceptibility but one that appears more like an amorphous region, Fewster (1992). We may then expect that the wave from the layer below will not couple into the layer above or at least not initially. We can then split the problem into the old wave-field and a new wave-field, figure 2.8.2. An additional factor will be the distortion around a dislocation at an interface and this will add to the scattering.

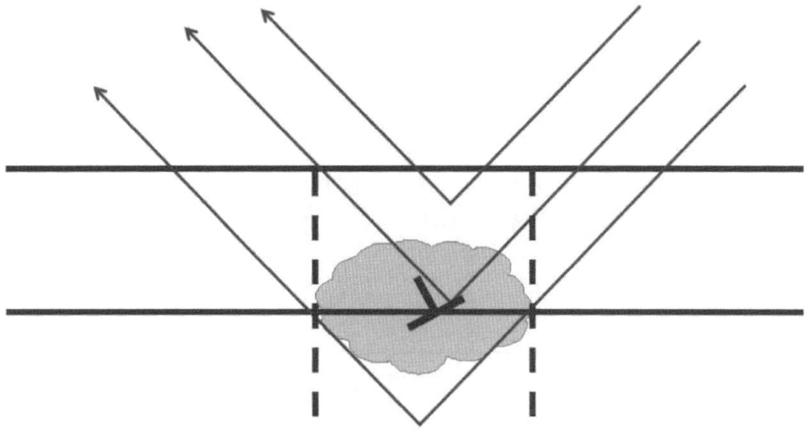

Figure 2.8.2 The different wave-fields that are created by the presence of a defect in the region of scattering.

The form of this scattering can be very complex, but first of all let us try and visualise the strain state close to the dislocation core, figure 2.8.3. In the figure it is clear that if we assume a dislocation locally relieves the strain then the planes of atoms normal to the interface either side of this dislocation cannot go through a step change, this only happens at the dislocation. The alignment must occur over a finite distance either side of the interface. If we assume that some form of elastic distortion is still applicable then the corresponding interplanar spacing parallel to the interface will take on a form given in figure 2.8.3. This simple analysis of the strain will now give us a strain variation as a function of depth and can be modeled using equation 2.4.17. We can make an assumption that the variation of strain given in figure 2.8.3 follows an exponential form and that the lattice plane spacing parallel to the interface is some simple geometrical mean.

$$\bar{d}_{INTERFACE} = d_{n-1} + \frac{(d_n - d_{n-1})Z_{n-1}}{Z_n + Z_{n-1}} \qquad 2.8.12$$

The subscript n refers to the layer number before distortion and Z_{n-1} and Z_n refer to the distance of the distortion below and above the interface respectively. The distorted region is divided up into a series of thin layers of constant strain whose interplanar spacing is determined by an exponential distribution with end points $d_{INTERFACE}$ and d_n. The interplanar spacing perpendicular to the interface can therefore be determined from an assumed local Poisson ratio for each thin layer. Of course the structure of any distorted interface is unlikely to be uniform laterally and therefore we can split the problem into a series of columns, figure 2.8.4.

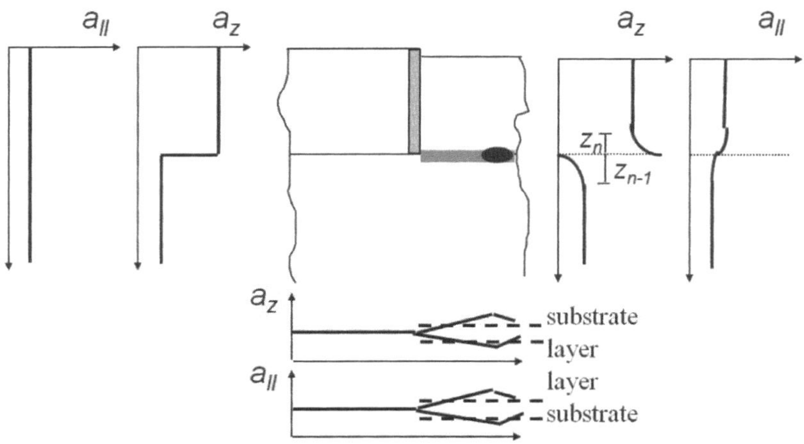

Figure 2.8.3 The variation of the distortion parallel and normal to the surface of a partially relaxed layer structure.

This column approach can be applied to good approximation provided that the lateral dimensions are large compared with the depth, although a further modification, see section 2.8.3, can accommodate exceptions to this. We should now consider the coherence associated with the possible wave-fields. As discussed at the beginning of this chapter we should consider that all possible paths of a photon are coherent if they can be emitted from the X-ray anode and arrive at the same point in the detector. Strictly therefore we should calculate all the path trajectories and bring them together coherently, however to reduce the calculation time we could bring in the idea of phase averaging. Phase averaging will lead to an effective incoherent addition provided that we model the major contributors to the scattering process. The intensity now becomes:

$$I(\omega) = [(X_a + X_b + ...)(X_a^* + X_b^* + ...) + X_c X_c^* + X_d X_d^* + ...]\left|\frac{\gamma_H}{\gamma_0}\right| \qquad 2.8.13$$

X_a and X_b represent the amplitude ratios from regions that scatter coherently with respect to each other and X_c and X_d represent regions that scatter independently and incoherently with respect to all other regions.

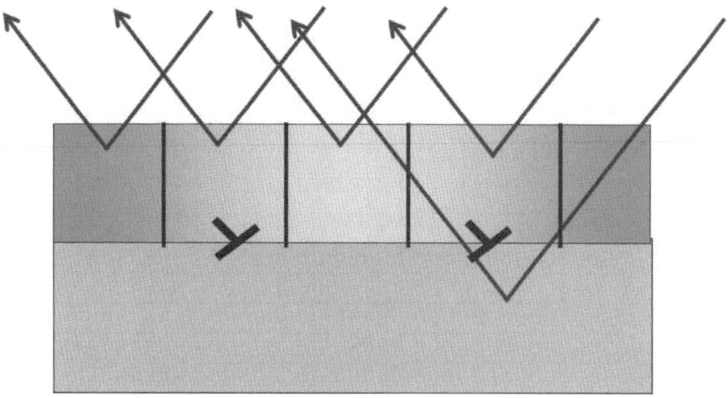

Figure 2.8.4 The procedure for generating the scattering from mosaic and distorted structures by splitting the problem into columns.

In this way we can now consider the scattering from layers below, within and above a distorted interface. In the opening argument we discussed that the susceptibility no longer becomes meaningful and effectively the region close to the dislocation becomes amorphous. The strength of the scattering is effectively negligible compared to diffracting regions except at very small angles in specular reflection. We must therefore grade the susceptibility as well as the strain. We could consider this as a very large static Debye-Waller factor, but that approach is only suitable for relatively small atomic misplacements (effectively within the harmonic limit of atomic vibrations). Here we are dealing with gross distortions.

Since we are concerned with columns that are relatively wide we have to average the structure within this width. If no correlation exist between these columns parallel to the surface then we can simply proportion the various contributions to the scattering, i.e.

$$I(\omega)=[(\Omega_a X_a+\Omega_b X_b+.....)^2+(\Omega_c X_c)^2+(\Omega_d X_d)^2+.....]\left|\frac{\gamma_H}{\gamma_0}\right| \quad 2.8.14$$

where

$$\Omega_a+\Omega_b+\Omega_c+\Omega_d+......=1 \quad 2.8.15$$

Ω represents the proportion of the area occupied by a given amplitude ratio. Taking this approach it is clear that we can start to model quite complex structures having both lateral changes in structure and in depth. We will take this a stage further although so far we have been taking a rather simple explanation of the scattering process as some pseudo one-dimensional effect. All we have been concerned with is the incident wave-field entering the sample being scattered somewhere and the detector producing a simple profile that we can model.

We shall now expand the method to account for the finite dimension of the regions of coherent scattering, this information is revealed experimentally with diffraction/reciprocal space maps. The definition of the inclination of the scattering plane with respect to the surface is defined as in figure 2.8.5. The scattering is now defined by two angles, ω and 2θ, therefore the experimental data collection is two-dimensional. We will see in the following chapters on X-ray diffraction instruments how this data is collected, but for the moment we will assume that we have an instrument that has an ideal detector acceptance and produces a perfectly well defined incident wave direction. Suppose now we have an ideal perfectly crystalline sample then the scattering will be confined to a line normal to the surface, this is realised through the equation 2.5.4, in section 2.5.

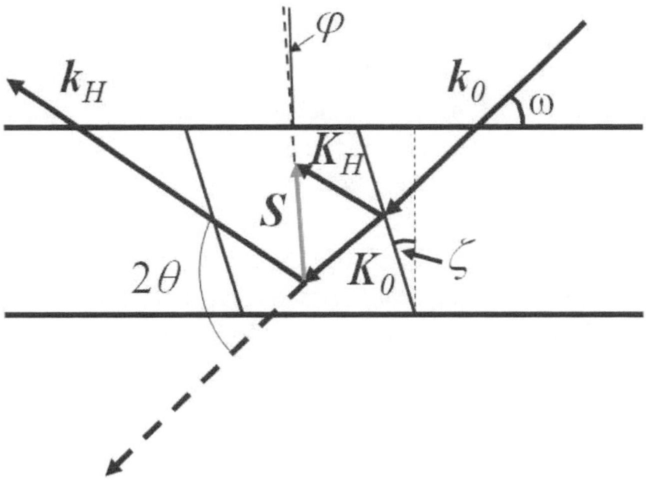

Figure 2.8.5. The components used to derive the relationships that satisfy the boundary condition for a simple case of a parallel-sided block.

For less than perfect structures where the crystallinity is not extended parallel to the sample surface as in figure 2.8.5 then the exit surface and for that matter the entrance surface to this crystal block may not be the sample surface. Effectively the incident beam passes through an absorbing medium, enters a finite crystalline block is scattered and the resultant wave exits through another surface and is absorbed before emerging at the sample surface. Since in general the crystal quality is less perfect, i.e. composed of mosaic blocks that are not all identical, we can relax the rigour somewhat and consider averages and fluctuations. From figure 2.8.5 we can work out the scattering angle for a wave entering a crystal block, by applying the usual boundary conditions we obtain:

$$2\theta = \sin^{-1}\{\sin(\omega - \zeta) - 2\sin\theta_B \sin(\varphi - \zeta)\} + \omega - \zeta \qquad 2.8.16$$

neglecting refractive index corrections. Angles φ and ζ represent the inclination of the scattering plane to the surface and the tilt of the exit and entrance surface of the mosaic block to the surface normal. This equation can be compared with 2.5.4. We have assumed for simplicity in this example that the block is parallel sided.

The broadening of the scattering profile due to the finite depth appears naturally in the calculations for the laterally infinite case, whereas the broadening due to the finite width is most easily estimated from the approximate description given in section 2.8.3. If we assume that the sides of the block are not parallel then the expression is essentially unchanged.

We can consider this by recognising that the incident wave enters the block and is subject to a small refractive index change, figure 2.3.5. The strength of the refractive index correction is dependent on the incident angle. The refracted wave, $\boldsymbol{K}_0$, will then undergo scattering according to equation 2.5.1, and create an internal scattered wave with a wave vector of $\boldsymbol{K}_H$. When this internal wave reaches the exit surface the boundary condition applies, which can be visualised by drawing the normal to the exit surface through the point of intersection of the refracted beam and the dispersion surface in figure 2.3.5. The intersection of this normal to the sphere centred on H of radius $|\boldsymbol{k}_H|$ will denote the external scattered wave direction (this represents the condition when $\{\boldsymbol{K}_H\}_{//} = \{\boldsymbol{k}_H\}_{//}$). We therefore see that the exit surface defines the broadening direction of the scattered wave.

To relate the coherence of the X-ray wave through different blocks we include changes in the phase due to the location. It is now possible to build very complex structures from an array of blocks that represent some average of the local strain and scattering strength. We therefore have an additional lateral dimension and position to include in our model as well as the normal input of interplanar spacing, orientation and scattering strength, etc. The local amplitude of a perfect layer based on dynamical theory can then be used as an approximate scaling parameter for the defect structure. Section 2.8.3 considered the case when the dimension of the lateral regions becomes significantly below the X-ray coherence length.

The approach given above is a numerical method that can be built in an almost atomistic way, and relies on an overall description of the strains and dimensions throughout the structure. This always depends on a proper description of how the structure responds to distortions. The principle is as always reliant on obtaining an approximation to the scattering process, by first describing the atom positions in a sample and their scattering strength and then conducting some Fourier transform to obtain a scattering pattern. Aspects of coherence of the X-ray source, instrument beam paths, and area over which the X-rays are correlated due to the sample should all be included.

Various authors have made good use of the work of Krivoglaz (1995), which relates some structural fluctuations, i.e. defects and distortions, and creates an analytical approach to form the diffraction patterns. The approach is discussed here briefly since it is used in a few examples, however for more detail this can be obtained from Krivoglaz (1995), Pietsch, Holý and Baumbach (2004), Kaganer, Köhler, Schmidbauer, Opitz and Jenichen (1997). The intensity can be represented in terms of a reduced scattering vector $\boldsymbol{s} = \boldsymbol{K_H} - \boldsymbol{K_0} - \boldsymbol{S}$, and therefore represents the deviation from the reciprocal lattice position, it is also complex since it contains the refraction and absorption components

$$I(s) = A\int_z\int_{z'}\chi_S(z)\chi_S^*(z')\Omega_x(s_x, z, z')\Omega_y(s_y, z, z')e^{-2\pi i\left(s_z z - s_z^* z'\right)}dz'\,dz \quad 2.8.17$$

where Ω represent the shape function, or in the case for modeling the diffuse scattering from dislocations, this will represent the deformation field due to the dislocations. So for example a dislocation parallel to the x-axis in the plane of the interface will result in a deformation field that is a function of its Burger's vector and elastic properties, the vicinity of other dislocations and the interface or surface above. This formula assumes the distribution of dislocations is random and there is no dynamical contribution. The equations become increasingly complex if the dislocations are not purely in the plane of the interface, or elastic anisotropy is included. It is interesting to note though that careful modeling of these relaxing structures is that the width of the layer profile is proportional to the square root of the dislocation density for relatively thick layers when the surface effects are minimal, which was the conclusion of the expression derived by Gay, Hirsch and Kelly (1953), and fits with equation 2.8.21 derived in the following section.

The assumption above is that each dislocation contributes individually to the scattering pattern and this is why it is in close agreement with the simpler expression of Gay et al. However, depending on the dislocation density, and certainly as it reaches certain values, perhaps when the separation is less than the layer thickness, then there will be some correlation and a further term is required in equation 2.8.17.

$$I(s) = A\int_{z}\int_{z'}\chi_S(z)\chi_S^*(z')\Omega_x(s_x, z, z')\Omega_y(s_y, z, z')e^{-2\pi i(s_z z - s_z^* z')}\left\langle e^{-2\pi i S.(u(r)-u(r'))}\right\rangle dz'\,dz \qquad 2.8.18$$

This additional term can be composed of correlated and uncorrelated positions for the dislocations. As may be expected the complexity increases and assumptions are needed, i.e. assuming either the longer or shorter length scales dominate or some mixture of the two. It is important to realize though that correlated dislocation positions will in general narrow the profile. Unfortunately relaxed structures are not always composed of misfit dislocations lying in the interface, but can also have a significant number of threading dislocation. Untangling different dislocation types has been investigated by Kaganer et al (2005), where it was found that the tails of the diffraction peaks can reveal some of this information, and by Daniš and Holý (2006). The details of this are best found in the original works, as these can be quite specific to particular problems, orientations and materials.

These integrals, equation 2.8.18 require an estimate of the distribution of dislocations, that defines the extent of the correlation and this can be estimated by statistically sampling. This allows significantly more complex distributions to be modeled and permits modeling of point defects, etc., Kaganer and Sabelfeld (2010). The correlation length is just an average, so in general as given by the Gay et al study and others this only gives some average assuming a normal distribution, that does not necessarily relate to the actual distribution and therefore a false description of the sample may result. However the integration can be performed by randomly selecting the correlation length within certain macroscopic bounds, e.g. overall elastic energy, sample thickness, and this will achieve a good estimate if sufficient samples are taken to indicate convergence. Kaganer and Sabelfeld (2011) have simulated various distributions of defects and illustrated how the peak broadening increases, for a fixed dislocation density, as the dislocations tend to distribute more unevenly, i.e. cluster. Clearly the single value estimate of dislocation density can be seriously problematic, and clearly needs an estimate of the distribution to obtain a definitive value

2.8.3. Scattering from mosaic structures

In this section we shall extend the arguments given in section 2.8.2 above to include lateral inhomogeneities into our scattering theory. Krivolglaz (1995) has developed an approach to include lateral correlation lengths by including what is effectively a Fourier transform of the length scales expected in the structure. The approach here yields similar results although Krivolglaz gives analytical expressions, whereas the approach given here is numerical. In effect the general emphasis is that the scattering pattern is the Fourier transform of the length-scales involved, with scattering strengths and some displacements due to the local strain distributions. However to make the derivation of lateral correlation lengths consistent with our arguments so far, based on the kinematic derivations of section 2.8.2, we will consider a phase difference approach.

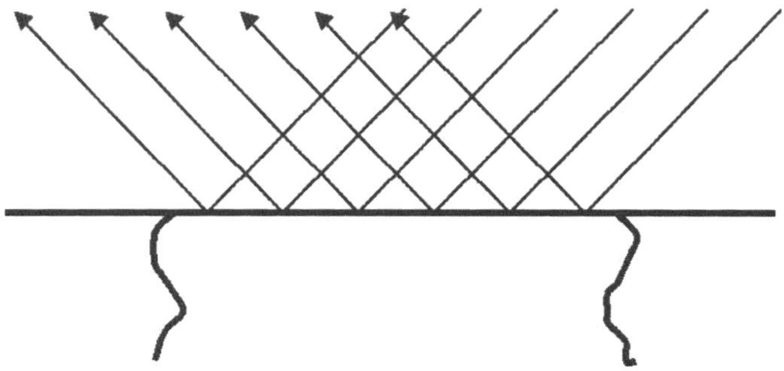

Figure 2.8.6. The scattering from a mosaic block is built by combining the amplitudes from an infinite number of possible paths.

Consider figure 2.8.6, where we have a mosaic block or some shape within our sample that can scatter along the whole dimension. If we consider the waves to be separated by some small dimension l then we can calculate the path difference between any two waves separated by l and derive the phase difference:

$$\Phi = \frac{2\pi l}{\lambda}\{\cos(2\theta - \omega) - \cos\omega\} \qquad 2.8.19$$

The amplitude is given by a sinc function as in equation 2.7.6, but in this case the dimension is considerably larger, matching that of the mosaic block. Of course ω relates to the incident wave-field direction and 2θ is given by equation 2.8.16. The dimension Nl is simply the dimension parallel to the surface plane of the sample. Now the separation of the individual wave-fronts contributing to this profile can be any small value and the shape is invariant. This is evident from the approximation that we can make for small phase angles:

$$A = a\frac{\sin N\Phi}{N\Phi} = \mathrm{K}\frac{\sin N\Phi}{\frac{2\pi}{\lambda}\{\cos(2\theta - \omega) - \cos\omega\}} \qquad 2.8.20$$

where $K = a/Nl$ and is constant since the scattered contribution from a region of length l can be considered proportional to the length involved.

Therefore we can add the influence of mosaic blocks to our scattering pattern. The amplitude scattered by the layer, assuming that it is perfect should now be smeared out such that the amplitude is distributed to give measurable scattering at 2θ for a range of incident angles:

$$A = a\frac{\sin\{\frac{2\pi L}{\lambda}[\cos\{\sin^{-1}\{\sin(\omega-\zeta)-2\sin\theta\cos(\varphi-\zeta)\}-\zeta\}-\cos\omega]\}}{\{\frac{2\pi d}{\lambda}[\cos\{\sin^{-1}\{\sin(\omega-\zeta)-2\sin\theta\cos(\varphi-\zeta)\}-\zeta\}-\cos\omega]\}}$$

2.8.21

for the case of a parallel sided block inclined at ζ to the sample surface. Clearly this approach can be extended to accommodate blocks of various shapes and different strain values.

2.9. A new theory that explains the diffraction from polycrystalline samples:

Any theory of diffraction should ideally be applicable to any crystalline material. The kinematical theory has been shown in section 2.7.1 to be the same as dynamical theory with a few assumptions, which relate to the crystalline quality. The conventional dynamical theory breaks down remote from the Bragg peaks along the crystal truncation rods, that is particularly evident in thin layers on a substrate section 2.4.4. The theory described in that section is based on the Fresnel approach and does not require the Bragg relationship. In all these models the sample is assumed to be aligned so that the incident and scattered beams are accurately coplanar with the scattering plane normal. A crystallite in a polycrystalline aggregate will in general not be aligned and this new theory takes account of this, so that the scattering can be modeled from a distribution of orientations. Firstly though we will discuss why we need to move away from the thought that the observed scattering is purely due to Bragg scattering or close to the Bragg condition. There are some problems associated with assuming the scattering from powders and polycrystalline aggregates depends on Bragg scattering that have been discussed in Fewster and Andrew (1999), Fewster (2000). The following discussion is based on Fewster (2014).

2.9.1. Which scattering theory should be used?

It is important to establish which theory, dynamical or kinematical, is appropriate for modeling the intensity from powders. The conventional approach is to combine three aspects; the magnitude in terms of the structure factor $|F_{hkl}|^2$ in the kinematical approximation, a profile shape that is a variable parameter (a mixture of Gaussian and Lorentzian form; Pearson VII, pseudo-Voight, etc.,), and a FWHM function that varies with 2θ, which is also a variable. That is, it is explicitly assumed that kinematical theory is appropriate; so this can be tested. The profile shapes are defined by variables that are fitted, so will contain little directly derivable physical information, although there is some evidence that a more Gaussian shape is more indicative of strain variations and a Lorentzian bias is indicative of crystallite size broadening, Louër and Langford (1982). Many authors have studied the profiles and associated the broadening with defect information within the grains, etc.: these analyses require considerable care and prior knowledge to extract reliable information. The variation in the FWHM across the diffraction profile is usually based on a 'tangent' quadratic function, first suggested by Caliotti, Paoletti and Ricci (1958) for neutron diffraction data. It is debatable whether it is suitable for high-resolution instrumentation, so modifications of this function have been given by Louër and Langford (1982) and others, and also to account for axial divergence, etc.

There is an implicit assumption that the peaks observed come purely from the Bragg condition, Bragg (1913), whereas the following explanation of scattering does not make this assumption, Fewster (2014). Suppose that the observed peaks are only from crystallites that satisfy the Bragg condition; then the profiles can be calculated based on dynamical theory, figure 2.9.1. The widths and intensities of two LaB_6 reflections (strong and weak) have been calculated as a function of the crystallite thickness, and also for Si *004*, based on a planar structure. An estimate of the differences between the kinematical and dynamical models is obtained by comparing differences in widths: this gives an underestimate in the width if the kinematical theory is assumed, figure 2.9.2.

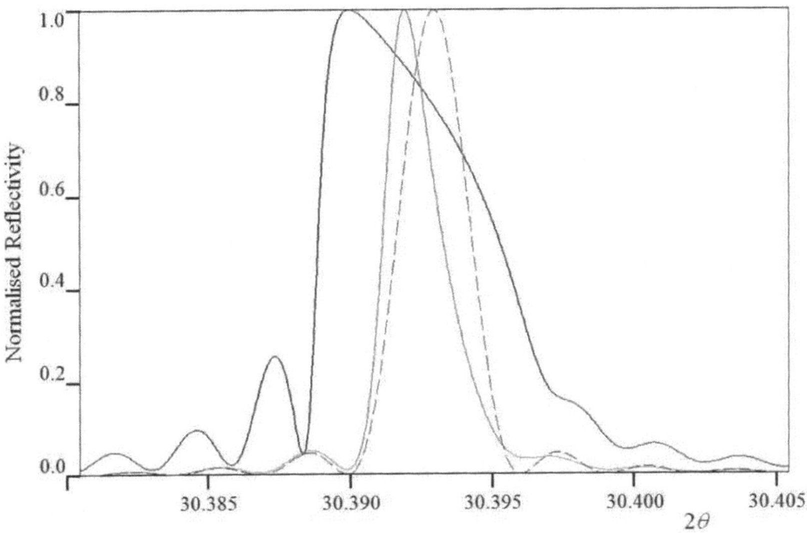

Figure 2.9.1. The *110* scattering profile for a 3 μm cube of LaB_6 calculated with dynamical theory in reflection mode (black line) kinematic theory including absorption (grey line) and kinematical theory with no absorption (dashed line). Note the significant broadening of the profile and importance of dynamical theory for this strong scattering and thickness. The peak positions in the kinematical simulations have been shifted from the Bragg angle to overlay the dynamical profile, and the peak heights have been normalized to 1.

Clearly kinematical theory grossly underestimates the width above ~0.2 μm for the *110* reflection of LaB_6, whereas for the *003* LaB_6 and *004* Si reflections this is less of a problem and the widths are ~2% and 2.5% under-estimated respectively at 1 μm. In kinematical theory the intensity is assumed to diminish only through absorption, however this is not necessarily the dominant process and losses due to scattering can be considerable and are the reason for these differences in peak widths. The distance over which the intensity falls to $1/e$ of its incident value due to scattering is called the extinction length and is the given by

$$\Lambda_R = \frac{\lambda}{|C|}\sqrt{\frac{\gamma_0 |\gamma_H|}{\chi_H \chi_{-H}}} \qquad 2.9.1$$

All these parameters have been defined previously. The equivalent length in the transmission case is called the Pendellösung distance. From this equation we can derive a simple rule that the extinction / Pendellösung distance should be significantly greater than the absorption length or, the crystallite size should be significantly below the extinction distance for the kinematical theory to give reliable results.

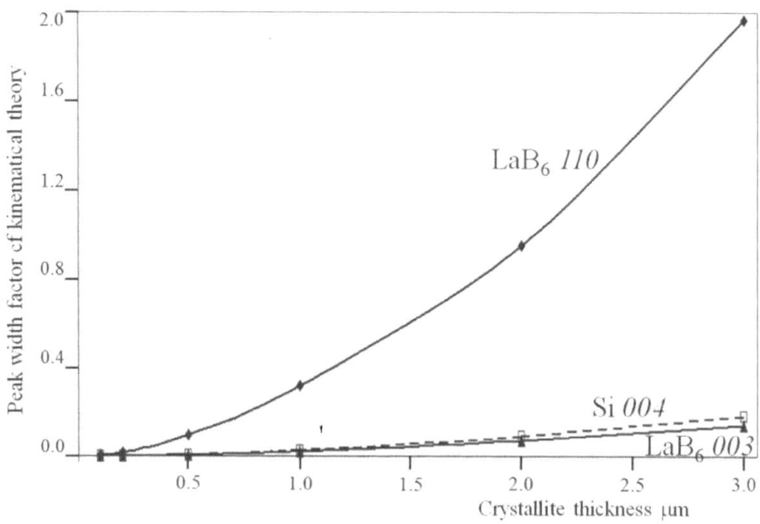

Figure 2.9.2. The fractional increase in peak width when using dynamical theory compared to using kinematical theory. For example, the *110* reflection from 3 μm LaB_6 crystallites will have a width assuming dynamical theory that is ~3× the width obtained when assuming kinematical theory. For reflections or materials with a longer extinction length this factor is reduced considerably, and at small thickness the peaks shapes from both theories match those of the interference function (figure 2.7.3) with identical widths and intensities.

For the examples given, LaB_6 obviously will have the same absorption length for both the *110* and *003* reflections, since it is a function of the bulk properties; the linear absorption coefficient and the density, equation 2.7.13. The absorption length, defined as the distance for the intensity to fall by $1/e$, for LaB_6 is 1 μm, whereas the extinction distances for the *110* and *003* reflections are 0.28 μm and 1.2 μm. The extinction length therefore dominates the width for the *110* reflection. For Si the absorption length is 63.7 μm and the extinction length for the

004 reflection is 1.2 μm, so the width in the kinematical model is only valid significantly below this level for this reflection. These calculations and widths in the figures are for the σ polarization, whereas the π polarization extinction lengths are 0.33 μm and 3.13 μm for the *110* and *003* for LaB_6 and 3.2 μm for *004* Si.

The corresponding analysis on the peak intensity shows that the kinematical theory will greatly over-estimate the value, figure 2.9.3. If the integrated intensity is used these over-estimates are approximately halved. Again this is a consequence of the extinction effect. The extinction or dynamical effects are particularly dominant at the Bragg peak, figure 2.9.1, and further down the tails of the peak the dynamical and kinematical theories become coincident and the intensities are equivalent. This is also observed in figure 2.7.1.

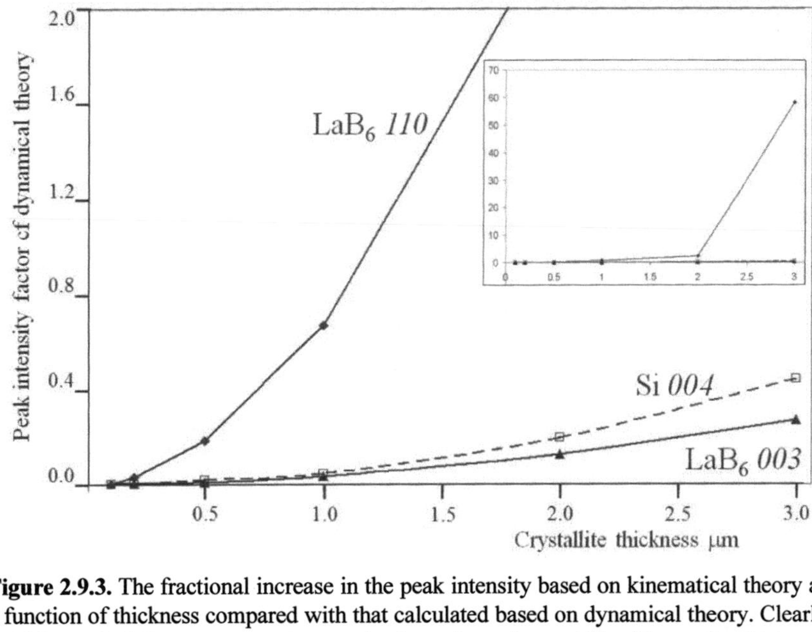

Figure 2.9.3. The fractional increase in the peak intensity based on kinematical theory as a function of thickness compared with that calculated based on dynamical theory. Clearly the kinematical theory over-estimates the intensity quite considerably as the thickness increases, whereas the dynamical theory is dominated by the extinction length.

Clearly if the scattering from powders assumes that the observed peaks are Bragg peaks, i.e. scattering from within the full-width-at-half-maximum, then the extinction distance becomes very important for making a decision on the appropriate theory. Since in general this information assumes that there is significant prior knowledge; this could prove problematic and can lead to considerable variation in the calculated and measured peak intensities. These calculations are based on dynamical theory in reflection mode and are illustrative. The same calculation in transmission mode gives greater differences.

2.9.2. Some important geometrical considerations

We shall consider a typical instrument used in powder diffractometry that will be described in greater detail in Chapter 4, section 4.4.1, the Bragg-Brentano geometry.

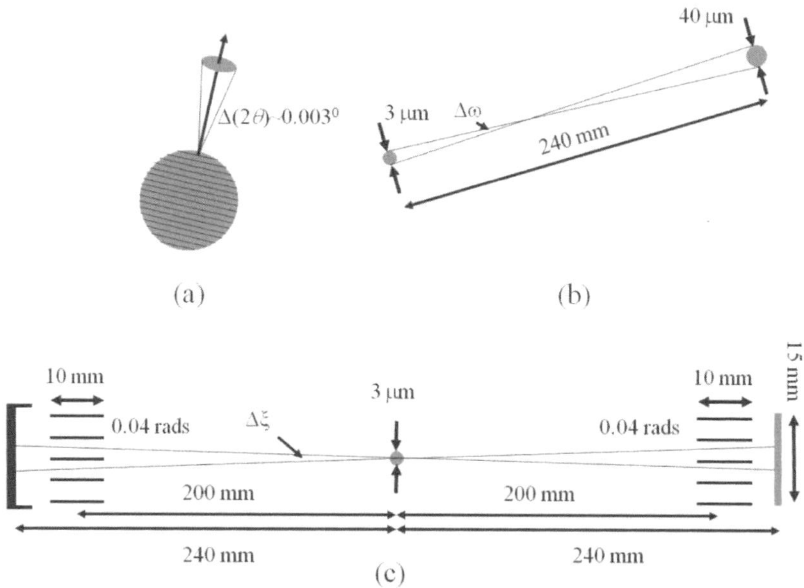

Figure 2.9.4. The geometry of the scattering from a single crystallite in the Bragg-Brentano geometry. (a) a 3 μm spherical crystallite will have an angle of acceptance for Bragg scattering within it intrinsic width of ~0.003^0. (b) the divergence experience by a crystallite in the scattering plane is ~0.01^0, and in the axial plane (c) is ~0.114^0.

A common X-ray source focus can be 400 μm × 15 mm, such that with a typical "take-off" angle the projected image of the source seen by the sample will be 40 μm × 15 mm, with the smallest dimension in the plane of the scattering experiment. The beam diverges, limited by a slit and axial divergence with Soller slits, to cover a sample area of 10 mm × 10 mm. The slit in front of the detector is placed at the same distance as the source from the sample to create a para-focusing geometry when both detector and source are symmetrically positioned and scanned. This creates good resolution and illuminates many crystallites.

If the crystals are ~3 μm in diameter and if the X-rays penetrate ~10 μm into the sample then there are ~37,000,000 crystallites illuminated, if they are perfectly packed. The angular spread of photons experienced by a single crystallite at a radius of 240 mm, will be = $2\tan^{-1}([0.04+0.003]/[2*240]) \sim 0.01^0$, figure 2.9.4b. A 3 μm cubic crystallite will have an angular acceptance given by the FWHM determined from the Scherrer equation 2.7.10 of 0.00295^0 at $\sim 30^0$ 2θ. However as mentioned earlier in this section, the true width is probably defined by dynamical theory at this thickness, e.g. for *110* LaB_6 that is a standard reference material the width is 0.0064^0 and the peak is not symmetric figure 2.9.1. This will not make a significant difference to the interpretation since the dominating influence is the divergence seen by the crystallite, so providing the scattering vector is within this 0.01^0 range then there should be some contribution to the scattering. The scattering vector of the crystallite has a degree of freedom in the axial plane (out of the scattering plane), and if the possible angular spread of trajectories passing through the diffractometer is calculated, this comes to a possible axial divergence of 0.114^0, i.e. = $2\tan^{-1}(L\tan[0.02]/[R_S])$ assuming the 0.04 radian Soller slit is 10 mm long (L) and the Soller slits are both at a distance (R_S) of 200mm from the crystallite, figure 2.9.4c. The probability of the divergence incorporating the scattering vector for any reflection is therefore

$$\frac{0.114\times 0.01}{4\pi\sin\theta}\frac{\pi}{180}\approx\frac{6.33\times 10^{-8}}{\sin\theta} \qquad 2.9.2$$

The $\sin\theta$ term is a geometrical factor to account for the increased range of capture for scattering by the axial slit (this will become clear further into this section). Therefore for scattering from a Bragg peak to occur requires ~ 100,000,000 crystallites, for a stationary sample with this geometry. Typically 20,000,000 crystallites can be illuminated with this geometry, however a full scattering profile can be observed without rotation. The intensity is clearly not dominated by the scattering within the main width of the Bragg peaks.

2.9.3. So what does a powder diffraction profile represent?

So far the above calculations suggest that the measured intensity would not be reliable because of the appropriate choice of theory is problematic, and also the geometry for typical samples will have very few or no crystallites in the diffracting condition. From the extensive published experimental work, the intensities follow the kinematical theory of scattering remarkably well and are reproducible; this is especially evident in structure solution. So the above arguments suggest that the profile is unlikely to be a result of Bragg scattering, but the process of generating a profile is much more subtle.

The scattering from a crystallite is similar to that from a mosaic block, but in this case the scattering throughout diffraction space will be considered, since we cannot simply assume that the scattering is from the 'Bragg condition' alone. It has been established that the intensities and profiles based on the kinematical and dynamical theories are coincident outside the 'Bragg condition', so it is easier to visualize the scattering based on a kinematical model outside this and apply dynamical theory when the Bragg condition is encountered.

Our interest is to be able to analyse the scattering contribution at a specific detector position from a crystal plane in any orientation. The dimensions of the crystal plane are taken to be L_x in the source-sample-detector plane, and L_y is the dimension of the plane orthogonal to L_x. Consider a line on a crystal plane that is in the same plane as the X-ray source and detector position, figure 2.9.5a.

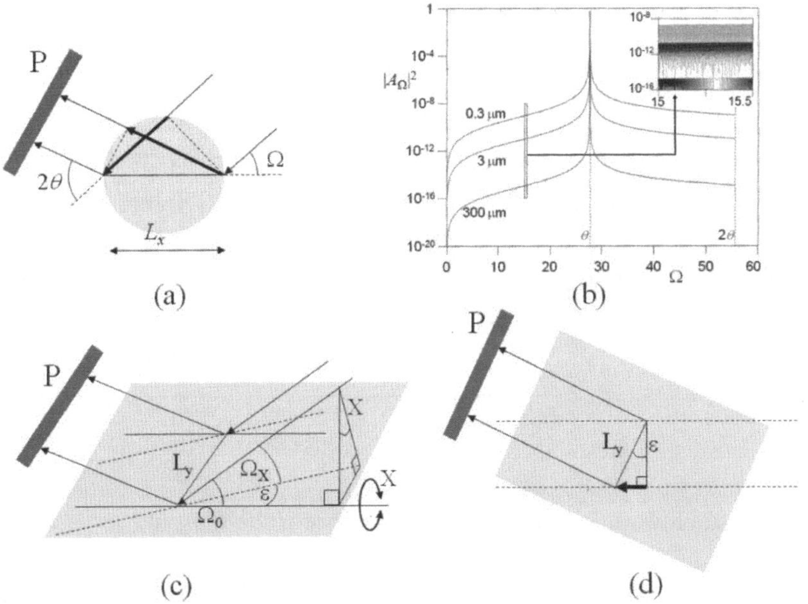

Figure 2.9.5. The different path lengths created by a parallel beam of X-rays impinging on the scattering planes of a crystallite: (a) as the crystallite is rocked in the scattering plane to change Ω; (b) is the resulting profile for various dimensions L_X. (c) and (d) show how the path length of the beam is changed by tilting the scattering plane normal out of line with the source-sample-detector plane.

We shall define two angles, Ω as the angle between the incident beam and the scattering plane, which is Ω_0 when the source, surface plane normal and detection point P are coplanar, and X as the tilt of this plane about this direction, figure 2.9.5c. The position of the detector is given by the angle 2θ that is relative to the incident beam. For a parallel beam of X-rays impinging on this crystal plane, we can determine the phase differences introduced when Ω_0 is varied (i.e. X = 0), figure 2.9.5a, and when X is varied at a fixed Ω_0 value, figure 2.9.5c and d. These phase differences, using the same arguments as in section 2.7.2 and 2.8.3 can lead to a sine cardinal (or 'sinc' function) for the amplitude when rocking in Ω_0:

$$A_\Omega = \frac{\sin\left(\frac{\pi L_x}{\lambda}\left(\cos(2\theta - \Omega_0) - \cos\Omega_0\right)\right)}{\left(\frac{\pi L_x}{\lambda}\left(\cos(2\theta - \Omega_0) - \cos\Omega_0\right)\right)} \qquad 2.9.3$$

The plot of $|A_\Omega|^2$ as a function of Ω is given in figure 2.9.5b. The amplitude resulting from the phase difference associated with changing X, with reference to figure 2.9.5c and d, is:

$$A_X = \frac{\sin\left(\frac{\pi L_y}{\lambda}\left(\sin\left[\tan^{-1}(\tan\Omega\sin X)\right]\right)\right)}{\left(\frac{\pi L_y}{\lambda}\left(\sin\left[\tan^{-1}(\tan\Omega\sin X)\right]\right)\right)} \qquad 2.9.4$$

The peak amplitude is assumed to be unity for the present.

Clearly if $X \neq 0$, then the incident angle Ω becomes a projected angle Ω_X, where:

$$\Omega_X = \cos^{-1}\left\{\frac{1 + \left(\frac{\cos\Omega_0}{\cos\left[\tan^{-1}(\tan\Omega_0\sin X)\right]}\right)^2 - (\sin\Omega_0\cos X)^2}{2\left(\frac{\cos\Omega_0}{\cos\left[\tan^{-1}(\tan\Omega_0\sin X)\right]}\right)}\right\} \qquad 2.9.5$$

This can then be substituted into equation 2.9.3 and combined with equation 2.9.4 to give the amplitude arriving at 2θ when the crystal plane is rotated and tilted to Ω and X, i.e.

$$A_{\Omega X} = \frac{\sin\left(\frac{\pi L_x}{\lambda}\left(\cos(2\theta - \Omega_X) - \cos\Omega_X\right)\right)}{\left(\frac{\pi L_x}{\lambda}\left(\cos(2\theta - \Omega_X) - \cos\Omega_X\right)\right)} \frac{\sin\left(\frac{\pi L_y}{\lambda}\left(\sin\left[\tan^{-1}(\tan\Omega\sin X)\right]\right)\right)}{\left(\frac{\pi L_y}{\lambda}\left(\sin\left[\tan^{-1}(\tan\Omega\sin X)\right]\right)\right)}$$

2.9.6

As the incident angle Ω_X is varied then the cross-section of the crystallite scattering plane to the beam is reduced, provided the beam is larger than the crystallite. Then the intensity available for scattering is simply given by:

$$P_{\Omega} = L_x \sin \Omega_{\mathrm{X}} \qquad 2.9.7$$

Another inclusion for estimating the intensity is the region of integration associated with each data point. This 'integration volume' or instrument function varies significantly with Ω and 2*θ*, and results from the oversampling of the scattering by the detector slit in the axial direction, shown in figure 2.9.6. As 2*θ* - Ω becomes small, the rotation in X will take an increasingly larger movement to take the scattering contribution outside the bounds of the detector. This results in a variation of oversampling with 2*θ* and a modification of the amplitude variation in Ω and X at each 2*θ*.

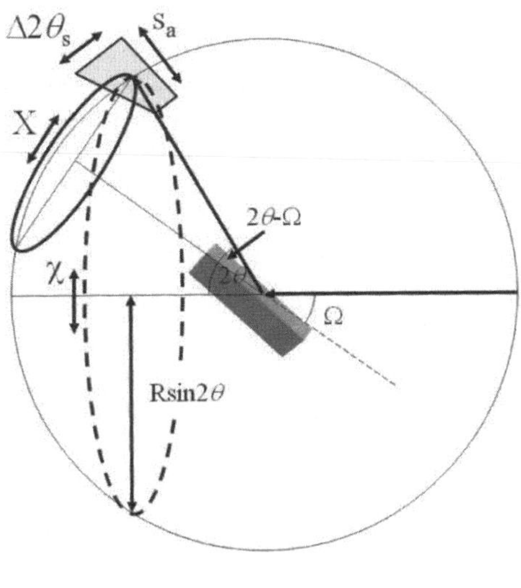

Figure 2.9.6. The exit angle (2*θ* - Ω) influences the angular acceptance of the scattered beam. $\Delta 2\theta_s$ and s_a represent the dimensions of the detector slit.

The distribution of intensity or $|A_{\Omega X}|^2$ at two different 2*θ* values is given in figure 2.9.7. Although the level of intensity outside the 'peak' region is low, the total intensity contribution is comparable to the intensity associated with the peak, for this case when the crystallites are ~10 μm. For smaller crystallites the 'non-peak' contribution becomes larger, and for larger crystallites the 'peak' contribution becomes larger.

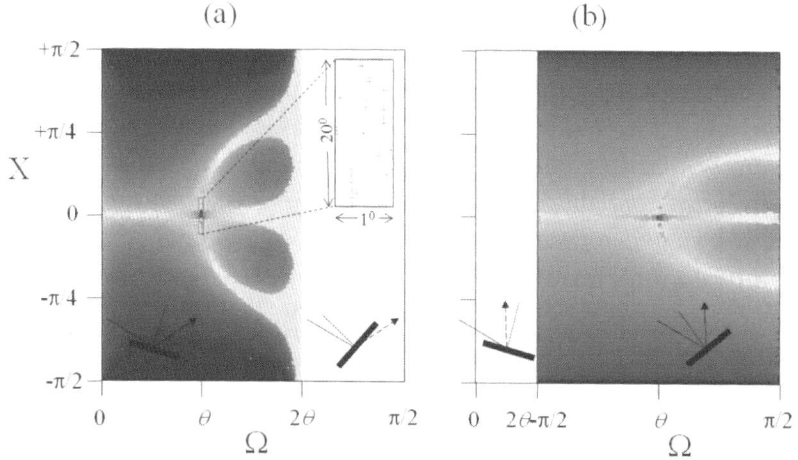

Figure 2.9.7. The distribution of intensity reaching the detector at 2θ for a crystallite as it is rocked in Ω and X, assuming the crystallites are cube shape. The figures are formed from sampling 10,000 randomly orientated crystallites with a typical instrument capture volume (as discussed in the text). A close up of the bounds of the individual capture volumes is given in the insert to (a). (a) 60^0 and (b) 110^0 2θ.

This amplitude from a single plane from a single crystallite will exist in 4π steradians, however for purpose of the calculations it is assumed that the dominant contributions occur when the incident beam and the diffracted beam are above the surface plane and on opposite sides of the surface normal, i.e. $0 < \Omega < 2\theta$ (or $\pi - 2\theta < \Omega < \pi/2$, if $2\theta > \pi/2$). Outside these bounds shown in figure 2.9.7a and b, the scattering is assumed to be weaker than in the forward scattering region, and to keep the subsequent calculation times reasonable these are assumed to be ~0; this has little effect on the mean, which is the total intensity calculated divided by the total area. This sinc function approach is an approximation to a full atomistic model, i.e. summing the amplitudes from every atom, however this is impractical at present for typical crystallite dimensions: these summations could involve $>10^{12}$ atoms per crystallite. Even this approach should include subsequent interactions as discussed in Fewster (2014), i.e. dynamical scattering effects. These approximations must be born in mind.

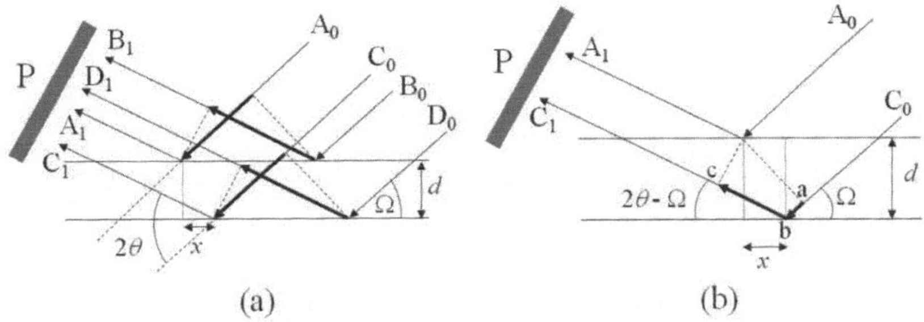

Figure 2.9.8. The path differences from plane to plane, in (a) the beam paths A_0A_1 and B_0B_1 interfere as in figure 2.9.5a, whereas the combinations of paths differences A_0A_1 and C_0C_1 can be equal to $2d\sin\theta$ depending on the value of x.

So far only a single crystal plane has been considered, and if there are N parallel planes separated by a length d then these contributions will have a varying phase relationship depending on d and the wavelength used, figure 2.9.8a. The interference between beams A_0A_1 and B_0B_1 give rise to the profile in figure 2.9.5b, and similarly for C_0C_1 and D_0D_1, and since all these beams can be combined with their appropriate phases there will always be beams, e.g. A_0A_1 and C_0C_1, which can constructively interfere from layer to layer. The path difference for the two beams A_0A_1 and C_0C_1 in figure 2.9.8b is given by:

$$\Delta = d(\sin\Omega + \sin(2\theta - \Omega) - x(\cos\Omega - \cos(2\theta - \Omega))) \qquad 2.9.8$$

The amplitude A_Ω is a maximum $\Omega = \theta$, from equation 2.9.3, which gives rise to a path difference of $\Delta = 2d\sin\theta$, and therefore the value of x that maintains this phase relationship is:

$$x = d\frac{\sin\Omega + \sin(2\theta - \Omega) - 2\sin\theta}{\cos\Omega - \cos(2\theta - \Omega)} \qquad 2.9.9$$

x/d varies up to values of the order of 0.1, well within the coherence length or crystallite size. Consequently we can define the phase relationship associated with N planes separated by d from each other, in a similar manner to a combination of N slits, as in section 2.7.2: the amplitude for this combination is given by:

$$A_{2\theta} = \frac{\sin\left(\frac{\pi d}{\lambda}(2\sin\theta) - n\pi\right)}{\left(\frac{\pi d}{\lambda}(2\sin\theta) - n\pi\right)} \frac{\sin\left(N\left\{\frac{\pi d}{\lambda}(2\sin\theta) - n\pi\right\}\right)}{\sin\left(\frac{\pi d}{\lambda}(2\sin\theta) - n\pi\right)} \qquad 2.9.10$$

The amplitude should go to a maximum when the phase difference, given by $\pi/\lambda(2d\sin\theta)$ is zero or equal to $n\pi$, where n is an integer. This condition is clearly identical to Bragg's equation: $2d\sin\theta = n\lambda$.

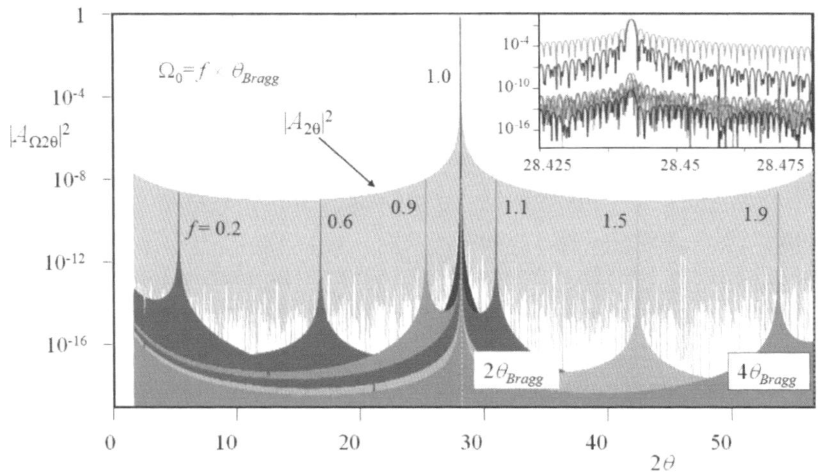

Figure 2.9.9. The profile along 2θ is given for different values of Ω_0, which shows a specular peak when $\Omega = \theta$ and enhancement at $2\theta_{Bragg}$ regardless of the Ω_0 value.

Equation 2.9.10 can now be combined with equation 2.9.6 to give:

$$A_{\Omega X 2\theta} = \frac{\sin\left(\frac{\pi L_x}{\lambda}(\cos(2\theta - \Omega_X) - \cos\Omega_X)\right)}{\left(\frac{\pi L_x}{\lambda}(\cos(2\theta - \Omega_X) - \cos\Omega_X)\right)} \frac{\sin\left(\frac{\pi L_y}{\lambda}\left(\sin\left[\tan^{-1}(\tan\Omega \sin X)\right]\right)\right)}{\left(\frac{\pi L_y}{\lambda}\left(\sin\left[\tan^{-1}(\tan\Omega \sin X)\right]\right)\right)} \qquad 2.9.11$$

$$\times \frac{\sin\left(\frac{\pi d}{\lambda}(2\sin\theta) - n\pi\right)}{\left(\frac{\pi d}{\lambda}(2\sin\theta) - n\pi\right)} \frac{\sin\left(N\left\{\frac{\pi d}{\lambda}(2\sin\theta) - n\pi\right\}\right)}{\sin\left(\frac{\pi d}{\lambda}(2\sin\theta) - n\pi\right)}$$

This is the amplitude from a set of N parallel crystal planes separated by d that are inclined at angles Ω and X, that arrives at the detector positioned at 2θ.

The consequence of this is that there is significant scattering remote from the Bragg condition with a distribution that is related to the crystallite shape, L_x, L_y and L_z, and the influence of N parallel planes. For a cube with scattering planes parallel to the surface there are streaks of scattering along the specular direction, i.e. $|A_{2\theta}|^2$ shown in figure 2.9.9 and along $|A_\Omega|^2$ shown in figure 2.9.7. For a random orientation of cubes, such that the planes are not necessarily orientated parallel to facets, the streaks will give an average distribution and contribute to the background, but the enhancement very close to $2\theta_{\text{Bragg}}$ will still occur. That is there will be scattering at the Bragg angle, but not necessarily at the Bragg condition. It is the addition of all these enhancements that make a very significant contribution to the intensity registered at $2\theta_{\text{Bragg}}$ and can explain the reason for presence of a complete, or near complete, diffraction pattern with very few crystallites.

As mentioned above the actual intensity measured by the detector is an integral over the captured region defined by the instrument at each position. The finite size of the focus and the dimensions of the crystallite in the scattering measurement plane will give a spread in Ω to be integrated, ΔΩ. Similarly the lateral dimensions of the axial divergence (e.g. dimensions of the Soller slits) and the inclination of the scattered beam to the crystal plane will change the bounds of the integral in X depending on 2θ and Ω. This leads to an intensity contribution at the position defined by Ω, X, and 2θ for each crystallite i given by:

$$I_{hkl}(2\theta) = f\left\{\left|F_{hkl}\right|^2\right\} pV \left[\int_{\Omega_i=\Omega-\frac{\Delta\Omega}{2}}^{\Omega_i=\Omega+\frac{\Delta\Omega}{2}} \int_{X_i=X-0.5\sin^{-1}\left(\frac{s_a}{2R\sin(2\theta-\Omega_X)}\right)}^{X_i=X+0.5\sin^{-1}\left(\frac{s_a}{2R\sin(2\theta-\Omega_X)}\right)} A^2_{X_i\Omega_i} P_{\Omega_i}\, dX_i\, d\Omega_i \right]$$

2.9.12

The parameters s_a, R, V, p and f will now be defined. s_a is the axial slit dimension and R is the radius of the goniometer; or alternatively the Soller slit acceptance angle is $\sin^{-1}(s_a/2R)$. This gives the minimum bounds of the tilt acceptance, ΔX. p is the polarization factor, but differs from that given in equation 2.3.22 that is solely related to the Bragg condition. From equation 2.9.11 it is clear that there is intensity everywhere and it all contributes to varying extents, therefore a more generalized description of parameters is required.

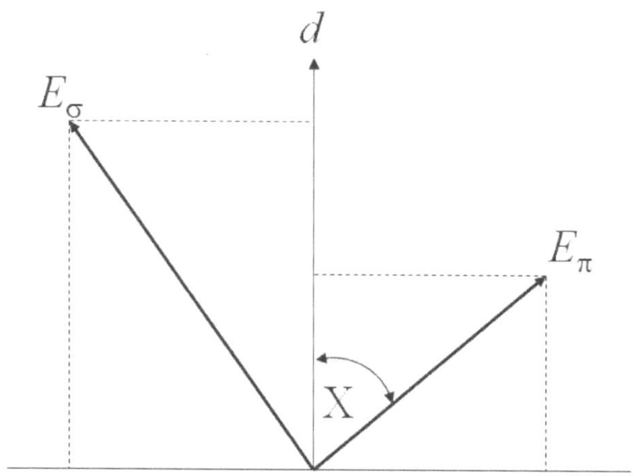

Figure 2.9.10. The resolved components of the electric field vector, with directions defined by the monochromator, as they appear on the inclined crystallite plane.

The polarization of the beam on scattering is given in figure 2.3.3, and the ratio of the two components of the electric field parallel (σ) and perpendicular (π) to the crystal plane surface are $1{:}\cos 2\theta$, provided there is no monochromator. If the instrument has a monochromator and its crystal planes and those of the crystallites are perpendicular to the scattering plane, then the scattering angle of the monochromator modifies this ratio to 1:$\cos 2\theta_m \cos 2\theta$. However for the more general case of the sample crystal planes inclined by X with respect to the crystal plane then this ratio is modified by the projection of these components onto the crystal plane normal of the crystallite, figure 2.9.10.

The components of $|E|^2$, i.e. the intensity parallel and perpendicular to the surface of the crystallite are given by:

$$E_{//}^2 = E_\pi^2 \sin^2 \mathrm{X} + E_\sigma^2 \cos^2 \mathrm{X}$$
$$E_\perp^2 = E_\pi^2 \cos^2 \mathrm{X} + E_\sigma^2 \sin^2 \mathrm{X} \qquad 2.9.13$$

E_σ and E_π are the components of the electric field scattered from the monochromator. If there is no monochromator then, the parallel and perpendicular components are the same for all X; in this case :

$$p = \frac{1+\cos^2 2\theta}{2} \qquad 2.9.14$$

p is the proportion of the incident intensity that can be scattered. If there is a monochromator then $E_\sigma^2 = E^2$ and $E_\pi^2 = E^2\cos 2\theta_m$, where E is the magnitude of the electric field vector incident on the monochromator and p becomes:

$$p = \left(\frac{\left[\cos^2 \mathrm{X} + \sin^2 \mathrm{X} \cos^2 2\theta_m\right] + \left[\sin^2 \mathrm{X} + \cos^2 \mathrm{X} \cos^2 2\theta_m\right] \cos^2 2\theta}{2} \right)$$

2.9.15

The parameter V in equation 2.9.12 refers to the integration volume in reciprocal space: this has been touched upon in terms of the influence of axial divergence, ΔX, and angular acceptance by the crystallite, ΔΩ, in determining the intensity at a specific 2θ, i.e. that component is included within the integral, of equation 2.9.12. The detector has to collect intensity over a range of 2θ, $\Delta 2\theta_s$, or the slit size. This is the third dimension of the integration volume, which is independent of X, i.e. Ω is orthogonal to X, and 2θ cannot be influenced by X. Referring to figure 2.9.11a, we can see the area in reciprocal space that is captured. The full explanation of reciprocal space will be covered in greater detail in section 4.2, but for now it can be seen that if a crystallite is rocked about an axis normal to the page then S will sweep through reciprocal space changing Ω. Similarly if the detector is rotated by changing 2θ, i.e. $\boldsymbol{k}_{\mathrm{H}}$ is rotated about the same axis, then the incident angle is unchanged and a region is mapped out defined by $\Delta\Omega$ and $\Delta 2\theta$ in the figure. The corners of this region in reciprocal space coordinates in figure 2.9.11 can be evaluated from

$$s_{xi} = \frac{1}{\lambda}\left(\cos\left(2\theta_i - \Omega_i\right) - \cos\Omega_i\right)$$
$$s_{zi} = \frac{1}{\lambda}\left(\sin\left(2\theta_i - \Omega_i\right) + \sin\Omega_i\right) \qquad 2.9.16$$

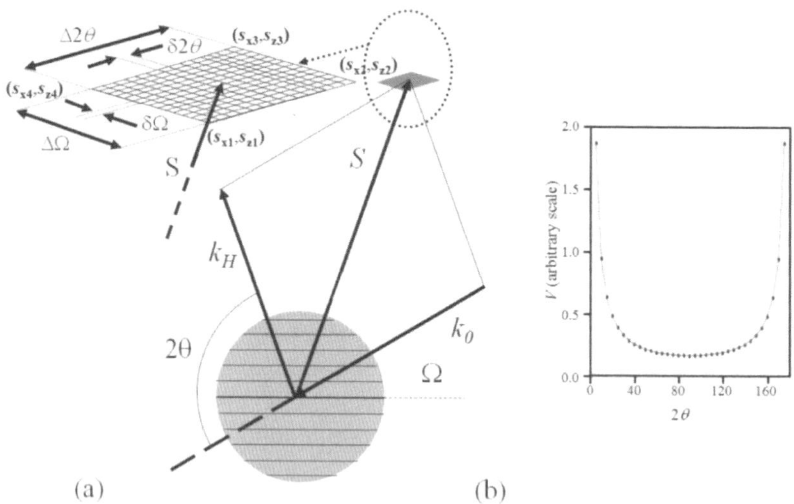

Figure 2.9.11. The region in reciprocal space captured for a spread in Ω and 2θ, which relates to the diffractometer resolution, (a). This contribution is purely a function of 2θ, as shown in (b).

From these four coordinates, the area of reciprocal space can be calculated. This region of capture is again an oversampling. This oversampling can be very small when angular high-resolution diffractometers are used, however it is the change in oversampling with $\boldsymbol{S}$ that is important for comparing intensities from different reflections. This is particularly relevant to structure determination and powder diffraction studies. The conventional approach to explain this oversampling is that given by Lorentz, which is referred to in Debye (1913) and Buerger (1940), and is discussed in terms of the time that a diffraction peak remains in the Bragg condition. This Lorentz factor has a $1/\sin 2\theta$ dependence, which is overlaid and scaled to the integration volume derived above, figure 2.9.11b. The results are identical therefore we can define V in equation 2.9.12 as $1/\sin 2\theta$. It is interesting to note that it is independent of Ω, and in evaluating the points in figure 2.9.11b, Ω was chosen randomly in the range from 0 to 2θ.

The parameter f, which precedes the square of the structure factor $|F_{hkl}|^2$, accounts for the dispersion of the scattering power. The scattering power, within this description, is not concentrated at the Bragg peak, but spread in Ω and X, as well as 2θ. From figure 2.9.7, it can be seen that this dispersion varies with 2θ, being most heavily dispersed at $2\theta = 90^0$ and when the crystallite size is small. In a normal experiment, not all of the intensity is captured and therefore the calculated intensity must be modified, this parameter f refers to the proportion that is captured.

Following the arguments in Fewster (2014), the dispersion D can be evaluated with sampling using:

$$D\left(\left|F_{hkl}\right|^2\right)=\left\{\int_0^{4\theta_B(or\pi)} A_{2\theta}^2 d2\theta\right\}\left\{\sum_{i=1}^{N}\left[\int_{\Omega_i=\Omega-\frac{\Delta\Omega}{2}}^{\Omega_i=\Omega+\frac{\Delta\Omega}{2}}\int_{\mathrm{X}_i=\mathrm{X}-0.5\sin^{-1}\left(\frac{s_a}{2R}\right)}^{\mathrm{X}_i=\mathrm{X}+0.5\sin^{-1}\left(\frac{s_a}{2R}\right)} A_{\mathrm{X}_i\Omega_i}^2 P_{\Omega_i}\, d\mathrm{X}_i\, d\Omega_i\right]\right\} \qquad 2.9.17$$

The parameter f is dependent on the range over which the intensity is calculated, i.e. how much of the intensity is captured in an experiment:

$$f = \frac{\left\{\int_{-\Delta 2\theta_B}^{+\Delta 2\theta_B} A_{2\theta}^2 d2\theta\right\}\left\{\sum_{i=1}^{N}\left[\int_{\Omega_i=\Omega-\frac{\Delta\Omega}{2}}^{\Omega_i=\Omega+\frac{\Delta\Omega}{2}} \int_{X_i=X-0.5\sin^{-1}\left(\frac{s_a}{2R\sin(2\theta-\Omega_X)}\right)}^{X_i=X+0.5\sin^{-1}\left(\frac{s_a}{2R\sin(2\theta-\Omega_X)}\right)} A_{X_i\Omega_i}^2 P_{\Omega_i} dX_i d\Omega_i\right]\right\}}{D\left(|F_{hkl}|^2\right)} \qquad 2.9.18$$

The experimental range in this case is from $-\Delta 2\theta_B$ to $+\Delta 2\theta_B$. The first term in D and f clearly accounts for the dispersion in 2θ, whereas the second term relates to the dispersion in Ω and X, and takes account of the oversampling in X.

The calculation of these Ω/X maps, figure 2.9.7, ideally should be performed at each 2θ value, however this is rather impractical computationally. The approach is to create a calibration curve that represents the integral of the intensities at Ω and X for various 2θ values. The resulting calibration curve is smooth and follows the same shape for a range of crystallite sizes given in figure 2.9.12. The integrals in this example are for 3 μm crystallites and are evaluated by a form of importance sampling, Fewster (2014). The first aspect to recognize is that the mean intensity for the full integral is only slightly above the integral of everything except the Bragg peak contributions. This simply means that the presence of contributions that are not associated with the Bragg condition give an underlying stable intensity, and superimposed on this are the far smaller number of contributions from the Bragg peaks. This results in a much more reliable intensity than the contributions from Bragg peaks alone.

For each diffracting plane we can determine the structure factor and its dispersion according to equation 2.9.18. The intensity profile is calculated at each 2θ value, as given in equation 2.9.13, for each crystal plane reflection within the bounds from $2\theta = 0$ to $2\theta = 4\theta_B$ or π, whichever is the smaller. Each profile at each 2θ is then scaled by interpolation with the calibration curve, which accounts for the dispersion of the scattering power (structure factor) in Ω and X. This profile includes the crystallite size normal to the crystal planes as Nd, and this dominates the peak width. The resulting profile is given in figure 2.9.13, i.e. all the peak shapes and background intensity associated with sample scattering are included.

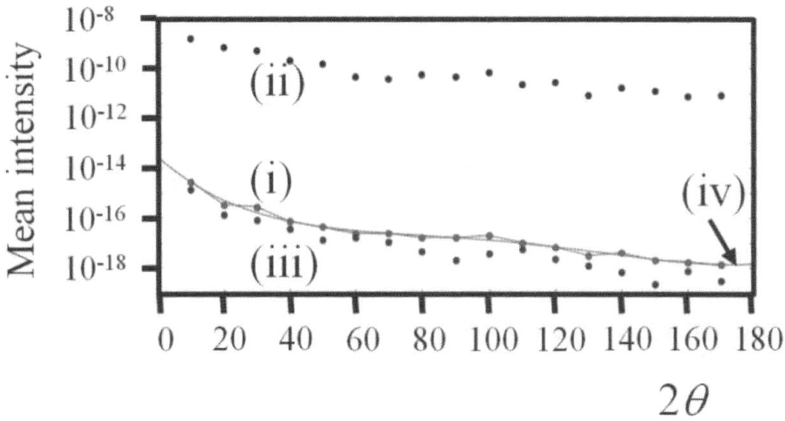

Figure 2.9.12. The mean intensities obtained by sampling from all the contributions (i), from the Bragg peaks (ii) and from the non-Bragg peaks only (iii). The calibration curve for interpolating the dispersion of the intensity is a polynomial fitted (iv) to the mean values from all the contributions.

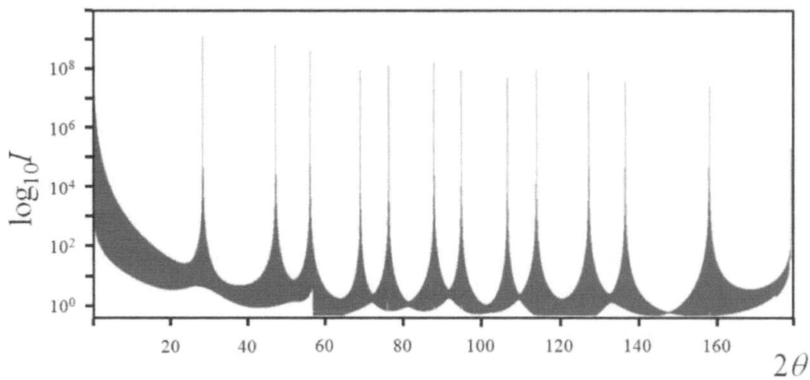

Figure 2.9.13. The intensity profile that has been calculated from a randomly orientated perfect silicon powder.

The inclusion of texture will define the sampling distribution for the Ω/X maps, which would then have to be evaluated for each reflection. For an assumed random distribution of orientations there is only one calibration curve and the sampling of the maps is uniform. An example of the experimental results and how they match with this theory are given in Fewster (2014).

Although the complexity is greater than in the conventional theory used to describe the scattering from powders; the result is that the intensity matches the experimental values much more closely, and account for features observed in careful experiments. An important consequence is that the scattering from very few crystallites can be explained. Suppose that figure 2.9.7 is at the Bragg angle, $2\theta_B$, then the map will represent the intensity that enters the detector from a crystallite misorientated from the Bragg condition. The proportion of the intensity that is not emanating from the Bragg condition but contributing, for a large number of 10 μm and 3 μm size crystallites is ~30% and ~35%, from calculation, assuming the crystals are perfect. This is significant and explains why the Bragg condition is not necessary to observe diffraction peaks, and why the intensities are reliable (because there is a large number of contributing intensities).

2.10. Optical theory applied to reflectometry

In the discussion on basic dynamical theory we showed that when one wave enters the sample and no scattered wave is produced the incident beam is refracted and a specular reflected wave is produced, equation 2.3.24 and figure 2.3.4. One advantage of the dispersion construction, figure 2.3.4 and 2.3.5 is that it becomes clear that each intersection of the surface normal with the dispersion surface should create a contribution. If we refer to figure 2.3.5 we can see that there are several intersections in the two-wave dynamical model. These arise from the normal wave-fields described in the basic dynamical model and a specular wave. The specular wave will therefore exist over all scattering angles but is generally very weak and insignificant except close to the condition of grazing incidence. Within the more complete dynamical theory, section 2.4.4, that uses the Fresnel equations, the principle of specular reflectivity is exploited and therefore includes the following results exactly. This section reiterates the methodology but simplifies it by assuming that the refractive index, i.e. polarizability, is uniform over the layer thickness, whereas in section 2.4.4 the refractive index is assumed constant over a very small dimension ~0.01×unit cell dimension. This macroscopic level makes it akin to optical theory approaches, and has the advantages of not requiring detailed structural knowledge.

Equation 2.3.25 indicates the effective refractive index for material is given by

$$n = (1+\chi_0)^{\frac{1}{2}} \sim \left(1+\frac{\chi_0}{2}\right) \qquad 2.10.1$$

Where

$$\chi_0 = -\frac{e^2\lambda^2}{4\pi^2\varepsilon_0 mc^2}\rho_e = -\frac{r_e\lambda^2 Z}{\pi m_p A}\rho = -\frac{r_e\lambda^2 N_A Z}{\pi A}\rho = -\frac{r_e\lambda^2}{\pi V}F_0 \qquad 2.10.2$$

from equations 2.3.7, 2.4.8 and 2.4.9. The electron density ρ_e has been replaced by the macroscopic density, ρ by transforming it with the number of electrons in a unit volume, *Z*, the atomic mass of these atoms, *A*, and the mass of the basic atomic building block, m_p of the hydrogen atom or Avogadro's constant N_A. This assumes that an average refractive index can be used is this case. At very low angles of incidence the X-rays are probing length-scales normal to the surface plane that are very large, i.e. $D \sim \lambda/2\sin\omega$ ($\omega = \theta$) from Bragg's equation, well in excess of any interatomic distance, *d*. Hence this assumption is valid since we are averaging the spatially varying electron density, $\rho(\boldsymbol{r})$, over these long length-scales this equates to the macroscopic average electron density, ρ, χ_0 is constant. Using a similar argument the absorption term must be some macroscopic average, which is included naturally as the imaginary component of F_0, whereas for the density it has to be included more explicitly. The phase of a wave at a position *r* along the wave-vector *K* into the sample will be given by

$$\Phi = \exp\{-2\pi inKr\} = \exp\frac{(-2\pi i\{1-\delta-i\beta\}r)}{\lambda} = \exp\frac{-2\pi i\{1-\delta\}r}{\lambda}\exp\frac{-2\pi\beta r}{\lambda} \tag{2.10.3}$$

where $n = 1-\delta-i\beta$, the refractive index is arranged into real and imaginary components. δ is the real part of χ_0 in this case. As before, we take the real part to contain the phase front and the imaginary part to include the absorption, figure 2.3.2. Now the loss of intensity due to the linear absorption coefficient is given in equation 2.7.13, and this is the same as the imaginary term of equation 2.10.3, and hence

$$\beta = \frac{\mu\lambda}{4\pi} \tag{2.10.4}$$

The extra factor of two arises from the equation 2.10.3 being associated with amplitude and equation 2.7.13 being associated with intensity.

The reflection coefficient for X-rays from the interface between two materials is related to the difference in the wave-vector normal to the surface. The parallel component, as described earlier is unchanged, section 2.3.1. We can therefore write the reflection coefficient at an interface between two layers j and j-1 as:

$$r_{j-1,j} = \left(\frac{K_{j-1} - K_j}{K_{j-1} + K_j}\right)_{\perp} \tag{2.10.5}$$

where $(\boldsymbol{K})_{\perp}$ is the wave-vector normal to the surface. This equation can be compared to equation 2.4.31.

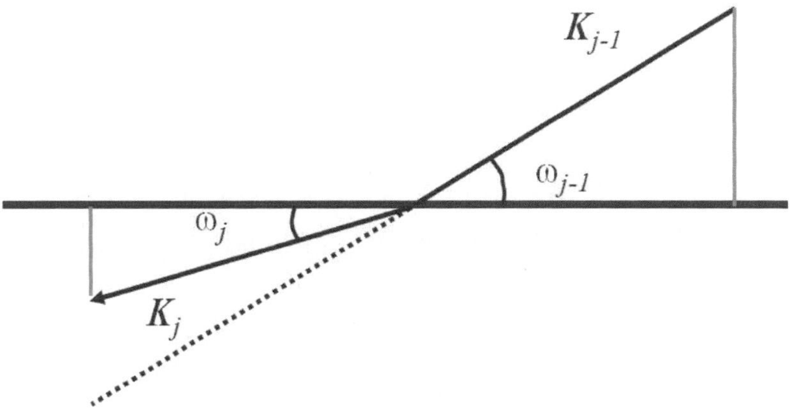

Figure 2.10.1. The influence of refraction on the incident beam to create a forward-diffracted beam.

Consider now figure 2.10.1 and include the conditions that we have stated

$$K_j \cos\omega_j = K_{j-1}\cos\omega_{j-1} = \mathrm{n}_j k_0 \cos\omega_j = \mathrm{n}_{j-1} k_0 \cos\omega_{j-1} \qquad 2.10.6$$

i.e. the parallel components of the wave-vector are equal and the wave-vector is changed in magnitude by the refractive index n compared with that in vacuum $\boldsymbol{k_0}$, equation 2.3.25. Now equation 2.10.5 becomes

$$r_{j-1,j} = \frac{K_{j-1}\sin\omega_{j-1} - K_j \sin\omega_j}{K_{j-1}\sin\omega_{j-1} + K_j \sin\omega_j} = \frac{n_{j-1}\sin\omega_{j-1} - n_j \sin\omega_j}{n_{j-1}\sin\omega_{j-1} + n_j \sin\omega_j} \qquad 2.10.7$$

Now from equation 2.10.6

$$\cos\omega_j = \frac{n_{j-1}}{n_j}\cos\omega_{j-1} \qquad 2.10.8$$

and hence

$$\sin\omega_j = \left(1 - \left\{\frac{n_{j-1}}{n_j}\right\}^2 \cos^2\omega_{j-1}\right)^{\frac{1}{2}} \qquad 2.10.9$$

Therefore the reflection coefficient can now be determined in terms of the individual refractive indices and the incident angle on the interface concerned where there is a change in the refractive index.

$$r_{j-1,j} = \frac{n_{j-1}\sin\omega_{j-1} - (n_j^2 - n_{j-1}^2\cos^2\omega_{j-1})^{1/2}}{n_{j-1}\sin\omega_{j-1} + (n_j^2 - n_{j-1}^2\cos^2\omega_{j-1})^{1/2}} \qquad 2.10.10$$

We now have to consider the magnitude of the incident and reflected electric fields, $\boldsymbol{D}_0$ and $\boldsymbol{D}_R$ respectively, which have the form given in equation 2.3.13.

$$\begin{gathered} D_{R,m}\exp\{2\pi S_m d_m/2\} \\ D_{0,m}\exp\{-2\pi S_m d_m/2\} \end{gathered} \qquad 2.10.11$$

This is the magnitude of the electric wave in the middle of the m th layer that has a thickness d_m. From figure 2.3.3 we can see that the change in the electric field component in the plane of the scattering (π polarisation) is insignificant for small incident angles whereas the electric field component parallel to the interface, the σ polarization is unchanged. We shall treat them equally. The wave-field components parallel to and at the interface (i.e. at a distance $d_{j-1}/2$ from the centre of layer j-1 and $d_j/2$ from the centre of layer j) are equal, therefore

$$(D_{0,j-1}e^{2\pi i S_{j-1}d_{j-1}/2} + D_{R,j-1}e^{-2\pi i S_{j-1}d_{j-1}/2}) = (D_{0,j}e^{-2\pi i S_j d_j/2} + D_{R,j}e^{2\pi i S_j d_j/2})$$

2.10.12

As discussed above the reflectivity is given by the change in wave-vector normal to the surface plane. The component of the amplitude of the wave normal to the interface will be modified in proportion to the scattering vector. The scattering vector $\boldsymbol{S}_j = 2\boldsymbol{K}_j\sin\omega_j = 2\boldsymbol{k}_0 n_j\sin\omega_j$, (i.e. twice the vertical component of the wave-vector) and since this is purely a specular reflection the scattering vector is always normal to the surface plane. For this to be continuous across the interface

$$(D_{0,j-1}e^{2\pi iS_{j-1}d_{j-1}/2} - D_{R,j-1}e^{-2\pi iS_{j-1}d_{j-1}/2})S_{j-1} = (D_{0,j}e^{-2\pi iS_{j}d_{j}/2} - D_{R,j}e^{2\pi iS_{j}d_{j}/2})S_{j} \qquad 2.10.13$$

Hence for a σ, π polarised or a circularly polarised incident beam the magnitude of these electric field amplitudes are in proportion, equation 2.10.12 and 2.10.13, and we can solve these two equations to give

$$X_{j-1} = (e^{-2\pi iS_{j-1}d_{j-1}/2})^4 \left\{ \frac{r_{j-1,j} + X_j}{r_{j-1,j}X_j + 1} \right\} \qquad 2.10.14$$

where

$$X_j = \frac{D_{Rj}}{D_{0,j}}(e^{-2\pi iS_jd_j/2})^2 \qquad 2.10.15$$

and

$$r_{j-1,j} = \frac{S_{j-1} - S_j}{S_{j-1} + S_j} = \left(\frac{K_{j-1} - K_j}{K_{j-1} + K_j} \right)_{\perp} \qquad 2.10.16$$

This is identical to equation 2.4.31, and so for completeness we include the transmission coefficient, t:

$$t_{j-1,j} = \left(\frac{2K_{j-1}}{K_{j-1} + K_j} \right)_{\perp} \qquad 2.10.17$$

Equation 2.10.15 is similar to our previous definitions, except that we have a phase term since we require the amplitude at the interface not at the mid-point of the layer j.

We now have a recursive expression for the reflected amplitude, X_{j-1}. Hence if we know the amplitude in the layer j, X_j (equation 2.10.15) and the reflection coefficient at the interface between the j-1 and the jth layers, then we can determine the amplitude in layer j-1 (equation 2.10.14). This is where we now need to apply a boundary condition since we do not know the amplitude in layer j. We can assume that the amplitude at the bottom of the sample is zero, i.e. there is no intensity scattered upwards from the bottom surface, then $X_N = 0$, therefore we can determine the amplitude immediately above this first interface, X_{N-1} from equation 2.10.14. The reflection coefficient at each interface has to be determined first, equation 2.10.10, and this is also recursive since we only know the angle of incidence at the surface, ω_0.

The reflected intensity or reflectivity of the sample is given by

$$R_0 = \frac{I}{I_0} = X_0^* X_0 \qquad 2.10.18$$

I_0 is the incident beam intensity and I the specular reflected beam intensity. The suffix 0 refers to the air (~ vacuum) layer above the sample.

It is clear to see that there are many assumptions that can be made, but without a full investigation it is difficult to assess their impact. Consequently the work described here will use the derived equation from the above approach.

2.10.1. Some general conclusions from this derivation

Consider equation 2.10.2, where we have two ways of describing the susceptibility. Either we can consider the value as a function of the electron density or in terms of a structure factor. The structure factor for $\boldsymbol{H} = 0$ is simply the sum of the electrons in the unit cell repeat and this is only applicable to crystalline solids. However the electron density can be related to the macroscopic density through Avogadro's number and the average atomic mass, equation 2.10.2, therefore we can apply this model to amorphous and poorly crystalline material. It is very important though to realise that amorphous and polycrystalline materials may not have the same density as their crystalline equivalents.

To obtain an indication of the form that this reflectivity takes, consider equation 2.10.10, for the case of a single homogeneous sample. We are now only interested in a single interface, the surface, hence

$$r_{0,1} = \frac{\sin\omega_0 - (n_1^2 - \cos^2\omega_0)^{1/2}}{\sin\omega_0 + (n_1^2 - \cos^2\omega_0)^{1/2}} \qquad 2.10.19$$

n_0 is of course unity (the refractive index of air) and n_1 is complex. However for incident beam angles where

$$\omega_0 < \cos^{-1}\{\mathrm{Re}[n_1]\} \qquad 2.10.20$$

the reflectivity is close to unity, although there is a small imaginary component that represents absorption into the sample. This penetration depth is very small. 'Re' represents the real part of the quantity. When the angle of incidence reaches a value such that

$$\omega_0 > \cos^{-1}(\mathrm{Re}[n_1]) \qquad 2.10.21$$

then the reflectivity begins to fall in value very rapidly. This angle where the rapid change in reflectivity occurs is called the critical angle. An indication of the shape is given in figure 2.10.2, as well as the variation in the penetration depth with incident angle.

As will be shown later in the examples, Chapter 5, the reflectivity from layer structures will exhibit oscillations characteristic of the thicknesses involved. In this case the penetration of the X-rays varies and is a maximum in the troughs of the oscillations and at a minimum at the peaks. The critical angle, when equations 2.10.20 and 2.10.21 become equalities, is clearly just a function of the material and the wavelength of the X-rays. From equations 2.10.1 and 2.10.2, we have

$$(\omega_0)_c \approx \cos^{-1}\left\{\mathrm{Re}\left(1 - \frac{r_e\lambda^2}{2\pi V}F_0\right)\right\} \approx \cos^{-1}\left\{\mathrm{Re}\left(1 - \frac{r_e\lambda^2 N_A Z\rho}{2\pi A} - i\frac{\lambda\mu}{4\pi}\right)\right\}$$

2.10.22

The derivation so far is based on an infinitely long sample that is homogeneous and abrupt at all its interfaces. The next section shall address some of the influences.

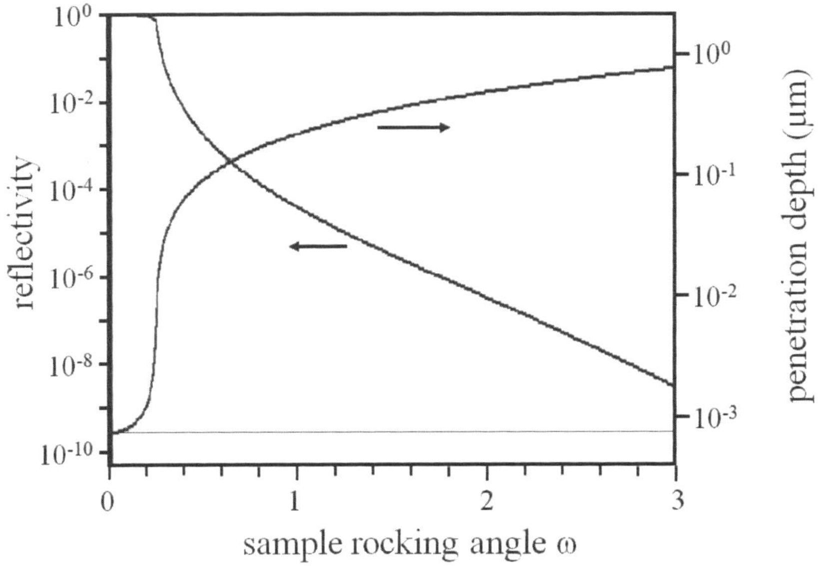

Figure 2.10.2. The variation of specular reflecting power of Si and its consequential penetration depth below the surface.

2.10.2. Imperfect interfaces

We can include imperfect interfaces in a number of ways. The simplest approach is to include a Gaussian smearing function to the reflectivity at each interface. The most robust method is to include many layers of different electron density close to the interface, just as in the case of the dynamical model of section 2.4. We can now calculate the influence of roughness on the shape of the specular profile for a single layer on a substrate, by varying the spreading of the surface and interface electron densities, figure 2.10.3.

The description of interfacial roughness considered above is based on the averaging of the density parallel to the interface and this is applicable for interfaces dominated by inter-diffusion and segregation. For laterally rough (jagged) interfaces we need to consider a different approach. This will be covered in the following section by the introduction of another theoretical approach.

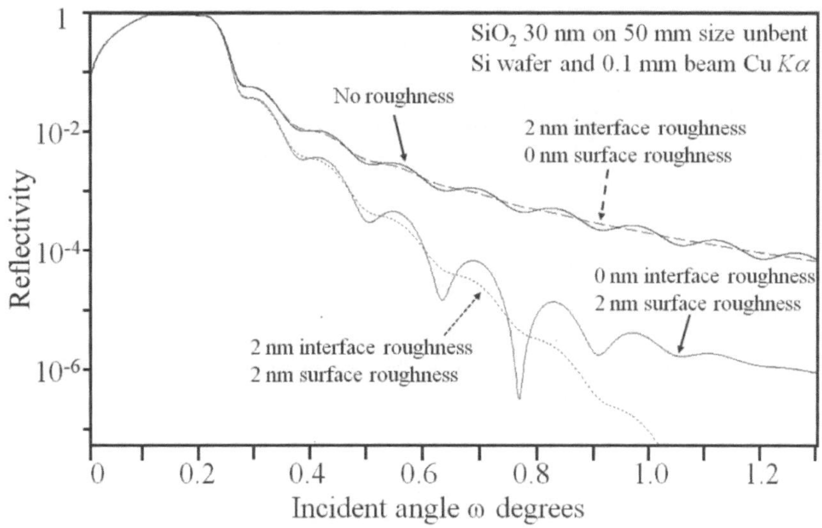

Figure 2.10.3. The simulated profile of Si with a SiO_2 layer with varying degrees of roughness. Note how the region of "total" external reflection is altered from the simulation of figure 2.10.2, because of the inclusion of finite sample and beam size.

2.11. The Distorted Wave Born Approximation:

The wave equation given in equation 2.3.16, is re-written here using electric fields E at a position r at time t:

$$\nabla^2 E = \frac{1}{c^2}\frac{d^2 E}{dt^2} \qquad 2.11.1$$

Since we are assuming that the frequency of the incident and scattered waves are the same, we are interested in the time-independent form of this equation (just as the time-dependent aspect falls out of the equations in dynamical theory). By separating the time and position dependent parts, $E(r,\mathrm{t}) = E(r)E(\mathrm{t})$ then:

$$\frac{\nabla^2 E(r)}{E(r)} = \frac{1}{c^2 E(\mathrm{t})} \frac{d^2 E(\mathrm{t})}{dt^2} \qquad 2.11.2$$

Since both sides are independent, but must always be equal, then they are equal to a constant if there is no scattering or a constant + scattering potential if in a scattering medium (e.g. a crystal); concentrating on the time-independent part (the left-hand side) we have the Helmholtz equation:

$$\left(\nabla^2 + k^2\right)E(r) = V(r)E(r) \qquad 2.11.3$$

∇^2 is the Laplace operator and k is the wave-vector in vacuo, defined in this section as $k = 2\pi/\lambda$, and $V(r)$ is the scattering potential.

The scattering from one point on the sample, $\boldsymbol{r}$, will create a spherical wave from the incident wave. The part of the spherical wave that arrives at $\boldsymbol{r}'$, will be re-scattered, some of which will travel along $\boldsymbol{r} - \boldsymbol{r}'$ and interact with the contribution directly along $\boldsymbol{r}$, figure 2.11.1. Thus the new field will be the original field plus the contributions from all the other re-scattered events:

$$E(r) = E_0(r) + \int G(r - r')E_0(r')V(r')dr' \qquad 2.11.4$$

The second term includes a correlation function and the scattering potential to give the field that is scattered from the point at r'. This re-scattering event can be further re-scattered at r'' leading to the Born expansion:

$$E(r) = E_0(r) + \int G(r - r')E_0(r')V(r')dr' + \int G(r - r')V(r')dr' \int G(r - r'')E(r'')V(r'')dr'' \qquad 2.11.5$$

Clearly this expansion can be extended indefinitely, however the contribution becomes weaker and weaker, therefore in this discussion only the first term will be considered, equation 2.11.4; the Born Approximation.

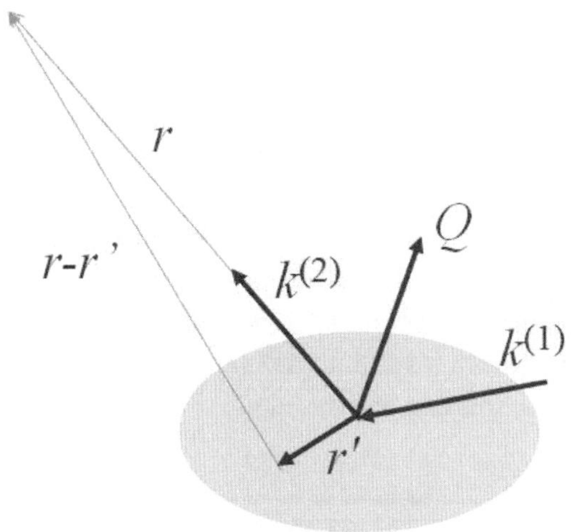

Figure 2.11.1. The incident wave with wave-vector $\boldsymbol{k}^{(1)}$ will produce scattering along $\boldsymbol{k}^{(2)}$ and will also scatter towards a position ***r'*** that will result in further scattering that will combine with $\boldsymbol{k}^{(2)}$.

From equation 2.11.3, $G(r - r')E(r)$ is simply $(\nabla^2 - k^2)E(r)$, which has the solution $-\exp(ik.r)/4\pi r$ using Green's function of this differential operator. Equation 2.11.4 can then be rewritten as:

$$E(r) = E_0(r) + \frac{e^{ik.r}}{4\pi r}\int e^{-2\pi i S.r'}V(r')dr' \qquad 2.11.6$$

The scattering potential can be expressed as $V(r) = -k^2\chi(r)$. $V(r)$ is strictly an operator of the scattering potential, $V(r) = \text{graddiv} - k^2\chi(r)$, however $\text{graddiv}E \sim 0$ for a transverse wave. For calculating the scattering close to the grazing incident and exit conditions, the scattering potential becomes $V(r) = -k^2\chi_0(r) \sim -2k^2(1 - n(r))$, where $n(r)$ is the refractive index at position r, from equation 2.10.1. This scattering therefore probes the density fluctuations. If it is assumed that the incident wave does not enter the detector during the experiment, then only the second term is relevant in equation 2.11.6. In addition, since we are assuming that the sample to detector distance >> *r*', which is the distance between scattering centres as shown in figure 2.11.1, the term $e^{-ik.r}/r$ can be assumed to be constant, i.e. $r - r' \sim r$. The whole of the second term is therefore equivalent to the structure factor, equation 2.3.9.

It is convenient to consider the covariance of the scattering from these various contributions: this can be considered as to how well the scattering from different contributions combine. For example if the scattering from all contributions combine perfectly in phase (regularly positioned) and their shapes are identical, then the intensity observed is high; whereas when the scattering contributions are random in size and phase the intensity is weak. A convenient route to determining the intensity is through the differential cross-section, i.e. the probability of finding scattering within a solid angle, and since there are 4π steradians in three dimensions, this can expressed as:

$$\frac{d\sigma}{d\Omega} = \frac{1}{16\pi^2}\operatorname{cov}(v, v) \qquad 2.11.7$$

If we define a disturbance as the impact that the incident wave $E^{(1)}$ has on the scattering potential V to create the scattered wave $E^{(2)}$, then the disturbance ν is given by:

$$v = \langle E^{(2)} | V | E^{(1)} \rangle \qquad 2.11.8$$

This approach allows us to easily add complexity by separating the scattering potential into two parts V_A and V_B, for example: these could refer to the scattering from a perfect structure and the disturbance, e.g. defects. This is a very useful and general way of analyzing dilute disturbances, so that there are sufficient regions of perfection that follow dynamical theory and small regions that follow kinematical theory. The perfect part can be considered to define the electric field at any position *r* in the structure and this can be used to give the field from which the imperfect regions scatter. This is the basis of the Distorted Wave Born Approximation, which is a very powerful approach for modeling the scatter from defects. This basic concept was used in section 2.8.2.

The following sections will give some examples of this methodology, including; analyzing interfaces close to the scattering conditions in reflectometry, in-plane scattering from quantum dots and periodic polycrystalline multi-layers, as well as grazing incidence small angle X-ray scattering. The scattering cross-section is different for each example, and so should give some insight into the possible applications of this approach and these methods.

2.11.1. The modeling of interface profiles close to grazing angles:

The reflectivity described in section 2.10 is from specular or 'mirror-like' reflections, and is insensitive to strain but sensitive to density variations. If there are local density undulations that are not parallel to the interface, figure 2.11.2, the specular reflectivity will average laterally. However lateral density fluctuations will create scatter away from the perfect specular condition that gives an indication of their extent. Considering the process of scattering from a density profile given in figure 2.11.2, we can immediately see the complications: multiple reflections can take place as well as transmission, and hence refraction, through jagged regions. Also the scattering will almost certainly be out of the general plane of scattering. However a number of authors have taken a more pragmatic approach by assuming that finite correlation lengths obtained from the diffuse scatter profiles can be directly attributed to a measure of lateral roughness. For two regions of different density separated by a distance $\boldsymbol{r}$, we can define some scattering function as the Fourier transform of a pair correlation function, $G(\boldsymbol{r})$.

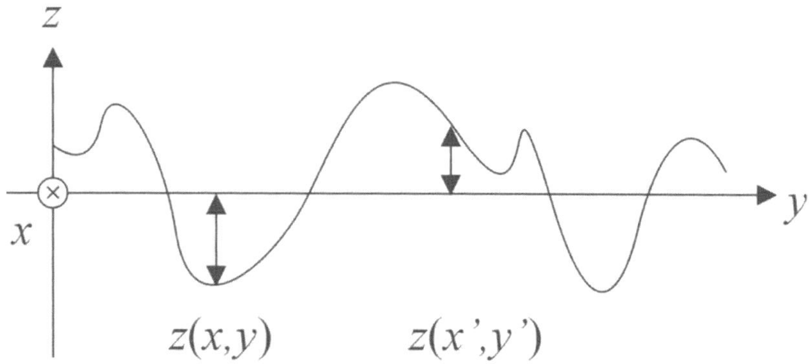

Figure 2.11.2. The imperfect interface can be characterized by a complex structure of intermingled densities.

The full manipulation of the formulae will not be included here, but the basics and results will be given. Now the disturbance v, given in equation 2.11.8 will be related to the density step between regions that are denoted by Δ and the scattering processes A. The scattering process is the amplitude associated with the wave scattered from the initial to final states and is described with reference to figure 2.11.3. In figure 2.11.3a, the vectors refer to the scattering vectors with time going forward; this is analogous to the incident beam or initial state and is denoted by the subscript (1) in the following explanation. Figure 2.11.3a will have time-reversal symmetry, i.e. if a wave can travel in one direction, it must be able to travel in the opposite direction; this is represented in figure 2.11.3b, which actually refers to the final vectors or final state denoted by the subscript (2). We can also understand this from the wave equation that is dependent on c^2 and therefore the solution is the same for $-c$ and $+c$. This results in an overall symmetry in the wave-vectors. This rather counter-intuitive thinking is explained by Feynman, Leighton and Sands (1966), Stokes (1903).

From the reasoning above and figure 2.11.3, we will have a transmitted and reflected wave for the initial and final states that leads to 4 amplitudes, A_a with $a = 1, 2, 3, 4$, that should be considered giving the following combinations; $T^{(1)}T^{(2)}$, $R^{(1)}T^{(2)}$, $T^{(1)}R^{(2)}$ and $R^{(1)}$ $R^{(2)}$. Each of these amplitudes have an associated reduced wave-vector, q, given by; $q_1 = K_0^{(2)}\text{-}K_0^{(1)}$, $q_2 = K_0^{(2)}\text{-}k_R^{(1)}$, $q_3 = k_R^{(2)}\text{-}K_0^{(1)}$ and $q_4 = k_R^{(2)}\text{-}k_R^{(1)}$ respectively. These can be visualized in figure 2.11.3.

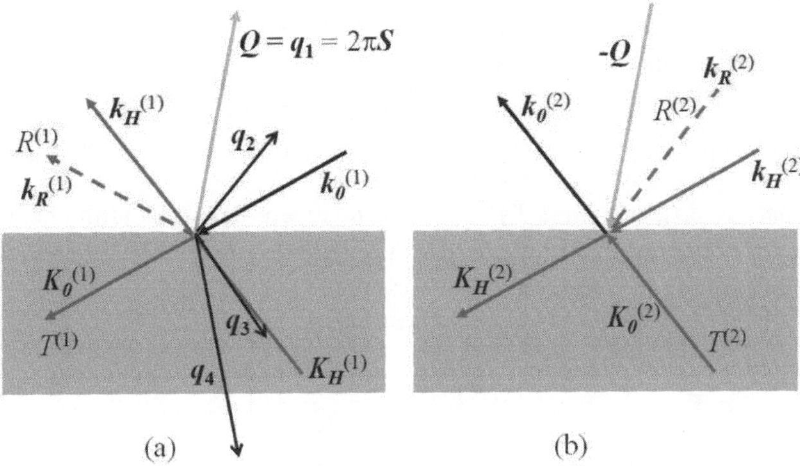

Figure 2.11.3. The wave-vectors associated with the scattering process initiated by the incident wave $\boldsymbol{k_0}^{(1)}$ given in (a). (b) is the time-reversed optical paths that relate to the final state and measured intensity along $\boldsymbol{k_0}^{(2)}$.

The covariance of the disturbance is therefore related to the density differences, the amplitudes of the scattering event and the nature of the interface:

$$\operatorname{cov}(v,v) = 4k^2 \sum_{i,j=1}^{N+1} \sum_{a,b=1}^{4} \Delta_i \Delta_j^* A_a^i \left(A_b^i\right)^* \frac{\mathrm{A}}{q_a^i \left(q_b^i\right)^*} F_{ab}^{ij} \qquad 2.11.9$$

'A' is the area of the sample that is illuminated and F defines the interface profile. The indices i and j refer to the interfaces, of which there are N, and a and b refer to the interactions given above. F is given by a phase term relating the coordinates of scattering points parallel to the interface plane, and the covariance of the scattering points at a distance above or below the interface reference plane:

$$F_{ab}^{ij} = e^{-iQ_{//}.(x-x')} \operatorname{cov}\left(e^{-iq_a^i \Delta z_i(x)} - 1, e^{-iq_b^j \Delta z_j(x')} - 1\right) \qquad 2.11.10$$

This function is summed over the irradiated sample area. Clearly the diffuse scattering associated with the interfaces, will only exist if there are density fluctuations laterally and vertically.

Consider figure 2.11.2; then the mean square deviation from the average interface position is given by:

$$g(R) = \left\langle \{z(x,y) - z(x',y')\}^2 \right\rangle = 2\left\langle z^2 \right\rangle - 2\left\langle z(0,0) z(x-x', y-y') \right\rangle \qquad 2.11.11$$

Where $R = ([x\text{-}x']^2 + [y\text{-}y']^2)^{0.5}$. The mean square deviation is the height-height correlation and $<z^2>$ is the mean square roughness, that we will call σ^2. This can be rearranged to become:

$$g(R) = 2\left(1 - \frac{C(R)}{\sigma^2}\right) \qquad 2.11.12$$

$C(R)$ is the in-plane correlation length = $<z(0, 0)z(x - x', y - y')>$. To obtain an understanding of this equation, if we assume that scattering features are so close together then it is unlikely that they will be correlated, i.e. $C(R) = 0$, then the height-difference correlation $g(R) = 2\sigma^2$. Whereas if there is any correlation in the plane of the interface $C(R) \neq 0$, making this an important parameter.

This general description will now be considered for several interface types given in figure 2.11.4.

2.11.1.1. The jagged interface:

The jagged interface of 2.11.4a, can be represented by a truncated fractal and for a single interface the following relationship can be considered to be a good approximation to the in-plane correlation function, Sinha et al (1988)

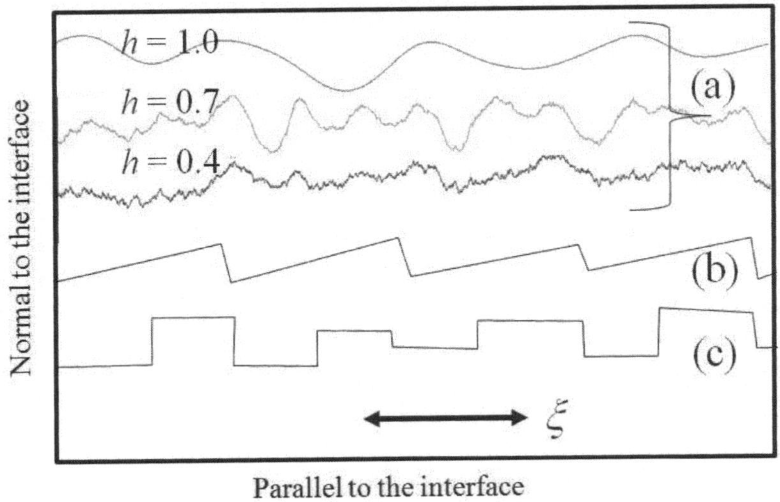

Figure 2.11.4. The three types of interfaces are simulated. (a) is the 'fractal' or jagged interface for several *h* parameters, (b) the staircase interface and the (c) the castellated interface.

$$C(x-x') = \sigma^2 \exp\{-\left(\frac{|x-x'|}{\xi}\right)^{2h}\} \qquad 2.11.13$$

where σ is the root mean square roughness (r.m.s.) and ξ is the lateral correlation length, which is in effect the maximum dimension for correlation. The parameter h is an additional variable that has a value between 0 and 1 and can be related to the fractal parameter D (= 3 – h). Following the derivation of Sinha et al we can determine the scattering function for his case as:

$$F_{ij} = \int\left\{ e^{q_z^i \left(q_z^j\right)^* C(x-x')} - 1\right\} e^{-iQ_{//}(x-x')} d(x-x') \qquad 2.11.14$$

$\boldsymbol{Q}$ (=2π$\boldsymbol{S}$) is resolved into components perpendicular and parallel to the interface. To help understand the general shape of this function, let us first assume that the pair correlation parameter $C(x - x')$ to be small, and since $\boldsymbol{Q}_{\perp}$ (i.e. q_z) is small, $\boldsymbol{Q}_{\perp}C(x - x') << 1$ then

$$F_{ij} = \int\left\{ q_z^i \left(q_z^j\right)^* C(x-x')\right\} e^{-iQ_{//}(x-x')} d(x-x') \qquad 2.11.15$$

Therefore when $h \sim 1$ the scattering function F_{ij} is Gaussian in shape and when $h \sim 0.5$, F_{ij} is Lorentzian shape. For $h \sim 0$ the pair correlation parameter is constant and represents a constant scattering function characteristic of very rough or jagged interfaces. The calculated profiles for a series of h values are given in figure 2.11.4a.

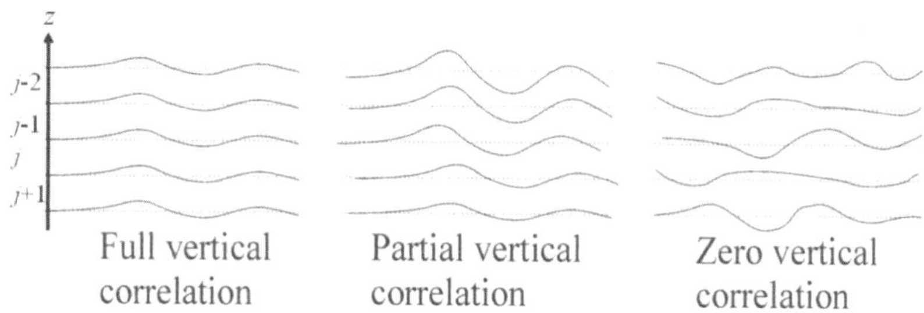

Figure 2.11.5. This indicates the shape of several interfaces with different levels of vertical correlation for the jagged or 'fractal' interface. It is easier to imagine that the longer length scales are more likely to be replicated from layer to layer.

We can use these expressions for roughness provided that the roughness is uncorrelated from layer to layer by considering each interface to contribute independently. The roughness contribution is therefore the sum of these contributions after taking into account transmission coefficients. The correlation parameter to take account of the roughness replication from layer to layer in approximate form can be expressed by

$$C_{ij}(x,z) = \frac{1}{2}\left(\frac{\sigma_j}{\sigma_i}C_i(x-x') + \frac{\sigma_i}{\sigma_j}C_j(x-x')\right)e^{-\frac{|z_i-z_j|}{\xi_v}} \qquad 2.11.16$$

where i and j are the layer indices and ξ_v is the vertical correlation length, Ming, Krol, Soo, Kao, Park and Wang (1993).

Now this expression is only valid when all lateral correlation lengths are correlated in depth through the structure, which will be an assumption. It is easier to envisage vertical correlation of long lateral correlation lengths, whereas the short lateral correlation lengths (high frequency components) are less likely to be replicated, figure 2.11.5. This can be incorporated by introducing an x^n dependence in the exponent.

The diffuse scattering intensity is therefore obtained by inserting this in-plane correlation length into equation 2.11.14 and combine with equation 2.11.7 and 2.11.9. An example of a jagged interface simulation is given in figure 2.11.6.

2.11.1.2. The staircase interface:

The form of the interface described as the staircase model is given in figure 2.11.4b. The parameters that describe this interface are the width l_n and the height h_n for the nth terrace, which can take on random values to give the crystal-plane to interface-plane angle of $\varphi = \sin^{-1}(h/l)$. From Holý and Fewster (2002) the equivalent function for equation 2.11.9 is:

$$F_{ab}^{ij} = V_{ij}q_a^i q_b^j \left(F\left(q_p q_a^i\right)F\left(q_p q_b^j\right)\right)^{0.5} \qquad 2.11.17$$

Where

$$F\left(q_p q\right) = \frac{4\pi}{kl\left|q_1 q_2\right|^2}\mathrm{Re}\left[\frac{\left(1-\left\langle e^{ih_n q_2}\right\rangle\right)\left(1-\left\langle e^{il_n q_1}\right\rangle\right)}{\left(1-\left\langle e^{ih_n q_2}\right\rangle\left\langle e^{il_n q_1}\right\rangle\right)}\right] \qquad 2.11.18$$

Where:

$$q_1 = \left(q_p\cos\varphi + q\sin\varphi\right), q_2 = \left(q\cos\varphi - q_p\sin\varphi\right)$$

and q_p is the projection of $Q_{//}$ onto the direction perpendicular to the terrace edges. V_{ij} is the vertical correlation that can take the form of the exponential part of equation 2.11.16.

2.11.1.3. The castellation interface:

This particular model describes the interface shape that may result from inadequate surface diffusion or incomplete deposition to achieve a uniform layer, despite being parallel to an atom layer. In this case the interface is described by step heights of h_n with lengths of l_{1n} separated by gaps of l_{2n}, which have mean values of l_1 and l_2 respectively. This interface, given in figure 2.11.4c, will result in an interface function for insertion into equation 2.11.9 of:

$$F(q_p q) = \frac{8\pi}{k(l_1 + l_2)|q_p q|^2} \mathrm{Re}\left[\frac{\left(1 - \left\langle e^{il_{1n}q_p}\right\rangle\right)\left(1 - \left\langle e^{il_{2n}q_p}\right\rangle\right)}{\left(1 - \left\langle e^{il_{1n}q_p}\right\rangle\left\langle e^{il_{2n}q_p}\right\rangle\right)}\right]\left(1 - \mathrm{Re}\left(\left\langle e^{ih_n q}\right\rangle\right)\right) \quad 2.11.19$$

As can be seen the complexity can be increased to represent a multitude of interface shapes, but in reality it is difficult to distinguish between the forms experimentally. Also it is worth noting that the castellation interface and the fractal (jagged) interface can provide reasonable fits to measured data, however the parameters l_1, l_2 can be very different from the lateral correlation length, chapter 5: section 5.7.6.1.

Examples of the simulation of these interface shapes are given in figure 2.11.6 and the reciprocal space maps in figure 2.11.7 for various interface shapes and vertical correlations.

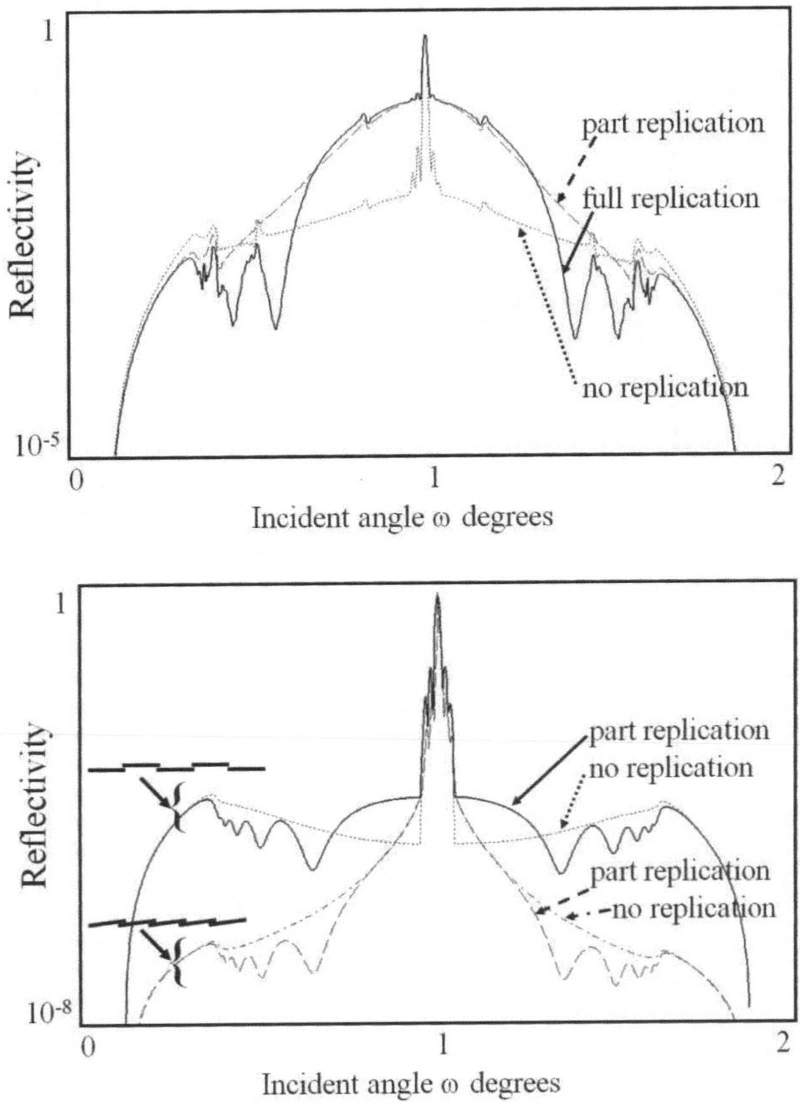

Figure 2.11.6. The calculated diffuse scattering profile for a laterally rough interface from a periodic GaAs/AlAs (15nm;7nm) multi-layer on GaAs, assuming a fractal model with a roughness of 3 monolayers, (top), assuming a half monolayer coverage on a flat surface (terrace width of 164nm) and monolayer steps created by a 0.1^0 inclined surface (terrace width of 164nm) (lower).

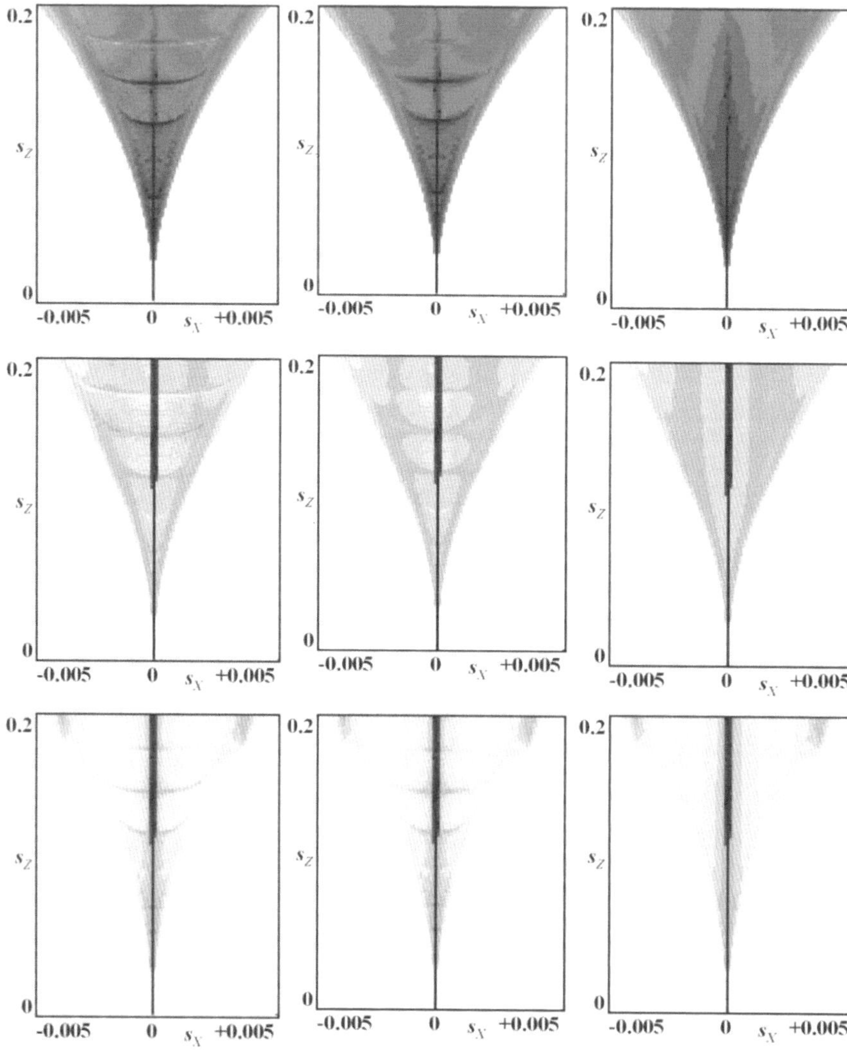

Figure 2.11.7. The calculated diffuse scattering patterns for the conditions described in the caption to figure 2.11.6, i.e. for fully correlated, partially correlated and not correlated roughness (left to right). (top) fractal model plotted over 6 orders in intensity, (middle) flat surface half-monolayer coverage plotted over 9 orders of intensity and (bottom) a monolayer stepped surface resulting from a 0.1^0 *<00l>* inclination to the surface normal. The observed lateral streaking is characteristic of vertical correlation and the specular scan is at s_x=0.

2.11.1.4. Some general observations on interface roughness:

A fractal interface cannot be the correct model for a nearly perfect interface, since the dimensions at the atomic level are discrete, therefore the example for GaAs (*001*) represents a surface with a minimum step height of a monolayer. Hence the interface roughness for the fractal model in figures 2.11.6 and 2.11.7 are for three monolayers, to give an approximate feasible fractal form. An almost perfect interface can be represented as a series of single monolayer steps distributed over the surface resulting from incomplete coverage. The diffuse scattering from this is different from that for a fractal surface and this becomes more pronounced as the replication from layer to layer becomes weaker. For a vicinal surface the correlated roughness is inclined at an angle of φ, this will have the correlated roughening of layer i displaced from layer j by a distance $(z_i-z_j)\tan\varphi$ therefore x becomes $x-(z_j-z_k)\tan\varphi$, etc., Kondrashkina, Stepanov, Opitz, Schmindbauer, and Kohler (1997). Again the effect can be quite dramatic, figure 2.11.6 and 2.11.7, indicating the sensitivity of the diffuse scattering to the interface model.

We immediately see that the complexity is growing rapidly and the number of parameters that should be refined is more uncertain; however this is where the instrument resolution or scan range can limit the spread of correlation lengths relevant to the calculation. As the resolution leads to smearing of the features the longer length scale limit is reduced and the possible scan range puts a lower limit on the measurable correlation length. Let us consider some approximate values. For Cu $K\alpha$ radiation and a divergence of the incident beam of 0.01° then when the sample is rocked about a scattering angle of 1° the lateral correlation lengths measurable range from 4.4nm to 440nm at best.

We shall now consider the intensity obtained from the equation 2.11.7 for the differential cross-section. If we assume the incident beam of area A and intensity I_0, subtends a solid angle $\Delta\Omega$ onto the sample surface, then the measured scattering is:

$$I_{diffuse} = I_0 \frac{\Delta\Omega}{\mathrm{A}} \left|\frac{d\sigma}{d\Omega}\right|_{diffuse} \qquad 2.11.20$$

From the figure 2.11.6, the intensity has an overall characteristic shape at low and high incident angles, with the former being higher than the latter. We also know that this diffuse scattering does not exist if the interfaces are perfectly smooth, i.e. the intensity will only be reflected; $I_{\text{specular}}=I_0|R|^2$. The enhancement in the diffuse intensity at very low angles, for a specific 2θ detector position, is a consequence of the incident beam being scattered by the rough surface as it tries to enter the sample and a contribution from the transmitted beam that is trying to penetrate the sample. This occurs above the critical angle when some penetration can occur. As the incident beam increases the incident beam will enter more successfully until the point arises when the scattered beam tries to exit the surface $\omega \sim 2\theta$. The exit beam will scatter from the rough the surface. These enhancements were first pointed out by Yoneda (1963).

As a general guide to the shape of the intensity of figure 2.11.6, the low and high regions are dominated by the 'Yoneda wings', the central strong intensity is dominated by the specular scattering and the broad envelope underneath is dominated by the interface roughness. In measuring and simulating the diffuse scattering, there are some cautionary aspects. Consider the scattering in the form of reciprocal space map, figure 2.11.7, where as can be seen the edges of the horizontal diffuse streaks there is considerable curvature. The scan in ω will rock through an arc centred on (s_x=0, s_z=0), therefore the scan profile will be very sensitive to the position of s_z. Experimentally this may be difficult and sometimes a reciprocal space map can be much more revealing.

2.11.2. In-plane scattering:

So far we have considered several theoretical models that will cope with the most perfect crystalline materials to those that are independent of crystalline form. These are all applicable to the types of analytical problems in semiconductors. As described in section 2.10, the X-rays barely penetrate the sample when incident below the critical angle and are almost totally reflected. However above the critical angle the X-rays penetrate and are available for diffraction and scattering. Scattering is a more general term that could be uncorrelated, correlated and in the limit could be a coherent sum of scattering from parallel crystal planes, i.e. diffraction. Scattering planes that can easily be brought into the correct angle for scattering are those roughly perpendicular to the sample surface normal, figure 2.11.8. Marra, Eisenberger and Cho (1979) first exploited this approach.

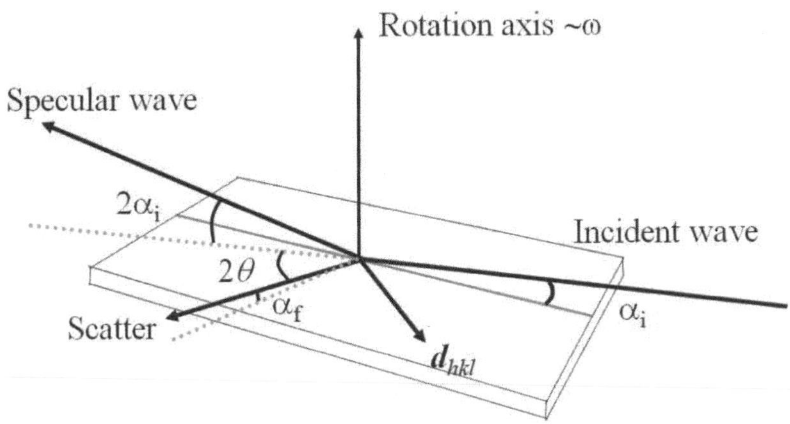

Figure 2.11.8. The geometry of the in-plane scattering experiment.

The penetration depth can be calculated and this could be a useful variable in the experiment. Below the critical incident angle, as we have seen in section 2.10, the X-rays are strongly externally reflected and so the flux available for scattering is that associated with the very top surface, figure 2.10.2 (reproduced with more detail in figure 2.11.9). As the incident angle is increased, the depth of penetration increases and the specularly reflected intensity falls, which results in more flux available for diffraction from suitably orientated planes. This is particularly marked close to critical angle, where for a simple flat surface of a crystal the diffraction from a suitably orientated plane becomes large and then gradually falls away with increasing incidence angle. At which point the penetration depth is very large and the depth sensitivity is lost. These general observations are based on the exit angle being ~$2\omega_c$. When the diffraction is measured with the detector set to only capture very low exit angles, the response is very sensitive to this angle, however for the $2\omega_c$ option the intensity response is almost indistinguishable from integrating the scattered intensity normal to the sample surface. This is helpful to know for experimental design.

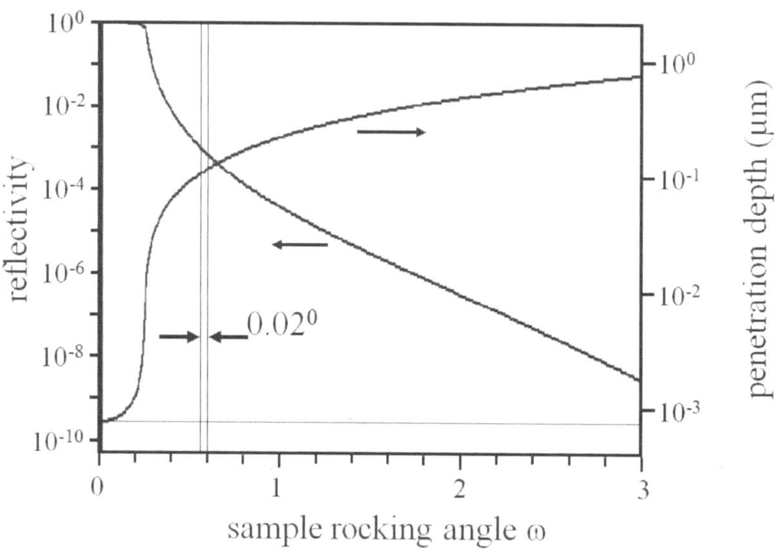

Figure 2.11.9. The variation of reflected intensity and penetration of X-rays is given for a Si wafer, as in figure 2.10.2. Superimposed is a typical incident beam divergence, to indicate the smearing of the information from this resolution.

This technique has advantages in that diffraction is possible from crystal planes approximately normal to the crystal surface, which is relevant for studying mosaic rotation or strain relaxation in multi-layers for example. These examples do not require depth sensitivity in general, however if this is required then this exists within a very small incident angle range, thus the angular resolution of the incoming X-rays has to be better than that required for the depth resolution, figure 2.11.9. The incident angle will also vary with crystal surface curvature as well as the roughness and this brings further limits on the resolution. It must also be remembered that the 'depth-resolution' is simply the depth at which the

X-ray flux has fallen to $1/e$ of its initial value, and so the scattered intensity is a complex combination of all the scattering with an available flux profile that is function of the transmission through the sample at the selected angle. Thus, obtaining quantitative information should be based on simulation, although peak positions and peak widths can be usefully interpreted.

The range of structures that can be simulated is too extensive to cover in this book, and so the concentration will be on the analysis of quantum dots buried below the surface and a polycrystalline multi-layer. The former relies on modeling the strain variation to obtain the information, and the latter uses standing waves to analyse the crystallite distortions as a function of depth. The following sections will give the underlying theory and in sections 5.4.7.1 and 5.7.6.2. The experimental method and some of the information that can practically be extracted will be discussed.

2.11.3. The analysis of Quantum Dots with In-Plane Scattering:

The role of quantum dots in semiconductor technology has shown to be of interest because it gives the opportunity to modify the bulk material band-structure by reducing the dimensionality and confining the charge carriers. This band-structure modification depends on the size and composition of the dots and this section will describe ways in which buried dots can be analysed. Dots on the surface require a different approach. As we have seen before the layer composition will create stresses, which we will become zero at the surface, so provided the surface is flat the strains can be evaluated. This is a reasonable assumption for buried dots, whereas dots on the surface will require strain modeling using finite element analysis, for example, before the scattering can be calculated. The following example is for quantum dots existing in a plane, or planes, parallel to the surface, where it is assumed that their strain-fields do not overlap, i.e. the strain-fields are independent, figure 2.11.10. If this is not the case then the calculation begins to become very complex, in a similar manner to the added complexity discussed concerning the correlated scattering from dislocations, section 2.8.2.

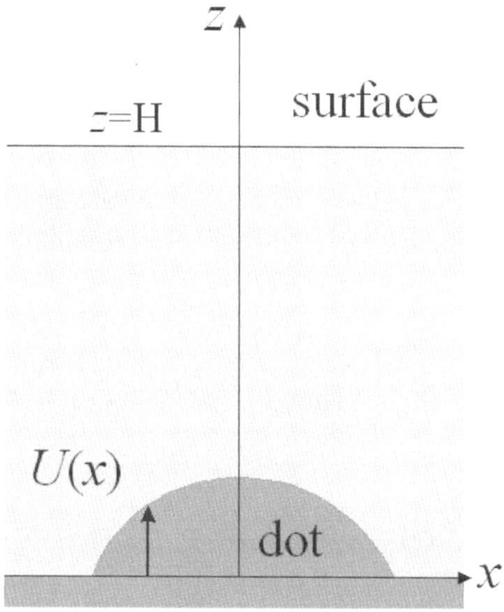

Figure 2.11.10. The description of the buried quantum dot, along with some of the information required for modeling. $U(x)$ defines the shape of the dot that gives rise to a displacement field $u(r)$.

The experimental configuration differs from the analysis of interface roughness, because in this case our concern is the calculation of the scattering associated with density changes, shape and strain. It is the latter parameter that requires us to scatter from planes approximately perpendicular to the surface, figure 2.11.8. The following analysis is the work of Holý, some of which is given in Fewster, Holý and Andrew (2001) and Fewster, Holý and Zhi (2003). We can write the intensity associated with the dots at a position defined by Q in reciprocal space as:

$$I(Q) = \left|t^{(1)}t^{(2)}\right|^2 \sum \delta(G - q_{//})\left|F_G\right|^2 \qquad 2.11.21$$

$t^{(1)}$ and $t^{(2)}$ refer to the transmission coefficients of the incident wave and scattered wave, equation 2.4.31 and equation 2.10.17. F_G is the structure factor associated with the column of quantum dots distributed in a 2-dimension array within the plane of the interface with the period G. The scattering makes the assumption that only the part of the structure that is influenced by the dot is calculated, or column of dots, the remainder of the structure is assumed to be amorphous, i.e. absorbing but not scattering. If there is no correlation between the positions of dots from layer to layer or there is only a single layer of dots then the delta-function is unity, and all dots scatter independently. The structure factor is given by:

$$F_G(q) = \int e^{-iGx} e^{-iq_z z}\left[\chi_h(r) e^{-h.u(r)} - \chi_h^{crystal}\right] dr \qquad 2.11.22$$

q is the internal wave-vector. The exponential function in z will accommodate the correlation of the dots from layer to layer, that occur when periodic multi-layer structures of dots are formed, and each plane of dots influence the location of the dots above. The term in brackets represents the structure factor (as the susceptibility) associated with the dots over and above that associated with the crystal matrix. The susceptibility associated with the dot varies with the three-dimensional position r, and $u(r)$ is the atomic displacement due to the presence of the dot. h is the scattering vector with respect to the reciprocal lattice point being mapped. The composition of the dot, compared with the crystal matrix, will create a strain because of the different covalent radii. The displacement parameter $u(r)$ can be deduced for a cubic material with an *001* surface from the following. This description will only give an outline of the method.

In the equilibrium, the following equation is valid for quantum dots in a crystal with anisotropic elasticity parameters:

$$\frac{\partial \sigma_{jk}}{\partial x_k} - \Delta({c_{11}}^{dot} + 2{c_{12}}^{dot})\frac{\partial \rho(\mathbf{r})}{\partial x_j} = 0 \qquad 2.11.23$$

where σ_{jk} is the stress tensor. The second term is the density of the volume force, where Δ is the lattice mismatch and $\rho(\boldsymbol{r})$ is the shape function of the dot; the density is assumed constant inside and is zero outside. This can be solved by applying the boundary condition

$$\sigma_{jz}\,|_{z=\mathrm{H}},\, j = x, y, z \qquad 2.11.24$$

i.e. these stresses are zero at the sample surface, where H is the distance of the surface from the dot. The full solution is too long and unnecessary to include here, but an appropriate route is to define the 2-dimensional Fourier transform of the displacement $u(r)$, which can be combined with the Fourier transform of the shape function. Unfortunately the shape function is in general difficult to solve analytically, unless of a simple symmetric form, and has to be done numerically.

If we assume the material is not anisotropic and we can introduce a Poisson ratio to describe the elasticity then the solution is considerably simpler. If we assume that the dot height is much smaller than the depth of the dot below the surface, the displacement $u(r)$ due to the whole dot column is

$$u(x,z) = \int U(x')\sum_{j=1}^{N} w(x - x',z,z_j)d^2x' \qquad 2.11.25$$

where $w(x - x', z, z')$ is given by:

$$w(x - x',z,z') = \frac{\Delta}{4\pi}\frac{1+\nu}{1-\nu}\left[\frac{h.R_1}{R_1^{\,3}} + (3-4\nu)\frac{h.R_2}{R_2^{\,3}} - 6z(z+z')\frac{h.R_2}{R_2^{\,5}} - \frac{2h_z}{R_2^{\,3}}((3-4\nu)(z+z') - z)\right]$$

$$2.11.26$$

ν is the Poisson ratio and Δ is the lattice mismatch of the dot with respect to the host crystal, and

$$R_1 = (x - x',y - y',z - z'), \quad R_2 = (x - x',z - z',z + z'),$$

The first term within the square brackets of equation 2.11.26 expresses the displacement field from a point source of stress within an infinite crystal matrix, whereas the second and third terms relate to the relaxation of the stress at the surface. The term exp($-ih.u(r)$) can be expressed using

Fourier coefficients of the dot shape, $U(\mathbf{x})$, i.e. the distance of the dot boundary from the interface, figure 2.11.10

$$U_G = \frac{1}{S}\int_S U(x)\, e^{-iG.x}\, d^2x \qquad 2.11.27$$

S is the surface area corresponding to a single dot, from which:

$$\exp(-i\,h.u) = \exp\left(-i\sum_G U_G W_G(z) e^{iG.x}\right) \qquad 2.11.28$$

where

$$W_G(z) = \frac{\Delta}{2}\frac{1+v}{1-v}\sum_{j=1}^{N}\left[e^{-G|z-z_j|}\left(-i\frac{h_{||}.G}{G} + h_z sign(z-z_j)\right) + \right.$$
$$\left. + e^{-G|z+z_j|}\left(i\frac{h_{||}.G}{G} - h_z sign(z+z_j)\right)\left(-(3-4v) + 2G\,z\,sign(z+z_j)\right)\right] \qquad 2.11.29$$

This is the two-dimensional Fourier transformation of the elastic Green function, where *sign*(A) is the sign of parameter A.

Equation 2.11.28 can therefore be substituted into equation 2.11.22, that in turn is substituted into equation 2.11.21 to calculate the diffuse scattering. The extreme sensitivity of the diffuse scattering to the distortions associated with the quantum dots, indicates that considerable information can be gleaned, however it also indicated in the analysis given in chapter 5, section 5.4.7.1, that for typical semiconductors that distort anisotropically the isotropic description has to be considered with care.

2.11.4. The analysis of Polycrystalline Multi-layers with In-Plane Scattering:

To overcome some of the complexities of obtaining depth sensitive information from multi-layers, Fewster, Andrew, Holý and Barmak (2005) made use of the standing waves that are created from periodic structures, to vary the position of the modulated electric-field flux to enhance or diminish the scattering strength from various layers. This adds complication to the simulation, however it is much more general and more reliable in extracting depth sensitive information compared with the simple relationship of relying on the transmission, figure 2.11.9.

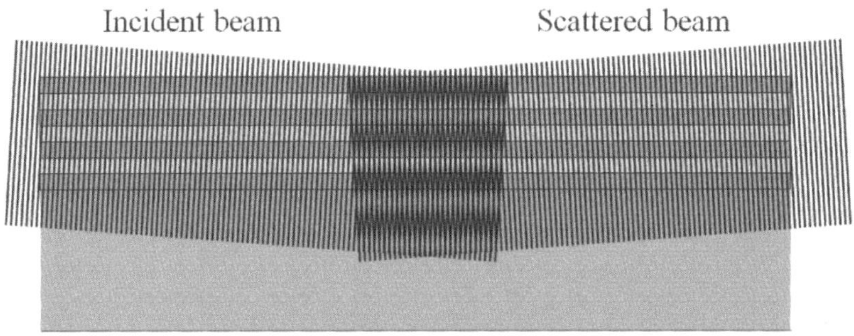

Figure 2.11.11. The generation of a standing-wave or Moiré pattern as the incident and specularly reflected waves interfere. At near grazing incident angles the standing–wave period can match typical periodic multi-layer thicknesses.

The geometry described in figure 2.11.11, shows the incident and scattered waves at low grazing angles will interfere and can produce a standing wave with a periodicity that could be comparable to the periodicity of a multi-layer. A periodic multi-layer is not an absolute requirement, but accentuates the features and so will be the concentration in the following description.

We can model the interaction of the reflected and transmitted waves in reflectometry for a periodic multi-layer with the following:

$$E_{Aj}^{(1)}(r) = e^{ik_{//}^{(1)}.r_{//}}\left\{ T_j^{(1)} e^{iK_{jz}^{(1)}.z_j} + R_j^{(1)} e^{-iK_{jz}^{(1)}.z_j} \right\} \qquad 2.11.30$$

$$E_{Aj}^{(2)}(r) = e^{ik_{//}^{(2)}.r_{//}}\left\{ T_j^{(2)*} e^{iK_{jz}^{(2)*}.z_j} + R_j^{(2)*} e^{-iK_{jz}^{(2)*}.z_j} \right\} \qquad 2.11.31$$

The parameters are best understood with reference to figure 2.11.3. (1) and (2) represent the initial and final states and $j = 0,1,\ldots N$, where N is the number of layers. These equations represent the electric field created from the undisturbed scattering potential V_A.

The crystallites will scatter to create the disturbed scattering potential, V_B, given by:

$$V_B(r) = k^2 \sum_{j=1}^{N-1} \sum_{n} \chi_{hj} \Omega_{jn}\left(r - R_{jn}\right) e^{ih_{jn}\cdot\left(r - R_{jn}\right)} \qquad 2.11.32$$

Ω_{jn} is the shape function, and R_{jn} the position vector of the crystallite n in layer j. The reciprocal lattice vector for this crystallite is h_{jn}. We have related the disturbance to the differential cross-section and hence intensity in equation 2.11.20. In this case the disturbance is given by a statistical average of the transfer of energy from the incident wave-field via the scattering potential to the scattered wave-field:

$$v_B = \left\langle E_A^{(2)} \middle| V_B \middle| E_A^{(1)} \right\rangle = \int E_A^{(2)*}(r) V_B E_A^{(1)}(r) d^3 r \qquad 2.11.33$$

So the intensity of the diffraction from the crystallites is obtained from the differential cross-section:

$$\frac{d\sigma}{d\Omega} = \frac{1}{16\pi^2} \left\langle \left| v_B \right|^2 \right\rangle \qquad 2.11.34$$

The angled brackets represent the statistical ensemble of random position and orientations. Now the diffuse scattering component of this differential cross-section is given by:

$$\frac{d\sigma}{d\Omega} = \frac{1}{16\pi^2} \operatorname{cov}\left(v_B, v_B\right) \qquad 2.11.35$$

Since we are assuming that the quantities v_B relate to random contributions and therefore $\operatorname{cov}(v_B, v_B) = \langle v_B v_B^* \rangle - \langle v_B \rangle \langle v_B^* \rangle$. This brings us to the full differential cross-section:

$$\frac{d\sigma}{d\Omega} = \frac{1}{16\pi^2} \sum_{j=1}^{N+1} N_j \left|\chi_{hj}\right|^2 \sum_{a,b=1}^{4} A_a^j A_b^{j*} \left\langle \Omega_j^{FT}\left(q_{ja} - h_j\right) \Omega_j^{FT*}\left(q_{jb} - h_j\right) \right\rangle \qquad 2.11.36$$

The amplitudes A have the same form as in equation 2.11.9, and q_{ja} is the associated diffraction vector for the irradiated crystallites in layer j. Ω^{FT} denotes the Fourier transform of the crystallite shape.

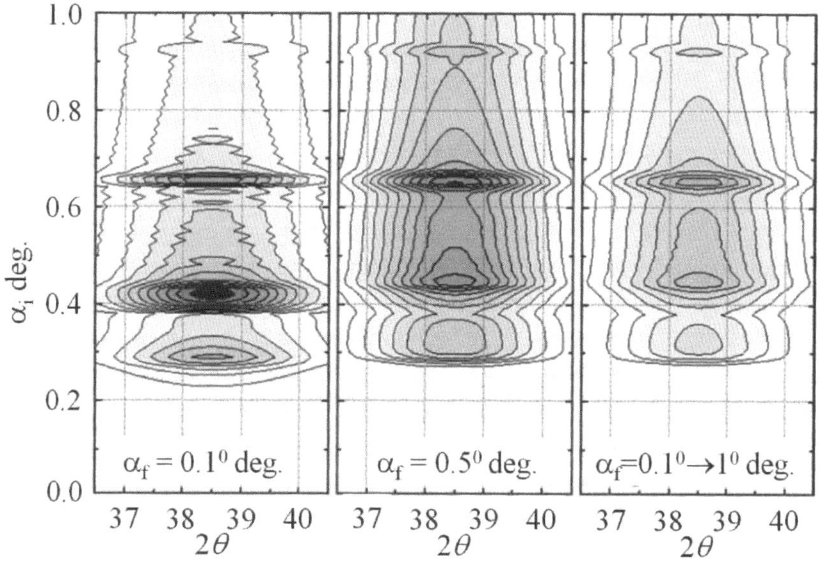

Figure 2.11.12. The enhancement and suppression of the diffracted intensity from a periodic multi-layer as the incident angle is varied, which changes the period of the standing waves of the electric field in and out of coincidence with the multi-layer period.

This approach will reveal the change in the scattering from the layers of scattering as the incident is varied, and for a periodic structure as discussed here the dominance of the scattering from specific layers can be enhanced or suppressed. This gives another variable in probing structures as a function of depth. An example of a calculation is given in figure 2.11.12, where it can be seen that the intensity for one of the bi-layers in a periodic structure oscillates as the incident angle and is varied. The distributions are given for different exit angles.

The assumption at this stage is that there is no roughness in the multilayer and this has been shown to have a significant effect on the intensities, Fewster et al (2005). Roughness can be included, but will not be reproduced here, except to say that the interactions now become more complex since the diffuse scattering from the roughness will contribute to the disturbance, v_B, in the following way:

$$v_B = \left\langle E_A^{(2)} + E_{A,rough}^{(2)} \middle| V_B \middle| E_A^{(1)} + E_{A,rough}^{(1)} \right\rangle \qquad 2.11.37$$

Clearly the level of complexity can be increased and the calculations become more involved, but the basic principles are the same. This should give an idea of the versatility of the Distorted Wave Born Approximation in obtaining information about the various structural features, interfaces, quantum dots and depth information, etc., from structures that are reasonably perfect.

2.11.5. Grazing Incidence Small Angle Scattering

In the discussion on the diffuse scattering close to the specular profile, section 2.11.1, the sample is rocked to change the incident angle. This can have limitations in accessing short length scales. As can be seen from figures 2.11.6 and 2.11.7 the lateral dimension in reciprocal space is restricted by the minimum incident and exit angle.

This can be overcome by rotating the detection point out of the plane of the specular beam, just as in in-plane scattering, but if this is kept to small angles the scattering will be sensitive to dimensions relevant to nano-structures, surface quantum dots, etc. The same principles discussed in section 2.11.1 are relevant, i.e. low incident angles are sensitive to the density fluctuations rather than atomic plane diffraction. The geometry related to the specular and diffuse scattering when rocking in ω, is given in figure 2.11.13 for a complex nanostructure, that is described in greater detail in section 4.3.8.

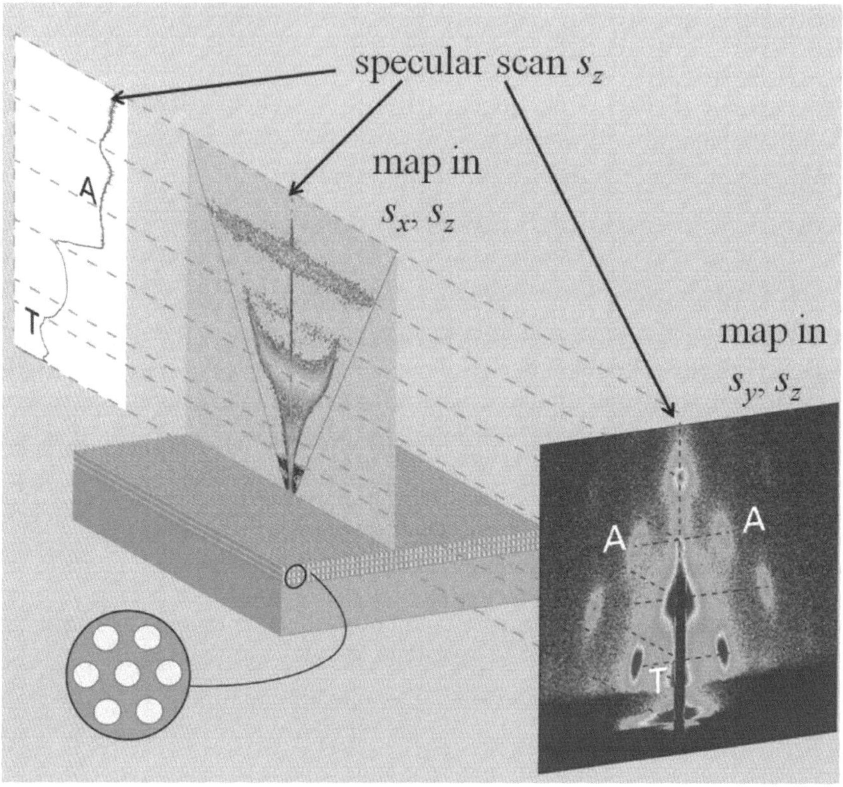

Figure 2.11.13. A diagram to show the relationship between the specular scan, the mapping of the diffuse scattering by rocking in ω, (s_x, s_z)and the in-plane mode (s_y, s_z), commonly termed GISAXS. The image was made by Patricia Kidd and the GISAXS data was collected by Milen Gateshki with a laboratory source (see section 4.3.8.2). The sample is a composed of cylindrical voids in SiO_2 forming an hexagonal arrangement.

The theoretical approach is exactly as in previous sections, except that there are no structure factor effects, just density changes associated with the shape of the features, although these can be correlated. For a distribution of similar shapes defined by Ω, distributed over the surface, these will have a differential cross-section of the form:

$$\frac{d\sigma}{d\Omega}=\frac{1}{16\pi^2}k^2|\chi_0|^2\left|\sum_{a,b=1}^{4}A_aA_b^*\Omega^{FT}(q_a)\Omega^{FT*}(q_b)\right|^2 \qquad 2.11.38$$

The amplitudes and scattering vectors take the form described as before. In this case we have assumed that the substrate below only contributes to the specular scattering and can therefore be ignored, i.e. this can be added as an incoherent contribution. If there is some ordering of the defects then this should include a correlation matrix of positions

$$G_{ab}=\left\langle\sum_n\sum_m e^{-i\left(q_a.r_n-q_b^*.r_m\right)}\right\rangle \qquad 2.22.39$$

r_n is the position of the nth defect. The specular contribution is often masked in an experiment. Since the method is strongly dependent on the experimental geometry, the examples have been deferred to chapter 4, section 4.3.8.6.

2.12. General conclusions

We have seen from the beginning of this chapter that; we have started with Bragg's equation, used it in the context of Maxwell's wave equation and derived a dynamical theory, it was then shown that Fresnel's equations can produce a full multi-wave dynamical theory not requiring Bragg's equation, and in the description of powder diffraction an optical based theory is applied. Clearly we are gradually getting closer to a more complete description of X-ray scattering.

The theoretical models described here cover many aspects of the scattering that we wish to interpret from semiconductor materials. The models discussed are generally applicable but the reader should be aware that we have largely concentrated on the interaction of a single photon with an initial pre-defined path. In general any source is divergent and we should consider the wave as having curved wave-fronts, Kato (1968a and b), although the Takagi theory partially overcomes this assumption. Kato approached this by making a superposition of plane waves and this is the basis of the approach used in Chapter 4 to take into account divergence in the X-ray source due to the instrumentation.

We have also not considered partial coherence in this Chapter due to wavelength spread, again we take a rather pragmatic approach and include it in the modeling of the scattering, i.e. add it as an instrumental effect, Chapter 4. However there are considerable possibilities in refinement, validating and extending as the materials challenge increases. As we shall see we cannot isolate all the components in an X-ray diffraction experiment and we have to combine the scattering theory, sample and data collection instrumentation as a whole. The next chapter will cover the instrumental aberrations associated with the beam conditioning components, which is then followed by the whole instrument sample integration in Chapter 4.

Bibliography

Authier, A (1996) in *X-Ray and Neutron Dynamical Diffraction: Theory and Applications.* pp 1–31 and 43–62. Editors Authier et al, Plenum Press: New York.
Bonse, U (1964) Z Phys. **177** 385
Bragg, W L (1913) Proc. Camb. Phil. Soc. **17** 43–57
Bragg, W H and Pierce,S E (1914) Philos. Mag. **28** 626
Broglie, L de, Recherches sur la théorie des quanta (Researches on the quantum theory), Thesis (Paris), 1924; L. de Broglie, Ann. Phys.(Paris) 3, 22 (1925). Reprinted in Ann. Found. Louis de Broglie 17 (1992) p. 22.
Buerger, M. (1940) *Physics,* **26**, 637–642
Cagliotti, G, Paoletti, A and Ricci, F P (1958) Nucl. Instrum. Methods Phys. Rev. **35** 223–228
Compton, A H (1917) Phys Rev 9 29
Compton, A H (1923) Physical Review **21** (5), 483–502
Daniš, S and Holý, V (2006) Phys. Rev. B **73** 014102
Darwin, C G (1914a) Phil. Mag. **27** 315
Darwin, C G (1914b) Phil. Mag. **27** 675
Debye, P (1913) Ann. Physik **348**, 48–92
Debye, P (1914) Ann. d Physik, **43**, 49
Debye, P (1915) Ann. d Physk **46** 809
Dederichs, P H (1971) Phys. Rev. B **4** 1041
Ewald, P P (1916a) Ann. d Physik **49** 117
Ewald, P P (1916b) Ann. d Physik **49** 1
Ewald, P P (1917) Ann. d Physik **54** 519

Fewster, P F (1986) Philips J Res. **41** 268

Fewster, P F (1987) *Thin Film Growth Techniques for Low Dimensional Structures* NATO ASI Series B:Physics 163, Ed: Farrow et al. New York: Plenum Press, pp 417-440

Fewster, P F (1992) J Appl. Cryst. **25** 714

Fewster, P F (2000) *Insight into polycrystalline materials with ultrahigh resolution and reciprocal space mapping* IUCr Commission on Powder Diffraction, Newsletter N0.24, December 2000 pp 17–19

Fewster, P F (2014) Acta Cryst. A **72** 257–282

Fewster, P F and Andrew, N L (1999) Reciprocal space mapping: and ultra-high resolution diffraction of polycrystalline materials IUCr series: Microstructure analysis from diffraction, Ed: Snyder, Fiala and Bunge, pp 346-364: Oxford University Press

Fewster, P F, Andrew, N L, Holý, V and Barmak, K (2005) Phys. Rev. B **72** 174105–1,11

Fewster, P F and Curling, C J (1987) J Appl. Phys. 62 4154–4158

Fewster, P F, Holý, V and Andrew, N L (2001) Material Science in Semiconductor Processing **4** 475-481

Fewster, P F, Holý, V and Zhi, D (2003) J Phys. D: Appl. Phys. **36** A217–A221

Feynman, R P, Leighton, R B and Sands, M (1966) *The Feynman Lectures on Physics* vol 2, chapter 20, London: Addison-Wesley.

Fock, V (1930) Zeit. F Physik **61** 126

Galihanou, M, Baumbach, G T, Marti, U, Silva, P C, Reinhart, F K and Ilegems, M (1993) Appl. Phys. Lett. **62** 1623–5

Gay, P, Hirsch, P B and Kelly, A (1953) Acta Metall. **1** 315

Hartree, D R (1935) Proc. Roy. Soc. **A 143** 506

Härtwig, J (1981) Acta Cryst. A 37, 802–804

Hedin, L (1974) Many Body Effects, edited by L V Azároff, chapter 5, X-ray Spectroscopy. New York: McGaw-Hill

Heisenberg, W. (1927), "Über den anschaulichen Inhalt der quantentheoretischen Kinematik und Mechanik", Zeitschrift für Physik 43: 172–198

Holý, V (1982) Phys. Stat. Solidi B **111** 341

Holý, V and Fewster, P F (2002) unpublished collaborative work, also see Holy, V, Pietsch, U and Baumbach, T (1999) *High Resolution X-Ray Scattering* Springer: ISBN 3-540-62029-X

Holý, V and Fewster, P F (2003) J Phys. D: Appl. Phys. 36 A5–A8

Holý, V and Fewster, P F (2008) J Appl. Cryst. **41** 18–26

Huang, K (1947) Proc. Roy. Soc. A **190** 102

Kaganer, V, Brandt, O, Trampert, A and Ploog, K H (2005) Phys. Rev. B **72** 045423

Kaganer, V, Köhler, R, Schmidbauer, M, Opitz, R and Jenichen, B (1997) Phys. Rev. B **55** 1793-1810

Kaganer, V and Sabelfeld, K K (2010) Acta Cryst. **A66** 703–716

Kaganer, V and Sabelfeld, K K (2011) Physica Stat. Solidi (a) **208** 2563–2566

Kato, N (1968a) J Appl. Phys. **39** 2225

Kato, N (1968b) J Appl. Phys. **39** 2231

Kato, N (1980) Acta Cryst. **A36** 763

Khrupa, V I (1992) Phys. Metals **11** 765
Kondrashkina, E A, Stepanov, S A, Opitz, R, Schmindbauer, M and Kohler R (1997) Phys. Rev. B **56** 10459
Krivolglaz, M A (1995) Theory of X-ray and Thermal Neutron Scattering by Real Crystals (New York: Plenum)
Laue M v (1931) Ergeb. Der exact. Naturwiss **10** 133
Louër, D and Langford, J I (1982) J Appl. Cryst. **15** 20–26
Marra, W C, Eisenberger, P and Cho A Y (1979) J Appl. Phys. **50** 6927
Ming, Z H, Krol, A, Soo, Y L, Kao, Y H, Park, J S and Wang, K L (1993) Phys Rev B **47** 16373.
Olekhnovich, N M and Olekhnovich, A I (1981) Phys. Stat. Solidi A **67** 427
Pavlov, K M and Punegov, V I (1998) Acta Cryst. **A54** 214
Pietsch, U, Holý, V and Baumbach, T (2004) High-Resolution X-Ray Scattering: From Thin Films to Lateral Nanostructures, 2nd Edition, Springer: New York.
Prins, J A (1930) Zeit. F Physik **63** 477
Reid, J S (1983) Acta Cryst. **A**39 1
Renninger, M (1937) Z Phys. 106 141–176
Scherrer, P (1918) Gott. Nachr **2** 98
Schevchenko, M (2010) Acta Cryst. A **66** 499–504
Sinha, S K, Sirota, E B, Garoff, S and Stanley, H R (1988) Phys. Rev. B **38** 2297
Stokes, G G (1903) Mathematical and Physical Papers Volume II p91
Takagi, S (1962) Acta Cryst. **15** 1311
Takagi, S (1969) J Phys. Soc. Japan **26** 1239
Tanner (1996) in *X-Ray and Neutron Dynamical Diffraction: Theory and Applications.* pp 147-166. Editors Authier et al, Plenum Press: New York.
Taupin, D (1964) Bull. Soc. Fran. Miner. Cryst. **87** 469
Weckert, E and Hümmer, K (1997) Acta Cryst. A 53 108–143
Yoneda, Y (1963) Phys. Rev. **131** 2010
Zachariasen, W H (1945) *Theory of X-ray Diffraction from Crystals,* ch III New York: John Wiley and Sons, Inc.
Zaus, R (1993) J Appl. Cryst. **26** 801

CHAPTER 3

COMPONENTS FOR MEASURING SCATTERING PATTERNS

This chapter presents the basic components, from X-ray sources to detectors including beam conditioners, monochromators, slits, etc., used in X-ray scattering experiments. An understanding of the components gives an indication of the options available to match the resolution and beam size to the experiment. However these should not be considered in isolation, since all the components including the sample can alter the achievable resolution, this latter aspect will be brought together in chapter 4. The emphasis on detectors will be on recent advances in photon-counting solid-state devices, since these offer improved opportunities with increasingly sophisticated electronics to optimize performance, and it is assumed that the use of proportional and scintillation counters will diminish.

3.1. General considerations

The instruments to obtain a diffraction pattern consist of a few basic features: an X-ray source, incident beam conditioning, sample stage and scattered beam capture. The nature of these features should be selected that best meet the needs of the material property to be analysed.

In achieving the ideal combination to analyse the material we have to balance diffraction pattern resolution, real space resolution, intensity and data collection time. These parameters are best understood by having a good grasp of the physics associated with each of the incident and scattered beam conditioners, as well as the X-ray source and detector electronics. Since we are primarily interested in extracting certain material properties from a diffraction pattern, it is important to know all those features that can influence this pattern. Clearly if we know little about the sample and wish to extract a lot of information then the instrument function has to be very well understood. If we already know a lot about the sample then perhaps we can perform a simpler experiment with a less well-defined instrument function. One very good feature concerning semiconductors is that we generally know a reasonable amount of information (approximate thickness and composition in each layer) and because they are amenable to a well-defined instrument function we can extract a real wealth of additional information.

A modern X-ray diffractometer is illustrated in figure 3.1.1, where the incident beam and scattered beam optics can be exchanged to suit the material problem to be analysed. The sample stage in this case is a precision goniometer with optical encoders on the axes, thus eliminating angular uncertainties on the ω and 2θ main axes, figure 3.1.2. The χ, tilt axis, allows for $\pm 90^{\circ}$ rotation and the ϕ rotation axis normal to the sample surface can rotate through 360°. The sample can also be positioned with an *xyz* stage. The 2θ angle can have a defined zero angle related to the direction of the incident beam, whereas the ω angle can be conveniently defined relative to the sample surface, if it is flat. With an instrument of this type the sample can be manipulated into any position and the incident beam and scattered beam optics chosen to suit the materials problem.

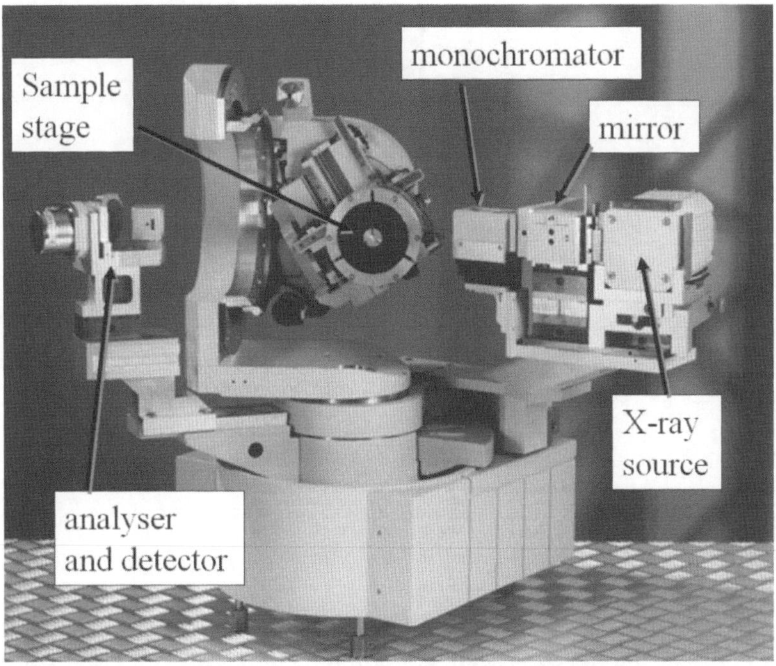

Figure 3.1.1. A modern versatile diffractometer from PANalytical. The X-ray source, monochromator, mirror and analyser assemblies are all pre-aligned and can be exchanged for alternative components. This allows the matching of the instrument to the sample problem to be analysed.

The occasional burst in development of optical elements for diffractometers has increased the number of options for beam conditioning. The principle of the instrument described above is for the various modules to be added resulting in an ever increasingly versatile base. The present day commercial instrumentation is becoming increasingly sophisticated as the demand rises, but also for ease of use and to meet strict safety requirements and tolerances. Therefore just as in the revolution in electronics the building blocks are getting bigger and the need for homebuilt instrumentation of this kind is reducing rapidly.

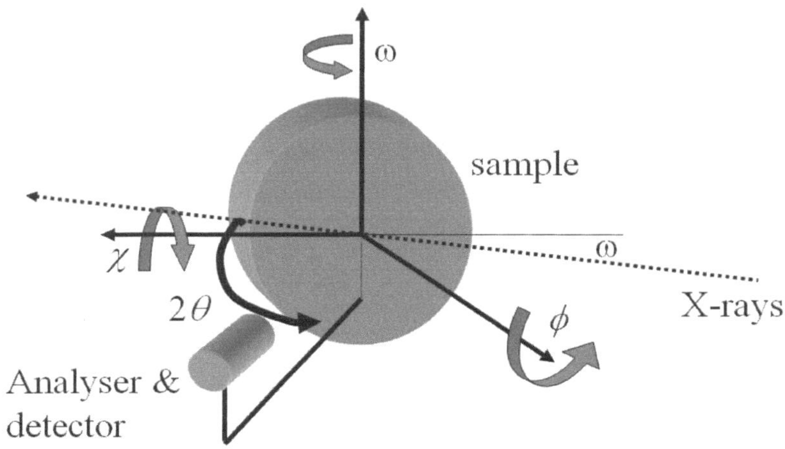

Figure 3.1.2. The angles associated with the diffractometer movements.

We have discussed the scattering processes in Chapter 2 and now we will consider the incident beam conditioning and scattered beam capture. As with all these approaches they fall into two main components, active and passive components. Active components consist of crystals or absorbers that can dramatically change the spectral characteristics and passive components consist of slits and such like that limit the beam that can pass.

3.2. X-ray Source

X-rays are generated by electron energy transitions in a solid (laboratory source) or in a confining magnetic field (synchrotron and free-electron laser sources). If the energy transition is well defined as in a solid (quantum transitions) then the energy of the emitted photon can be very well defined. X-rays are generated from transitions to the innermost electron orbitals and are characteristic of the atom concerned. The common laboratory sources are composed of copper, molybdenum, cobalt, iron and silver anodes for diffraction with most experiments favouring copper. The emission lines arise from excitations that transfer sufficient energy to remove an inner electron and allow the more loosely bound (higher energy states) to transfer to the vacant inner states. For example an electron in the *L* shell transferring to the *K* shell will create a $K\alpha$ photon and an electron transferring from the *M* shell to the *K* will create a $K\beta$ photon, etc. Of course there are more transitions than this and the $K\alpha$ line is a doublet representing transitions between $2p^{3/2}$ to $1s$ ($K\alpha_1$) and $2p^{1/2}$ to $1s$ ($K\alpha_2$) states. The $K\beta$ lines come from the $3p$ states and again the lines are not single valued. Basically we should consider that the intensity from an X-ray source is a complex function of wavelength that has been very closely measured and is essentially invariant for any given tube current.

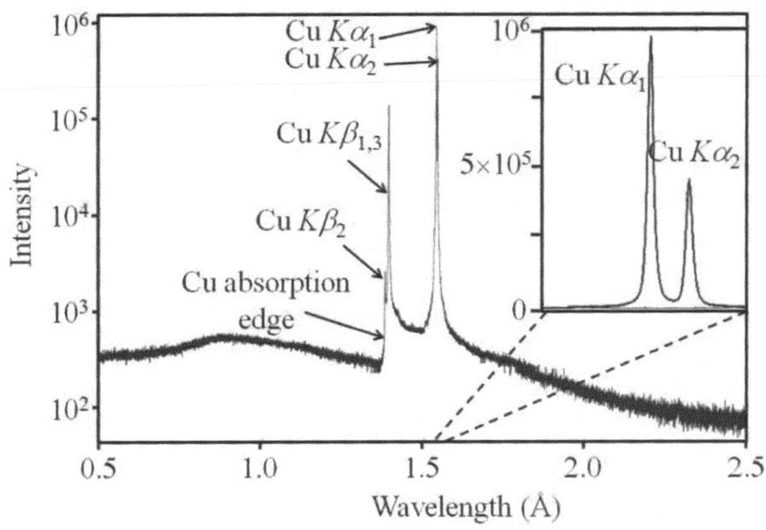

Figure 3.2.1. A typical spectral distribution from a sealed laboratory X-ray source.

The electrons that do not excite these lines will loose energy and emit a continuous background giving a total distribution of intensity with wavelength as given in figure 3.2.1. The controls we have are the tube voltage and tube current and we should optimise these for the maximum intensity and stability. In general, the characteristic intensity increases linearly with tube current and as the nth power of the voltage above the critical excitation voltage for the wavelength emission, where n can be between 1.5 and 2 up to ~4× the critical voltage when it becomes linear. The continuous background or Bremsstrahlung, increases as the square of the voltage setting. Increasing the voltage too high will create some self-absorption, i.e. the X-rays are generated on average deeper into the anode and are partly absorbed. The simplest method of determining the optimum tube voltage and current is by experiment. For a 2kV X-ray Cu anode tube the typical values of 45 kV and 40 mA are close to optimum when considering all aspects of focus stability.

The X-rays are generated in a confined area; generally termed the focus of the anode and its size is a function of the tube geometry. The thermal load from high-energy electrons impacting on the anode is very high and efficient cooling is essential in good X-ray tube design. A balance is therefore always necessary between focus size and the power available. The filament, figure 1.4.1, is a small linear coil and its dimension gives an approximate rectangular shape to the focus. The X-rays are emitted in all directions, and the projected size of the focus seen by the incident beam optics or sample, can be manipulated by altering the take-off angle, figure 3.2.2. There are two directions that are used; a line focus and a point focus, with typical sizes of 12 mm × 40 μm and 400 μm × 400 μm.

As the focus size is reduced the anode can more easily extract the heat, i.e. a small area surrounded by high thermal conducting material (copper). So it appears as though there is some gain in that the photon emission per unit area can be greatly increased, however the total number of photons available compared with a typical focus size is considerably less. Although tube design is never static, an approximate indication is that if a sub-100 μm sample area is probed using a pinhole, then there are advantages with a micro-focus (~10 μm dimensions) tube. Some applications or incident beam optics require a small focus and there is no alternative, e.g. focusing mirrors in two orthogonal planes, and some geometries for high-resolution topography. Again we have to compromise depending on the nature of the experimental problem to be resolved.

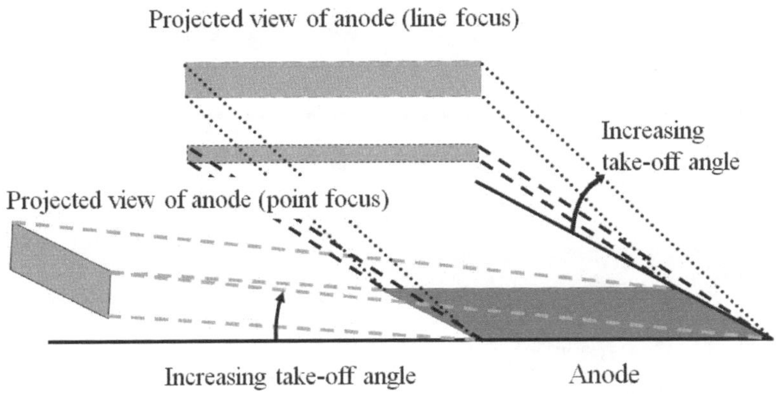

Figure 3.2.2. The different projections available from a sealed laboratory X-ray source.

3.3. Incident beam conditioning with passive components

Passive components in this context are defined as those that just restrict the beam spatially and in angular divergence. The wavelength distribution is unchanged unless it has a spatial distribution.

3.3.1. *Incident Beam Slits: Fixed arrangement*

Inserting slits into the beam path reduces the divergence of the X-ray source and the beam size arriving at the sample. We shall firstly consider the divergence in the scattering plane, since this has the largest influence on the scattering pattern, then consider the importance of axial divergence out of the scattering plane. We cannot simply state a divergence in the scattering plane without qualifying the shape of the distribution.

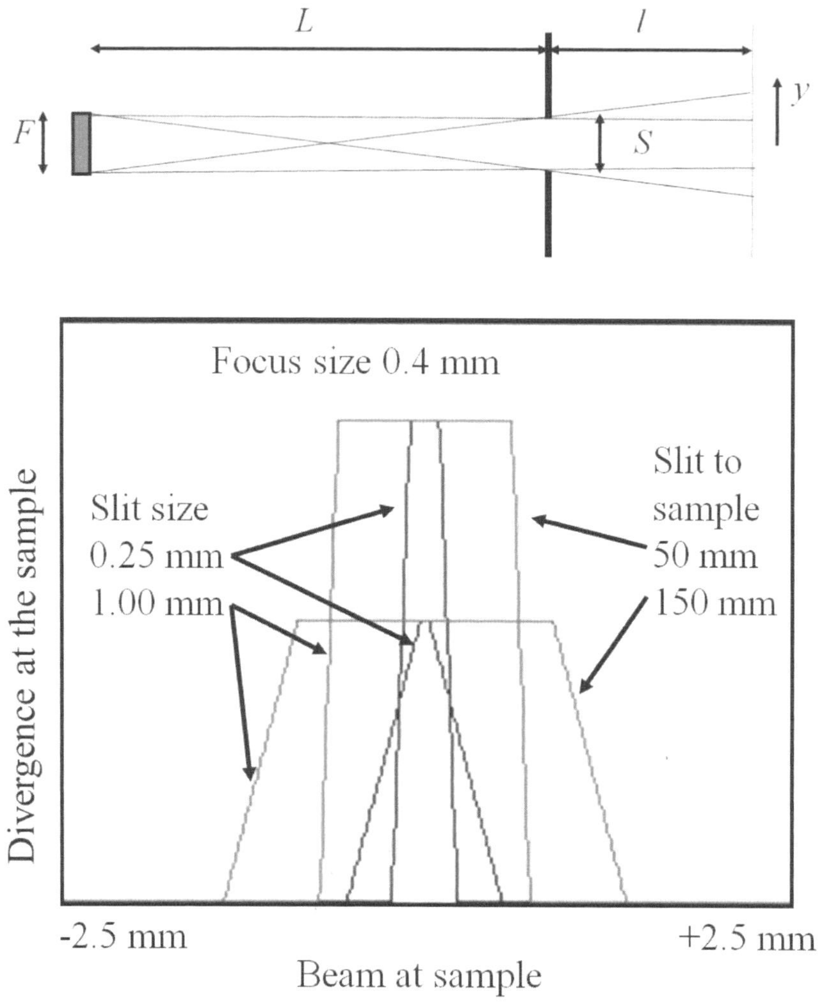

Figure 3.3.1. (a) The geometry of a single slit and (b) the variation in divergence on the sample for several combinations and a fixed focus size for $L+l = 320$ mm.

We will firstly consider the profile of the beam from the simplest configuration of slit and source, figure 3.3.1a. If we assume that the generation of X-rays is radially uniform, so that the X-ray flux reaching the slit is uniform then the profile can be expressed by:

$$I \propto \frac{|\Phi_2 - \Phi_1|}{2\tan^{-1}\left\{\dfrac{F}{2(L+l)}\right\}} \qquad 3.3.1$$

where $|\Phi_2 - \Phi_1|$ is the subtended angle at the position y and

$$\Phi_2 = \tan^{-1}\left\{\frac{y+F/2}{L+l}\right\} \qquad 3.3.2$$

or

$$\tan^{-1}\left\{\frac{y+S/2}{l}\right\}$$

or

$$\tan^{-1}\left\{\frac{y-S/2}{l}\right\}$$

depending whether the slit intercepts the view of the focus from the position y.

$$\Phi_1 = \tan^{-1}\left\{\frac{y-F/2}{L+l}\right\} \qquad 3.3.3$$

unless the slit intercepts this beam, following the arguments above. Expression 3.3.1 relates the view of the focus from any position y at a distance l from the slit. The slit and focus are assumed to be co-axial and therefore the normalisation factor gives the intensity that would be registered on this axis at the determined distance from the source.

Figure 3.3.1b gives the shape of this distribution as we change some of these parameters. It can be seen that a simple divergence angle is insufficient to characterise all these shapes. This shape varies with distance from the slit, therefore the position of the sample with respect to the slit and source will affect the scattered profile. From these profiles we can observe very useful details. Firstly the peak intensity diminishes with distance from the source, because we are capturing a smaller distribution of photon trajectories, this is important for limited sample sizes. The profile closely resembles a triangular, rectangular and more rounded shape as we progress from narrow combinations to larger dimensions. Clearly there is no simple description of the divergence.

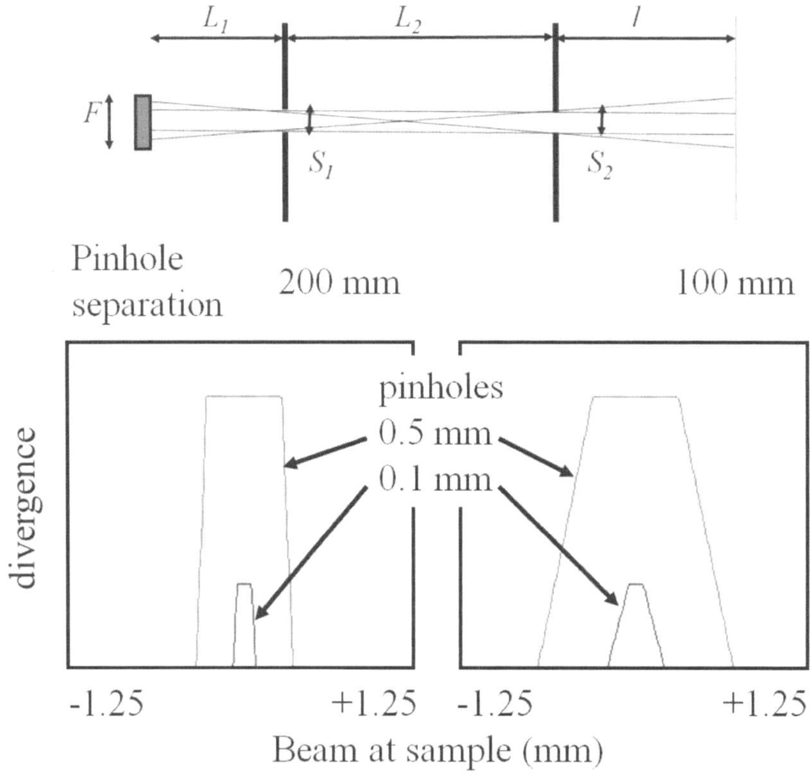

Figure 3.3.2. (a) The geometry of a double slit arrangement and (b) the resulting divergence on the sample for L_1 = 70 mm and a total focus to sample dimension of 320 mm.

Since these profiles are spatial distributions, each point represents a subtended angular divergence of the source and hence the angular divergence varies along *y*, normal to the axis. This angular divergence will be modified depending on the size of the sample and its projection onto the beam axis normal. It should be apparent that the complexity builds up very rapidly with errors in alignment, variable flux density across the focus and the inclusion of air scattering and absorption.

However as the source size is reduced the intensity distribution becomes more rectangular.

The addition of an extra slit to create a double slit collimator creates an extra degree of freedom and complexity, figure 3.3.2a. The spatial distribution of intensity arriving at the second slit will be as described in figure 3.3.2b, however because this has a varying divergence across this distribution the influence of this second slit will in general create a more triangular profile. This triangular profile is indicative of a rapid variation of angular divergence across the emerging beam. If the sample is smaller than the beam size then the divergence will vary with incident angle. If the sample is inhomogeneous then the varying spatial distribution in the divergence could also add complications and it is important to be aware of the variation. The divergence can be made more uniform with a line focus (i.e. a small focus in the scattering plane) however this increases the axial divergence normal to the scattering plane. The axial divergence can be contained as discussed in section 3.3.3.

Although these are relatively simple expressions, the addition of the focus emission can be included. These shapes can be quite complex and are a function of the tube design, tube voltage and current. For the purposes of this approximation this will not be included here. To give an indication, the X-ray emission from the focus follows the emission characteristics of the cathode. For a circular coil the emission can be higher towards the edges, such that the shape is not an exact rectangle in point focus mode. Some of these effects are reduced when the take-off angle is changed. Also as the tube current is increased the higher number of electrons will result in some spreading from repulsion and change the focus profile, which is more severe at low voltages. Clearly this brings in a vast number of additional parameters that may be important for the most precise analyses.

If the sample is inhomogeneous or we wish to maintain a constant sampled area then we have to control the illuminated area automatically with incident angle. This is discussed in the next section.

3.3.2. *Incident Beam Slits: Variable arrangement*

The illuminated area can be maintained constant by varying the width of the slits. This can be done very effectively under computer controlled data collection, figure 3.3.3a. The mechanical tolerances are very tight and the motion is controlled in general by rotation of the upper and lower parts of the slit. The corrections to the measured intensities to convert them to equivalent fixed divergence and an infinite sample are purely geometrical, depending on the radius of the slits, sample to slit and slit to X-ray source focus distances. The geometry of this arrangement requires a small focus to maintain a controllable divergence and therefore the line-focus option is preferred (figure 3.2.2).

As we could see from the explanation of the control of the divergence for a single slit the source to slit and slit to sample distances are all important. High mechanical precision is required for an automatic divergence slit, since the divergence and hence illuminated length for both upper and lower slits are different. The shape of the slits is rather critical to create a good edge to limit scatter from all working openings. Remember of course from our previous section, the divergence varies over the angular scan range so again we cannot define a simple instrument function.

Suppose that all the X-rays arriving at the sample undergo the same scattering probability, as would be the case for a perfectly random distribution of crystallites (the perfect powder sample). The intensity for an equivalent "fixed" slit arrangement would then relate to the measured profile for this "variable" slit arrangement, with the following expression:

$$I = I_{measured}\left\{\tan^{-1}\left(\frac{\sin\omega}{2\dfrac{L+l}{x}-\cos\omega}\right)+\tan^{-1}\left(\frac{\sin\omega}{2\dfrac{L+l}{x}+\cos\omega}\right)\right\}^{-1} \qquad 3.3.4$$

($L+l$) is the sample to source distance and x is the length of the sample illuminated. The variation of the intensity is presented in figure 3.3.3b. It is clear from this variation that the intensity on the sample is heavily reduced close to grazing incidence. At higher incidence angles this intensity increases to a maximum at normal incidence and then declines, although going beyond normal incidence is not relevant to typical geometries. The divergence of the incident beam in the scattering plane approximately follows this intensity variation. Therefore any experiment making use of these slits will produce a response that is dependent on the sample, since the divergence is rapidly varying with incident angle, i.e. can the sample make use of this increasing divergence?

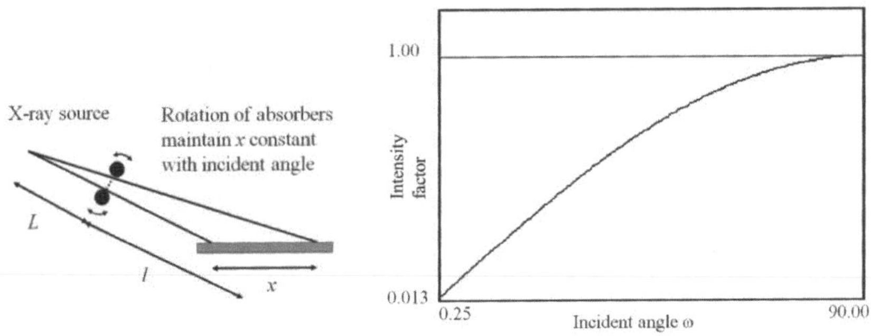

Figure 3.3.3. (a) The mechanism for obtaining constant area of illumination of X-rays on a sample and (b) the variation of intensity incident on the sample.

From Chapter 2 we could see that the specularly reflected intensity falls very rapidly with increasing angles of incidence and therefore this configuration can be used as a very powerful way of compensating for this effect. However since each beam is specularly reflected we must consider the scattered beam acceptance very carefully. This will be covered briefly in chapter 4, section 4.5.1.

3.3.3. *Incident Beam: Parallel Plate Collimators*

Another method for controlling the divergence of an extended X-ray source is illustrated in figure 3.3.4. The slit is composed of a series of thin parallel plates and the divergence is calculated in the same way as for a double slit arrangement. The advantage here is that the divergence can be maintained at low levels whilst still maintaining the intensity by using a large X-ray source.

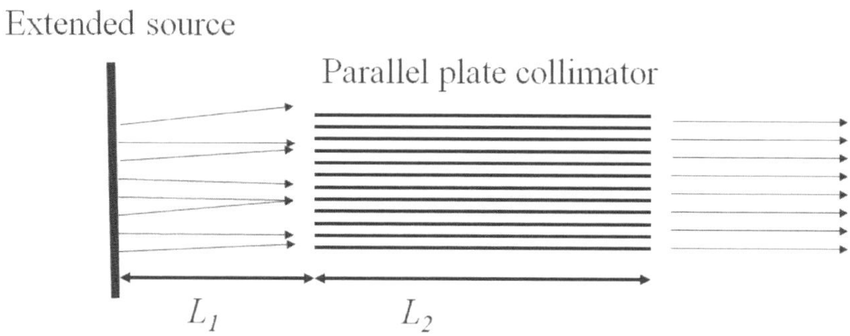

Figure 3.3.4. The parallel plate collimator does offer a quasi-parallel beam from an extended source.

The mechanical tolerance cannot be as good as a double slit arrangement and therefore the restriction on the divergence is rather limited. So far we have considered controlling the divergence in the scattering plane, however this arrangement can be used to control the axial divergence (normal to the scattering plane). The parallel plate collimator is rotated through 90° about the X-ray beam axis (Soller slit). This is a very common method used in conventional powder diffraction methods because significant axial divergence introduces a large asymmetric tail on the low scattering angle side of Bragg peaks $2\theta < 90^0$ and on the high side for $2\theta > 90^0$, leading to additional complications in fitting the shape and position. Typical Soller slits can restrict the divergence from 0.16 radians down to 0.02 radians.

Most applications of this parallel plate collimator considered in this discussion are as an analyser in the scattered beam, whereas the Soller slit is used on either or both. An example of the use of parallel plate collimators in the incident and scattered beam is given in chapter 4, section 4.5.3, and chapter 5, section 5.7.6.2.

3.3.4. *General considerations of slits*

The optimum slit configuration to control the axial divergence will depend on the size of the X-ray source and how much intensity is required for the experiment. The double slit combination given in figure 3.3.2, can be a double pinhole combination (defined equally in all directions) and when used with a point source will produce equal divergence in all directions. When good angular resolution is required in the scattering plane and a line focus is used then the axial divergence is best limited by Soller (parallel plate) slits figure 3.3.4, since the projection of the source will be extended in this direction.

Various slit combinations therefore create very different beam profiles and divergences. However the full spectral distribution of the X-ray source is passed and clearly the width of even the strongest characteristic energies can severely broaden the scattered beam and the angle over which the sample will scatter. However we can select a narrow wavelength band by inserting a crystal or periodic multilayer that scatters a specific wavelength band at a specific scattering angle. We shall firstly consider absorbing filters that are frequently used to modify the spectral distribution.

3.4. Incident beam conditioning with active components

An active component is defined as one that changes the spectral distribution. The distribution is changed by removing certain bands of wavelengths by scattering or through the use of absorbing filters. Inevitably the divergence and intensity will also be modified.

3.4.1. *Incident beam filters*

In Chapter 2 section 2.3 we discussed the absorption of X-rays due to resonance, equation 2.3.4. What we find is that various elements will absorb X-rays very strongly on the high energy (short wavelength) side of a resonance condition, but above this the absorption is dramatically reduced. By placing an absorbing material of an appropriate thickness, that has an absorption edge very close to the characteristic radiation of the X-ray tube, we can change the spectral distribution quite dramatically. The spectral distribution of the X-ray source is given in figure 3.2.1, where we can see that there are several characteristic peaks due to the $K\alpha$ doublet and a complex of $K\beta$ lines of lower intensity. Having several characteristic lines adds complications to any diffraction pattern. The energy difference in the doublet is small and high-energy resolution is required to separate these contributions, however we can select an elemental material with an absorption edge just on the low energy side of the $K\beta$ line. For Cu anode tubes, nickel fits the requirement and for molybdenum tubes zirconium fits the requirement. The spectral distribution is changed so that the $K\beta$ line is almost completely eliminated, but also the broad "white" radiation is reduced and a sharp absorption edge is seen, figure 3.4.1.

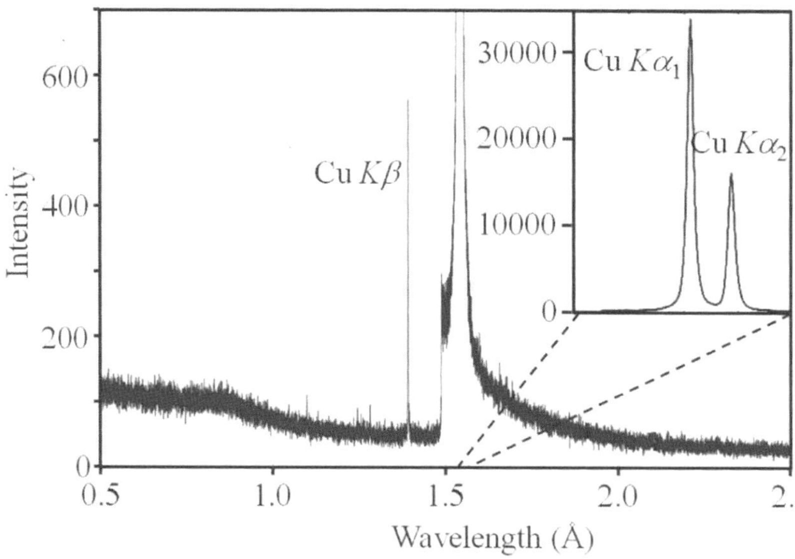

Figure 3.4.1. The distribution of intensity as a function of wavelength is modified with the inclusion of a Ni filter. The Ni absorption edge is between Cu $K\alpha$ and Cu $K\beta$, therefore Ni suppresses the latter with little effect on the former. The profile is also a function of the source, sample (perfect Si) and the solid state detector response.

For low-resolution measurements of generally weakly scattering material this degree of energy selection may well be adequate, since only the characteristic peaks create measurable scattering. The $K\alpha$ doublet is then accepted and predictable and for some small crystallites (creating diffraction broadening) or heavily strained material the doublet may not even be resolved.

3.4.2. *Incident beam single crystal conditioners*

The spectral distribution can be modified in a simple way by a slit and a crystal, figure 3.4.2. The slit limits the angular divergence, $\Delta\xi$, as described above and therefore from Bragg's equation:

$$\frac{\Delta\lambda}{\lambda} = \cot\theta\Delta\xi \qquad 3.4.1$$

where $\Delta\xi$ is the angle of divergence defined by the slits and the X-ray source size. The magnitude of the wavelength band-pass, $\Delta\lambda$, is defined by the slit and X-ray source, but the actual centre value of the wavelength, λ, is determined by the chosen Bragg angle, θ. If we limit the divergence to low values, then it is possible to obtain the spectral distribution, e.g. figure 3.2.1, with this geometry when rotating the crystal.

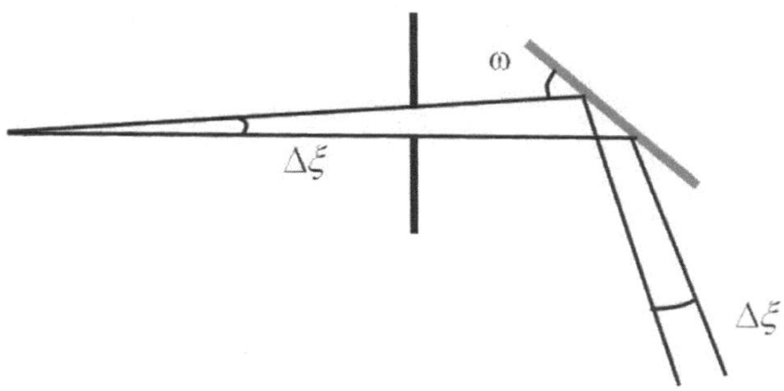

Figure 3.4.2. The combination of a slit and crystal to limit the wavelength spread.

The resolution of the spectral profile is increased as the divergence is reduced. However the divergence cannot be reduced much below significant fractions of a degree and therefore the wavelength band pass is still large and can dominate the scattering widths.

If however the crystal is kept stationary then a defined wavelength band is selected and each wavelength scatters at an angle defined by Bragg's equation. The double-crystal diffractometer that will be described in chapter 4, section 4.3.1 uses this property. The finite source size creates a blurring of this band-pass because it effectively increases the range of incident angles for each wavelength. However this blurring only occurs over the finite diffraction width and averages out the spatial distribution of wavelengths. The scattered beam from this perfect crystal is hence characterised by spatially distributed beam paths for each wavelength. Each wavelength has a clearly defined direction that is parallel to within the diffraction width of the perfect crystal.

3.4.3. *Single crystal groove conditioners*

From figure 3.4.3 we can see that the profile of a single reflection from a perfect crystal is sharp and narrow for reflectivities above about 10%. Below this value there are long tails of intensity that would be very evident for experiments using very large intensity ranges. This instrumental artifact can introduce considerable aberrations to an experimental profile. These "tails" can be reduced with a grooved crystal, where the X-rays are scattered many times, figure 3.4.3. From the dynamical scattering model we can calculate the scattering profile from a perfect crystal, as figure 3.4.3 for a single wavelength (each wavelength will have a similar profile, but will be shifted in angular position, ω, for wavelength differences within ~1%). The scattering from the second reflection in the groove will have an identical profile to that of the first reflection, provided that the crystal planes are parallel, i.e. from the same crystal block.

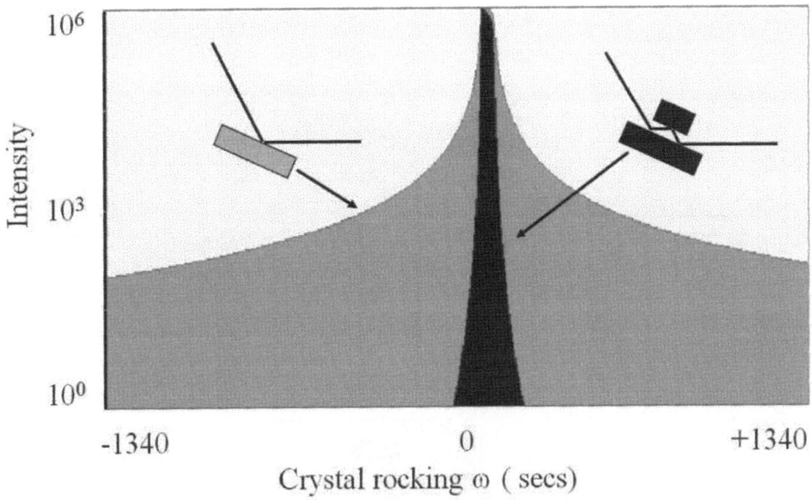

Figure 3.4.3. The simulation of the scattering from a perfect Ge crystal (*220* reflection) for a single reflection and three reflections.

The intensity is a function of the scattering angle and the scattering angle for any part on the curve of figure 3.4.3 will be the same for both crystal faces, if the crystallographic planes are parallel to the surface. The resultant profile is therefore the product of all the profiles. Since the scattered intensity varies from a value close to the incident intensity to a fraction less than 2% within about 20" of arc for the single crystal reflection, the equivalent fraction for N reflections is $(2\%)^N$. This is a useful way of producing very sharp scattering peaks and reducing the instrumental aberrations without significantly reducing the peak intensity.

As more reflections take place, the tails are reduced further to values that are completely unobservable. However this is only valid for perfect crystals and no crystals are perfect. Defects resulting from surface damage, point defects and dislocations will create diffuse scattering and broaden the tails of the profiles. The advantages of many reflections can therefore be lost beyond about $N = 3$. Also the maximum reflectivity is slightly less than unity and the overall intensity and shape reduces the intensity quite rapidly, especially if the crystals are not absolutely perfect.

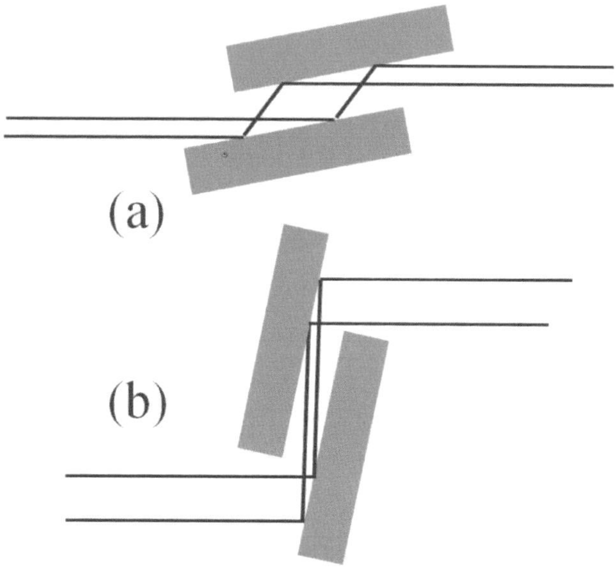

Figure 3.4.4. Some possible methods used for increasing the acceptance and divergence of the beam (a) and decreasing the acceptance and divergence (b) resulting in an increase and decrease in intensity respectively.

The approach above is very useful and can of course be extended to many combinations including crystals cut with non-parallel sided grooves and reflections from planes not parallel to the groove sides. Let us consider the case when the scattering is from planes not parallel to the crystal surface (or groove side). From figure 3.4.4a we can see that X-rays incident at a grazing angle will arrive at the crystal surface over a large area, resulting in a broad scattered beam after the first reflection. Also the intensity profile over the "rocking angle" is broadened compared with the equivalent reflection from planes parallel to the surface. If the X-rays approach the surface at higher grazing angles and scatter at small angles to the surface then the profiles are narrowed, figure 3.4.4b. For the condition of low angles of incidence, the angular acceptance of the X-ray beam is enhanced and more intensity is passed. This can be understood from the broadened "rocking angle" profile; although the crystal is not rocked the divergence incident on the crystal is far larger than this broadened "rocking angle." For the geometry of figure 3.4.4b the intensity passed is reduced. For a grooved crystal with scattering planes inclined to the surface, we again follow this reasoning through the system, although now the options are increased immensely.

To calculate the profile of the intensity through this groove is a little more complex since we have to account for the beam paths rather carefully. For an incident wave-vector arriving at the first crystal at an angle to the surface of ω_1 this will create an incident angle at the second crystal of $\cos^{-1}\{\cos\omega_1 + 2\sin\theta\sin\varphi\}$, equation 2.5.4. This in turn will create a scattered wave-vector travelling in a direction parallel to the incident wave-vector on the first crystal. Therefore there is no angular deviation of the X-rays passing through this groove crystal. However the angular spread of the X-rays after the first reflection is reduced and the acceptance of the X-rays on the second crystal matches this acceptance. The beam leaving the second reflection is as divergent as the beam accepted by the first crystal; therefore the intensity is improved at the expense of angular divergence. A factor of 5 in intensity is easily obtainable, compared with an equivalent arrangement when the scattering planes are parallel to the surface. However the crystals have to be larger to capture the same size beam from the X-ray source. A monochromator based on these principles will be discussed in the next section. X-ray beams passing through a groove with non-parallel crystal faces will experience different refraction peak-shifts and this must also be considered.

3.4.4. Multiple Crystal Monochromators

The monochromators described so far rely on the wavelength band being selected by the angular acceptance of the slit. For a typical slit with a divergence of 0.1^0 and the crystal centred on a Bragg peak at $2\theta = 60^0$ with Cu $K\alpha_1$ radiation then the wavelength spread is $\sim 1.55 \times 10^{-4}$ nm. The broadening of the profile from this wavelength spread varies very strongly depending on the sample scattering geometry. To obtain a versatile diffractometer it is important to have a method of restricting the angular divergences and the wavelength band-pass so the resolution varies only gradually throughout the range where we wish to carry out our experiments. The angular divergence can be kept roughly constant throughout the whole region where we collect data, but the wavelength contribution will always create a varying instrument smearing that increases with scattering angle. This is clear if we rearrange equation 3.4.1

$$\Delta\xi = \tan\theta \frac{\Delta\lambda}{\lambda} \qquad 3.4.2$$

Clearly however small the band-pass is as θ approaches 90^0 the broadening approaches very large values. For examining high quality crystals and to achieve a versatile instrument it is important to reduce the wavelength band-pass.

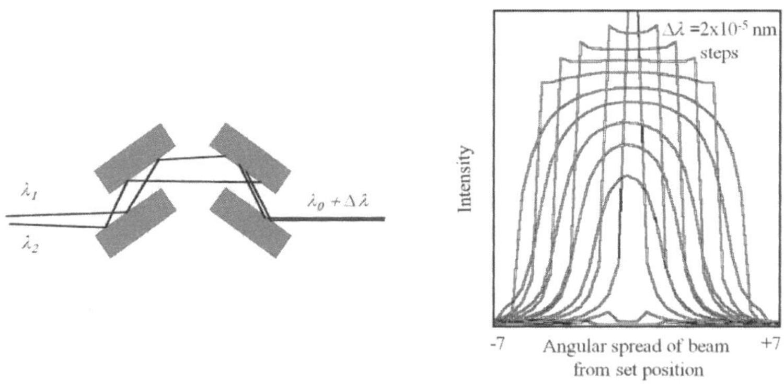

Figure 3.4.5. (a) The 2-crystal 4-reflection monochromator in symmetrical reflection geometry and (b) the resulting complex divergence variation in arc seconds with wavelength for Ge *220* reflections for the most intense beam (changing the offset can reduce the divergence and intensity).

Consider the combination of crystals illustrated in figure 3.4.5a and the associated X-ray beam paths for two wavelengths. If we followed Bragg's equation very strictly (i.e. in the kinematical approximation), then only one wavelength can be scattered by both crystals. But because the crystals have finite scattering widths and the beams are divergent then there is a finite wavelength band-pass. This principle is used in the 2-Crystal 4-Reflection monochromator first proposed by DuMond (1937) and built for a laboratory source diffractometer by Bartels (1983). However the intentions of DuMond to produced narrow profiles was not realised and the widths predicted by Bartels were too excessive. Further calculations by others still over-estimate this value. DuMond's narrowing of the profiles is easily achievable and the monochromator performance can be shown to exceed what Bartels had predicted. The main problem with many calculations is that they are based on convolutions of the angular distributions and wavelength contributions. A more exacting analysis is obtained by considering the angular and spatial distributions of the beam paths.

The approach used here is to calculate the absolute beam direction and position as it passes through the monochromator for each wavelength. Each wavelength from each part of the X-ray source is spatially separated on passing through the first grooved crystal, but the size of the source tends to give an approximately even distribution of wavelengths across the beam in the scattering plane. Let us assume that we have rotated the first U-shaped crystal block to scatter the mid-point of the Cu $K\alpha_1$ spectral line, figure 3.4.5a. The second U-shaped block is then rotated to capture and scatter as much of this distribution of wavelengths and angular divergences as possible. Now any beam of a specific wavelength scattered from the first crystal block will have its own divergence corresponding to the intrinsic scattering profile from reflections in this first block. The direction of this beam will be spread from its central value of $\omega_0(\lambda)$, and we can consider the path of a beam deviated $\Delta\omega(\lambda)$ from this angle.

Let us suppose that the second crystal block will accept and scatter X-rays that are incident on it within the intrinsic scattering angle and the mid-range of this corresponds to an angle $\omega_0(\lambda)$. The path of our deviated beam will impinge on the second block at an angle $\omega_0(\lambda)+\Delta\omega(\lambda)$, when ideally it should arrive with an incident angle of $\omega_0(\lambda)-\Delta\omega(\lambda)$. Therefore the optimum scattering angles are moving away from each other at twice the rate of the deviation in angle, i.e. $2\Delta\omega(\lambda)$. The width of the intensity profile emerging from the second block will therefore be approximately half that of the intrinsic scattering width. If we now consider a similar analysis for a different wavelength then the second block acceptance angle $\omega_0(\lambda')$ will be offset from $\omega_0(\lambda)$. Wavelengths other than the chosen value will either be deviated rapidly from the scattering condition or go through an optimum position before being filtered out. If we include all these factors then we obtain a family of curves that represent the divergence associated with each wavelength. The resultant distribution of angular divergences for each wavelength centred on the $K\alpha_1$ peak is given in figure 3.4.5b. It is abundantly clear that we cannot consider a wavelength distribution and a common angular divergence in isolation, but rather each wavelength having its own divergence and intensity distribution. Another important point here is that the alignment of the crystal blocks is quite crucial to maximise on the chosen wavelength and to optimise the angular divergence.

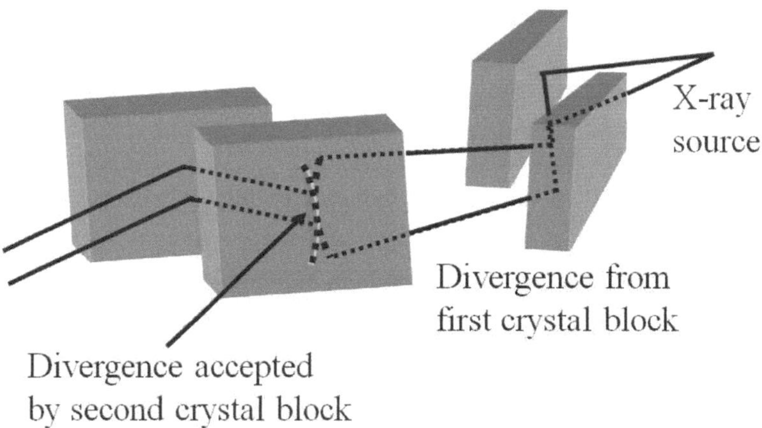

Figure 3.4.6. A schematic of how the axial divergence is limited on passing through a 2-crystal 4-reflection monochromator.

We have only considered the scattering in the plane of the diffractometer, yet there will be axial divergence that will also influence the profile shape from our experiments. If we consider the path of a single wavelength then it will only be scattered from a crystal surface if it approaches it within the angular acceptance (intrinsic scattering profile width). Therefore if this wavelength is optimised for scattering in the diffractometer plane and the crystal planes are perpendicular to this, then those paths out of the plane will be limited to the projection of the intrinsic scattering width. Consider figure 3.4.6, where the locus of the optimised X-ray beam path is drawn for a single point source at a distance x from the crystal. The beam trajectory is unchanged on passing through the first crystal block in terms of the axial divergence. The projection, θ_1, of the optimum scattering angle, θ_0, for any X-ray beam path will be increased as its trajectory moves out of the plane of the diffractometer. This change in the projected incident angle, $(\theta_1 - \theta_0)$, varies as the square of the axial divergence angle out of the plane, χ, equation 4.3.1. The X-ray beam therefore arrives at the second crystal block with a projected incident angle of $(2\theta_0 - \theta_1)$. Since the projected angle for the optimum scattering of an X-ray beam following this trajectory should be θ_1, the deviation of the actual to optimum incident angle is $(2\theta_0 - 2\theta_1)$. Hence the axial divergence is contained to small values by this monochromator.

$$\Delta\chi = \{[2\theta_0 - 2\theta_1]\cot\theta_0\}^{\frac{1}{2}} \qquad 3.4.3$$

This is discussed on a purely geometrical argument but for the scope that we wish cover in this book it does explain that the axial divergence is constrained, however for a fuller description we must include all the wavelengths and the alignment of the monochromator, etc. If the source is larger than a point, then the axial spread of the beam will be larger, but the angular divergence will still be constrained.

3.4.5. *Multilayer Beam Conditioners*

A laboratory source is divergent and most of the X-rays generated are lost, especially when the divergence acceptance of the monochromator is small or reasonable resolution is required from slits. Ways of capturing this divergence have been discussed for grooved crystals when using crystallographic planes inclined to the surface. However we can make use of a very simple principle of a parabolic reflector. Consider the geometry of figure 3.4.7. The divergent source is now reflected and concentrated into an almost parallel beam. Of course with X-rays this is not so simple, since they are not reflected efficiently at all angles, section 2.10. This loss of reflectivity is recovered by constructing the parabola as a periodic multilayer whose repeat changes along the surface, Arndt (1990). The satellites from a multilayer composed of strong and weakly scattering materials can give very high reflectivites (~70% or more). To satisfy this condition of maximum conversion of a divergent beam into a nearly parallel beam requires exact manufacture so the period varies to satisfy the Bragg condition for the wavelength of the source. The interface quality between the layers should also be of high quality since any imperfections will reduce the reflectivity.

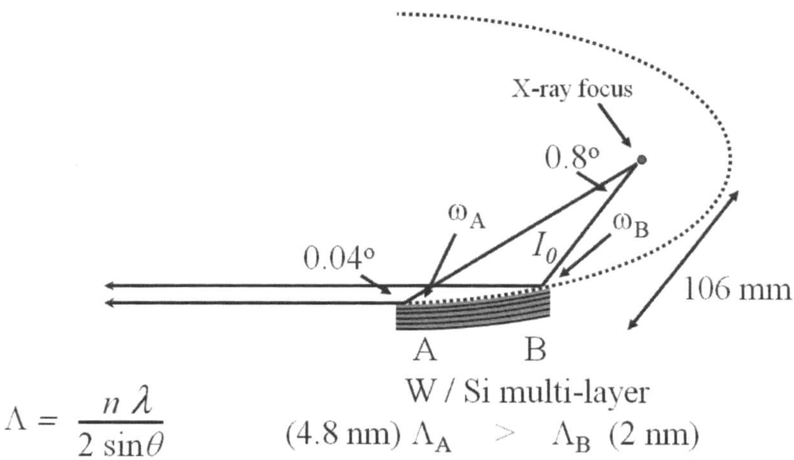

Figure 3.4.7. This illustrates the principle of the multi-layer parabolic X-ray mirror. Λ_A and Λ_B are the periods at the extreme ends of the multi-layer mirror.

An estimate of the gain in intensity can be obtained by including; the captured divergence, $\Delta\xi$ (0.8^0 in figure 3.4.7), the degree of parallelism achievable, $\Delta\xi'$ (0.04^0 in figure 3.4.7) and the reflectivity, R, for the chosen satellite reflection. The gain in X-ray flux (intensity per unit angle) is

$$Gain = \frac{\Delta\xi}{\Delta\xi'} R \qquad 3.4.4$$

For typical values of $R = 0.70$, $\Delta\xi = 0.8^0$ and $\Delta\xi' = 0.04^0$ the flux gain is ~ 14. Clearly as the periodic quality declines, the period variations laterally are not precise and the interface quality is poor the gain is diminished. Also the focus should be ideally point-like: this can be achieved with a line focus with a suitable take-off angle, to obtain a 40 μm dimension. The X-ray mirror, figure 3.4.7, can be large in the axial direction (normal to the plane of the figure) to make full use of the line focus. Alignment of the mirror is clearly rather critical but using the prefix arrangements of modern diffractometers, e.g. figure 3.1.1, this is very stable and realignment after removal and return is unnecessary.

As always the advantages of the mirror are not without some disadvantages. We have converted a divergent source into a parallel source and this is not ideal for some applications, since the beam is expanded and is less advantageous for diffractometers collecting data by position, e.g. slits discussed in section 3.5 and the following chapter.

There is some energy selection but this does not compare with single crystal reflections at higher angles. The mirror utilises a low angle satellite for improved reflectivity and therefore the wavelength band-pass is broad, equation 3.4.1, and Cu $K\alpha_1$ cannot be separated from Cu $K\alpha_2$ without great difficulty, but Cu $K\beta$ can be virtually eliminated. The axial divergence is increased and therefore the resolution can be made worse, especially in high resolution reciprocal mapping when compared with point focus tube settings. We also have to consider whether the sample can make use of the additional size of the beam to benefit from the increased intensity.

The mirror is therefore another optional component that helps in some analyses, but is not suitable for all. It is though a very useful additional component used in conjunction with slit geometries and multiple crystal monochromators. Increasing the intensity per unit angle has a significant advantage for the latter but now the axial divergence is less controllable and the resolution declines, however for many applications this is a very useful combination for speed. The resolution can be improved with a Soller slit to limit the axial divergence with some loss in intensity.

Although the X-ray mirror described here is for creating a close-to-parallel beam, the same principle can be used to produce a focusing (elliptical) mirror. These can have advantages in powder diffraction and some high-resolution studies: but this will require careful engineering to ensure the detector or sample (depending on the application) is accurately placed at the focus.

Toroidal mirrors can also be used to contain the axial divergence (similar to a mirror in scattering and axial planes) but these require a small X-ray source in both planes, i.e. a microfocus source, Arndt, Duncumb, Long, Pina and Inneman (1998). Again we see the intensity can be gained but at the expense of increased beam size. If we use the toroidal mirror to focus onto a small spatial region on the sample then we find the divergence is increased. Some further discussion on this is given in section 3.4.8.

3.4.6. *Beam Pipes*

X-rays in general are very difficult to guide unless they are diffracted as above, because the refractive index in all materials is so close to unity. Limited focusing can be achieved with Fresnel zone plates and these have ring diameters and separations to maintain the phase relationship and acts rather like a lens. However to condense an X-ray beam by guiding has been successful with a refractive lens, Snigirev et al (1998), which are more applicable to highly intense sources, i.e. synchrotrons, and with using glass capillaries, Komakhov and Komarov (1990). The latter procedure relies on the total external reflection of X-rays and gradual curvatures.

Consider a bundle of tapered glass capillaries, figure 3.4.8, where the capillaries are pointing towards the point focus of the X-ray tube. Those X-rays that fall within the angular spread of the capillaries have a good probability of entering them (some will be absorbed by the glass), since the X-rays will be entering roughly along the capillary axis. If they arrive at the walls at a low angle they will have a high probability of being specularly reflected. Higher energy X-rays have a lower critical angle and therefore there will be some energy filtering in favour of low energy X-rays. In general the capillaries close to the centre of the bundle (where the curvature is less) will transmit all energies and towards the edges the energy distribution will favour the lower energies.

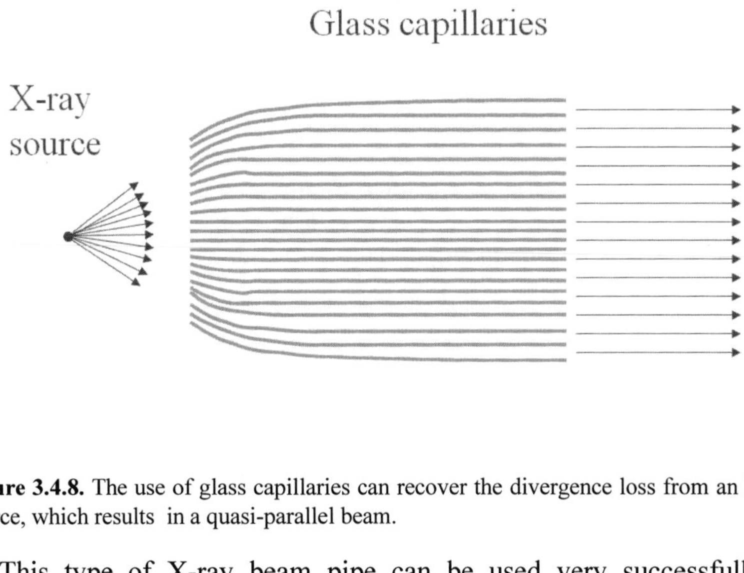

Figure 3.4.8. The use of glass capillaries can recover the divergence loss from an X-ray source, which results in a quasi-parallel beam.

This type of X-ray beam pipe can be used very successfully to capture a divergent source and convert it to a quasi-parallel source. This can prove very useful for low resolution X-ray diffraction studies with large intensity enhancements in terms of flux at large distances from the source. The size of the beam is controlled by the diameter of the capillary bundle, which are generally circular in cross-section giving an even resolution in the scattering plane and axial direction. The divergence is governed by the critical angle and will differ slightly for each X-ray wavelength; however as an approximate guide the divergence is twice that of the critical angle.

In some cases a single capillary can be of benefit, either as a straight capillary or one that is tapered to try and induce some spatial focusing. The same principle applies as above, but in this case this beam pipe is useful for transporting intensity with smaller divergence losses than a simple double pinhole for example. The capillaries can be made with diameters less than 100μm.

As with the multi-layer mirror, the glass capillaries can be brought back to a focus. Although the beam shape can be controlled to an extent, the divergence is largely defined by the critical angle of the glass, or its coating. So typically glass will have a critical angle of ~0.2^0, making the overall divergence spread in the region of 0.04^0.

3.4.7. *Bent crystal monochromators*

Consider the possibilities with the geometry shown in figure 3.4.2., which is used for selecting a wavelength range with a slit, a crystal and a small focus. In that case the wavelength band from the crystal is diverging. For some applications, especially requiring focusing optics, the emerging beams would be much more useful if they are made parallel or focusing. This has been achieved by bending the crystal, Johansson (1933), figure 3.4.9. If the crystal planes are parallel to the surface then the focus point is at the same distance from the crystal as the source, so the crystal is bent to the radius of the focusing circle. That is the crystal surface, the X-ray source and the focusing point all lie on the circle. Sometimes it is convenient to introduce an asymmetry, as in figure 3.4.9, then the crystal planes must be inclined to the crystal surface.

This approach is used in powder diffraction to isolate the $K\alpha_1$ characteristic line from the $K\alpha$ wavelength doublet, with high quality Ge crystals and *111* reflection. The curvature needs to match the divergence of the source, so that the Bragg condition is maintained across the whole crystal. Very careful and exact curvatures are required to match the geometry of the instrument: the focus point could be considered as a pure $K\alpha_1$ source for a subsequent geometry, e.g. as in figure 3.5.3. If the curvature is inexact and gives an imperfect focus then this can be compensated for by placing a slit at the focus. Geometries that rely on bringing the beams to a focus at the detector are more problematic, since these cannot be slit down so easily, and require very careful control of the curvature. Some of the focusing geometries making use of these ideas are described in chapter 4, section 4.4.1.

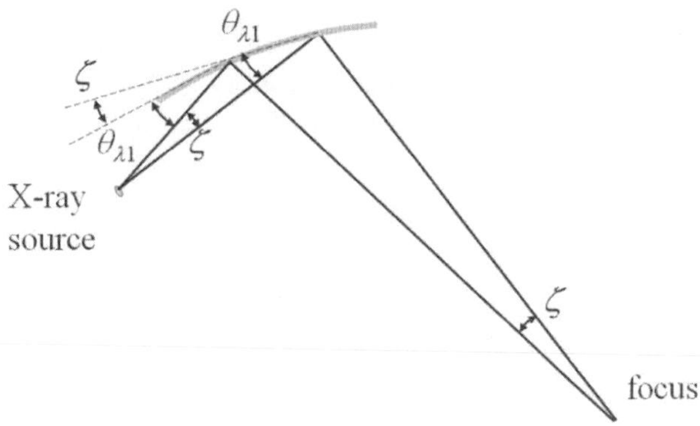

Figure 3.4.9. The bent crystal focusing geometry. The crystal is bent and positioned so that the crystal planes bring the divergent beam to a focus. If the focus point and the source are at different distances from the crystal then the crystal surface has to cut at an angle to the crystal planes.

Applications that can accept a larger band-pass, are able to use graphite crystals that can be mosaic (the angular spread can be from 0.3^0 to 0.5^0). This mosaic spread allows $K\alpha_1$ and $K\alpha_2$ to pass and has applications that are less concerned with the profile shape, but do want to reduce the spectral range: this reduces the background. The intensity can be kept high and suitable for measuring integrated intensities in single crystal studies.

3.4.8. *Double focusing mirrors*

Conditioning the incident beam in both directions, scattering and axial planes, will reduce the intensity unless some of this divergence is recovered. Exploiting these ideas will need a small focus in both directions, i.e. a micro-focus X-ray tube, and very careful alignment of the mirrors. Combining two good-quality single crystals orthogonal to each other will reduce the intensity passed, because of the very small divergence accepted, however two mirrors (figure 3.4.7) at right-angles to each other is a good option. The alignment becomes a significant challenge, requiring good source focus stability as well as good mechanical tolerances.

The number of possibilities is large and have been exploited at synchrotrons and then copied for laboratory sources. The ideas are not recent, but come from ideas proposed for X-ray telescopes; for example the work of Kirkpatrick and Baez (1948) and Wolter (1952). These options require the same graded multi-layer mirror technology described earlier, section 3.4.5, and because of the closeness of the optics to the source, the best performance is achieved with elliptical-shaped mirrors. As with all attractive options there are caveats, and in this case the use of micro-focus X-ray tubes. Although the emission per unit area from the anode of a micro-focus tube can be very high, the overall number of photons is low because of the limited power loading possible, section 3.2: so a well conditioned beam may not have the desired intensity and is another consideration.

3.4.9. *General comments on active incident beam components*

Crystals and mirrors create a large number of possible methods to condition the incident beam. These components should be considered in conjunction with the X-ray source. They do have the advantage over the slit-based components, in that the wavelength range can also be selected. The setting-up and alignment is more complex, although modern commercial instruments are engineered to ensure a stable environment once aligned. The only variable is the X-ray tube because of the heating of the focus (the inherent inefficiency of producing X-rays), and the inevitable movement through expansion until it achieves equilibrium with the coolant.

X-ray mirrors that recover the divergence will obviously expand the beam, and this should be a consideration whether this larger beam can be used. If a small beam is required then it is possible to focus the beam, but then it is convergent. As always these considerations always results in a compromise.

3.5. Scattered beam conditioning with passive components

The principles of the scattered beam conditioning are identical to that for the incident beam. However the source will now be equated to the sample and the output of the component will most likely go straight into a detector.

3.5.1. *Scattered beam slits: Fixed arrangement*

A single slit position will be associated with a measured scattering angle, $2\theta_{meas}$, figure 3.5.1. However as it can be seen from the figure, the reliability in determining the actual scattering angle from a sample for a finite size source is rather poor. The 2θ associated with the sample is with respect to the incident beam direction, and the measured angle is only correct with respect to its rotation axis. Figure 3.5.1 shows a series of incident beams that are all scattered through the same 2θ angle, and in this case the slit, with a detector behind, is set to capture this scattering, i.e. $2\theta_{meas} = 2\theta$. It can be seen that several of these rays associated with the scattering angle 2θ, are not captured when they emanate from a different region of the source, whereas some beams impinge on the sample at a different incident angle that do not scatter at 2θ but will be measured. These effects blur the measured profile with a range of sample information that cannot easily be resolved. The situation can become further confused if there is significant inhomogeneity across the sample.

The extracted information from the measured profile will be confused further if the sample is not on the rotation axis of $2\theta_{meas}$. This will lead to systematic 2θ displacements in the measured angles.

Suppose the source is very small, and the sample is accurately placed on the rotation axis for $2\theta_{meas}$, then the situation is improved. However there will still be intensity contributions from other 2θ angles that scatter from different positions on the sample. Basically three defining positions are needed to make this work, i.e. small sample, another slit (this is considered next) or by coupling the sample and $2\theta_{meas}$ rotations (chapter 4, section 4.4.1).

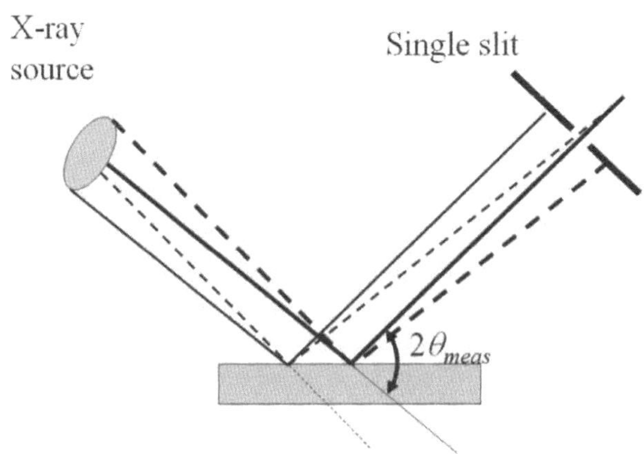

Figure 3.5.1. The loss in reliable information resulting from a finite size source and single slit in front of a detector. All the incident beams scatter through the same angle.

The advantages of the double slit configuration are clear to see in figure 3.5.2. This configuration defines the area of the sample from which the scattering is coming, and depending on the slit size will discriminate in 2θ. As the slits are narrowed the angular resolution improves but the intensity declines; this compromise is a regular feature in X-ray diffraction. A combination of double slits on the incident beam side (section 3.3.1) as well as on the secondary side will define an area of scattering in the depth dimension. This method is well suited to very high energy X-ray or neutron scattering, where the penetration depths are significantly larger than the dimensions of the scattering region, otherwise the loss of intensity is rather prohibitive.

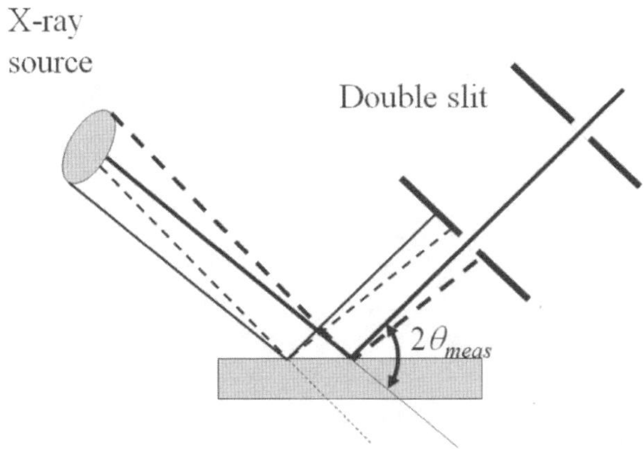

Figure 3.5.2. The double slit improves the resolution by limiting the area on the sample from where the scattered beams come.

3.5.2. *Scattered beam slits: Variable arrangement*

This option is generally used in conjunction with variable incident beam slits and a small source size. This combination ensures that the whole of the illuminated area, defined by the source and incident beam slits, is seen by the scattered beam slit. To make this configuration work well a 'receiving' slit is required, that has similar dimensions to the source and is at the same distance as the source from the sample, figure 3.5.3. If the receiving slit and the source are at the same distance from the sample, and the sample normal bisects this angle between the incident and scattered beams, then this symmetrical para-focusing geometry only allows those beams that scatter through $2\theta_{meas}$ to pass.

This para-focusing geometry is covered in chapter 4, section 4.4.1. It is not specifically used with variable slits, in fact it is more often used with fixed slits, or with just an incident beam slit and receiving slit or small detector. From this latter point, it is clear that the second variable slit has little impact because the region of sample scatter is already defined: however the scatter from slits by the broad X-ray spectrum is less likely to reach the receiving slit so that the profile has less spurious scatter.

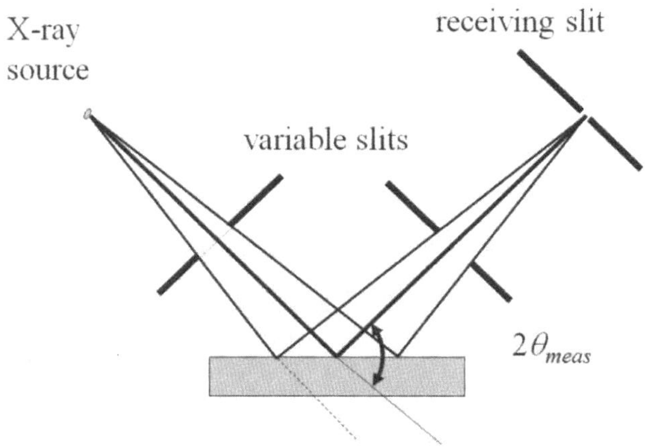

Figure 3.5.3. The combination of small source size and narrow receiving slit can be combined with automatic slits to illuminate large areas and maintain good resolution.

3.5.3. *Scattered beam slits: Parallel Plate Collimators*

Parallel plate collimators allow the capture of scattering from large regions on the sample, figure 3.5.4. These are used in combination with very low angle of incidence probing beams for analysing thin films, and as a Soller slit, when rotated through 90^0, in powder diffraction studies. The principle is obviously the same in just accepting those scattered beams that lie within the accepted divergence. The separation between the plates relates to the accepted divergence, just as in the calculation for the double slit configuration. The length of a parallel plate collimator is usually much greater than for a Soller slit to give a much smaller angular acceptance. Typically for a Soller slit used in the laboratory this will have an angular acceptance ranging from 1.14^0 to 4.58^0, whereas a parallel plate collimator varies from 0.09^0 to 0.27^0.

Although each individual pair of plates will have a similar characteristic intensity distribution to a double slit, figure 3.3.2b, the combined profile can be much more complex. Close to the exit of the collimator the intensity distribution across the beam will have a rectangular wave-form, with a width matching that of the separation between the plates and a region of no intensity corresponding to the width of the separating plates. The contribution from each collimated beam will begin to overlap and smooth this wave-form shape as the detecting point is moved further from the end of the collimator (or Soller slit). In general the intensity at any one angle-setting is integrated and assigned a value.

These scattered beam collimators are best suited to powder diffraction experiments, where the intensity scattered by the sample is quite even across the beam. A typical configuration is given in figure 4.4.2. These collimators are used in the scattering plane and limit the cross-fire, just as Soller slits achieve in the axial plane, and when used in conjunction with single slits or crystal optics can limit the stray scattering associated with angles not intended to be measured.

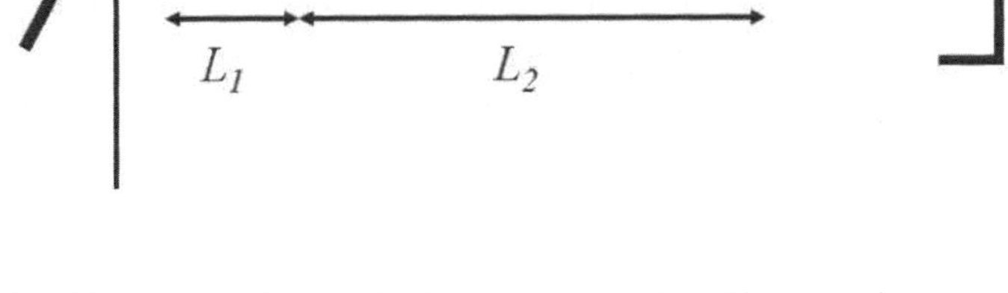

Figure 3.5.4. The parallel plate collimator used as a scattered beam analyser.

3.5.4. *General consideration of slits in the scattered beam*

Slits in the scattered beam are required to make a compromise, if good 2θ angular resolution is wanted, then the slits need to be narrow at the expense of intensity. Since 2θ is defined as the angle between the incident beam and the scattered beam, the resolution of a scattered beam component can never be considered in isolation. The parallel plate collimator is useful as a passive component for capturing the scattered beams by angular selection and not by position. This has the advantage that the position of the scattering point on the sample is less relevant, provided the incident beam direction is defined. This makes this configuration less susceptible to sample setting errors.

Single slits can work effectively, for example: to measure the 2θ angle to high precision, this will require the source, sample and scattered beam slit to be aligned accurately. If the source and the scattered beam are equidistant from the sample rotation axis, and the sample is flat, then the scattered beam slit will measure one angle from all positions across the sample, i.e. 2θ. However since the incident angle varies across the sample; it is especially useful for powder diffraction studies, since the crystallite scattering planes are not generally aligned with respect to the surface. This again emphasizes that it is the combination of components for various sample types that defines the resolution, rather than them being considered in isolation.

The use of crystal optics restricts the angular acceptance defined by the intrinsic scattering width and does not require a small beam. The compromise though, is that in general this angular acceptance is very narrow. This is discussed in the following section.

3.6. Scattered beam components: active components

In section 3.4., it was shown how crystal optics can be used to; select wavelengths, select a small angular range from the source and methods of shaping the beam. Those approaches are just as valid in the scattered beam. There are some differences though, in that stray radiation, for example fluorescence from the sample and scatter from slits, can be excluded from reaching the detector.

3.6.1. *Wavelength range selection in the scattered beam*

The use of filters as discussed in section 3.4.1 can be used equally well in the incident or scattered beam. There can be advantages though, to select the wavelength range using a crystal in the scattered beam. An example of this is to reduce the fluorescence from the sample that cannot be achieved in the incident beam. A single high-quality crystal will require good mechanical tolerances to maintain the correct orientation as the crystal / detector arm is rotated.

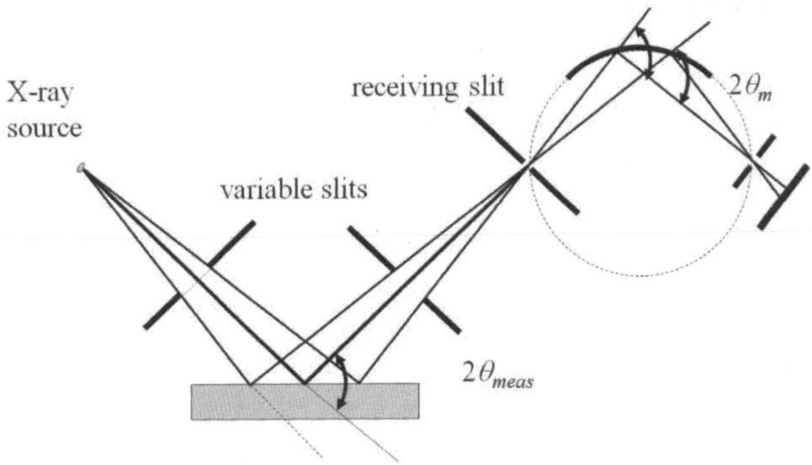

Figure 3.6.1. The extension of the geometry shown in figure 3.5.3, achieves high resolution and wavelength selection, but requires high mechanical stability. The curved analyser crystal is bent to the focusing circle.

Lang (1956) has described curved crystal monochromators, placed on their own focusing circle, which can be coupled to the geometry (as illustrated in figure 3.5.3); this is shown in figure 3.6.1. Since all common 2θ beams come together at the receiving slit, only those that diffract strongly at the fixed $2\theta_m$ will come to a focus at the detector.

Since $2\theta_m$ is defined for the crystal with a known interplanar spacing d, the wavelength can be selected:

$$\lambda = 2d \sin \theta_m \qquad 3.6.1$$

This will select a specific characteristic line. If, however, a graphite crystal is used then this will not discriminate the $K\alpha_1$–$K\alpha_2$ characteristic lines, but will remove the $K\beta$ lines. This configuration with the graphite crystal can also be used to select the $K\beta$ line: this geometry has been used by Kharrak and Cox (1977) and Fewster and Andrew (1999). The former was used in molecular structure refinement and the latter for analyzing texture from reciprocal space maps, to improve the peak shape. The loss in intensity, approximately a factor of 5, compared with the $K\alpha_1$ is not that significant, when compared to a bent incident crystal or a filter to suppress the $K\beta$ line. Because of the limited quality of the graphite there can be some evidence of $K\alpha$ contamination, however this can be easily suppressed with a mask or slit in front of the detector.

3.6.2. *Single flat crystals in the scattered beam:*

To improve the wavelength resolution of the parallel plate collimator a large flat single crystal can be placed after it: this can be visualised as the reverse geometry of figure 3.5.4. The beam direction onto the crystal is defined and the same principle as illustrated in figure 3.4.2 apply. The wavelength resolution is defined by the divergence passing through the collimator, and the centre wavelength depends on the crystal scattering angle. Generally graphite is used since the highest resolution is rarely needed with the parallel plate collimator: alternatively $LiNbO_3$ and Si have been used, but the intensity passed is often too low for typical applications with this geometry. For the collimator with the smallest accepted divergence, mentioned in section 3.5.3 of 0.09^0, the wavelength range is 0.01 Å, which is twice the separation of the $K\alpha$ doublet for Cu radiation, when using a perfect crystal, e.g. $LiNbO_3$ and Si. Ideally the crystal should have an acceptance angle that matches the collimator to optimize the intensity: $LiNbO_3$ and Si are too narrow and graphite is too broad. So reducing the slit size acceptance too far is unnecessary with the latter.

The attributes of a single flat perfect crystal are that it accepts only certain wavelengths impinging on its surface at a specific angle. Therefore if the incident beam is monochromatic and passes over the axis of the 2θ arm incorporating the crystal and detector, then it will discriminate with high-angular resolution. The resolution is determined by the intrinsic scattering width of the crystal. If the sample is placed on the axis then the 2θ angle can be measured accurately, but is independent of the beam position or size, just the difference in angle between the direct incident beam and the scattered beam. This discrimination by angle rather than position is a significant advantage of crystal optics in the scattered beam compared to purely slit-based systems, for the accurate measurement of angles.

If a crystal is combined with a slit then the wavelength range accepted can be chosen, as in figure 3.4.2. This can be useful for analyzing the scattering from a source that is not monochromatic: however, the selection of a narrow wavelength range will have significant impact on the intensity. This, as will be described in the next chapter, can be overcome by matching the scattering angle of the analyser to that of the sample.

3.6.3. *Grooved crystals:*

A grooved crystal as illustrated in figure 3.4.3, will make the angular selection, and therefore wavelength spread, accepted much narrower. The number of reflections within the groove reaches a limit to the improvement of the long diffraction tails, because of the intrinsic defects present in the 'nearly perfect' crystals used. Another important aspect of consideration is whether the number of reflections is odd or even. If the number is even then the beam path is only displaced a few millimeters from the beam entering the groove, which can cause partial transmission of intense beams through the crystal or unwanted forward scatter from imperfect surfaces to contaminate the signal at the detector. The use of three bounces as in figure 3.4.3, has the advantage of requiring the detector to be out of line with this unwanted scatter.

3.6.4. *Multilayer mirrors in the scattered beam:*

Any of the optics discussed regarding the incident beam can be applied to the scattered beam. For example, a large beam can be captured with a multilayer mirror, such that when a narrow slit is placed at its focus the angular acceptance angle is ~0.04^0 within the spectral range for the mirror. That is, it can be considered as just the reverse of figure 3.4.7.

3.6.5. *Beam pipes in the Scattered beam*

From section 3.5.3, the control of the accepted divergence has been discussed for a large beam from the sample. However this restriction to controlling the accepted divergence is either in the scattering plane or the axial plane, so to combine them a bundle of glass capillaries appears attractive, figure 3.6.2.

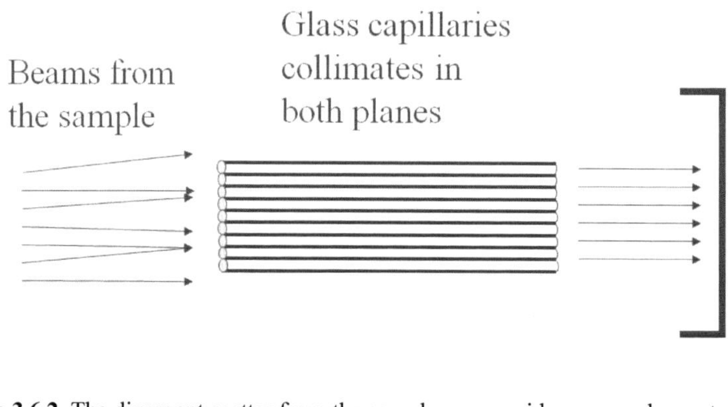

Figure 3.6.2. The divergent scatter from the sample over a wide area can be captured into scattering angle ranges of ~2×critical angle for glass capillaries in both scattering and axial planes.

There is no need to focus in this case, as in figure 3.4.8, so at first sight the manufacturing tolerances appear to be reduced. However, with strong scattering from the sample, or use of penetrating X-ray energies, the beam will not necessarily be well-contained unless they are coated. This adds to the complexity of manufacture. The resulting limit of the degree of collimation is determined by the critical angle of the glass or any coating that may be on the glass. For glass this gives an angular acceptance of ~ 0.4^0, and for materials of greater density, and hence able to contain the beams, the angular acceptance increases. This makes the capillaries less attractive for analyzing the scattered beams, but much more suitable for incident beam modification, when the beam is used with a small point source.

3.6.6. *Some comments on scattered beam optics:*

Apart from the use of glass capillaries, most incident beam optics can be used in the scattered beam in reverse. The significant advantage of using X-ray reflections in the scattered beam is that the size of the beam is less restricted, the centering of the sample is less critical. It is, though, the combination of the incident beam optics, with the sample properties to be analysed in conjunction with the analyzing optics that significantly increases the possibilities. That is, no single component should be considered in isolation.

3.7. X-ray detectors

The main emphasis here will be on the solid state detectors, however the gas-filled proportional counters and scintillation detectors will be covered briefly since they are still extensively used. There are some basics that we should discuss. The ideal requirement of any detector is to record every X-ray photon and produce a measurable signal proportional to the number of photons arriving per second over a large flux range. To maintain the linearity in the response of X-ray photons arriving at the detector to the recorded signal we have to reduce the contribution of any residual noise for weak signals and compensate for any time when the detector is inactive due to high fluxes. Most detectors work on the basis of the X-ray photon creating electron-ion (or hole) pairs either in a gas or a solid.

3.7.1. *The proportional detector*

One of the most reliable detectors is the proportional counter, which is simply a sealed cylindrical outer electrode with an inner coaxial wire kept at a large positive voltage with respect to the outer electrode, figure 3.7.1(a).

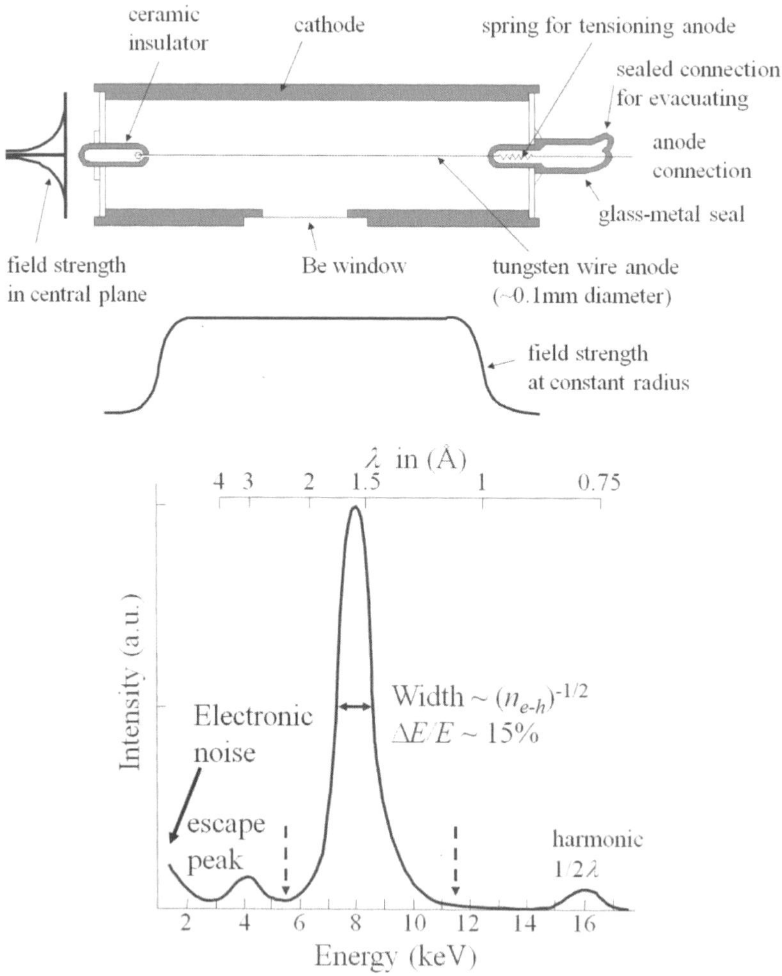

Figure 3.7.1. (a) The proportional X-ray detector. (b) The energy distribution for discriminating Cu $K\alpha$ X-rays with a proportional detector.

The region between the electrodes is filled with an inert gas that acts as an insulator at high electric fields. To obtain a highly efficient counter we require all X-rays to be absorbed within the gas, this partially determines the choice of inert gas. Argon is a low absorber and therefore not ideal, Krypton favours radiative absorption processes and is hence unsuitable for creating a significant current signal. Xe creates Auger electrons and gives about 93% and 45% absorption efficiencies for Cu and Mo X-rays respectively and with careful design can produce a high quality detector. The gas pressure can also influence the sensitivity to different energies and allows for some further optimisation. An X-ray transparent window (Be or mica) is placed at the side of the cylinder. X-rays entering the window then ionise the gas and the intense electric field accelerates the electrons towards the wire creating further impact ionisation events, more electron-ion pairs, and a recognisable current pulse that can be amplified and detected. This gas amplification, $\sim 10^4$, is important for creating a large signal compared with any residual noise. The pulse is further amplified with a low noise high quality linear amplifier mounted close to the counter to reduce the possibilities of additional noise. The highest electric field is very close to the wire and this is where most of the impact ionisation takes place. The electric field must be chosen carefully to give a strong pulse but not increased too much otherwise the plasma created will modify the local electric field significantly and the proportionality will be lost.

The energy of the X-ray photon determines the number of electron-hole pairs and therefore influences the strength of the current pulse in the detection circuit. The favoured absorption mechanism for reliable counting is the Auger process, which is non-radiative and ejects electrons. Xe is therefore the best choice of inert gas. The energy to eject an electron from a Xe atom is ~20.8eV, hence an 8.04 KeV Cu $K\alpha$ photon has enough energy to create ~386 electron-ion pairs. The residual energy from the first ionisation event is converted into the momentum of the ejected electron that in turn creates further ionisation. This aspect gives the energy discrimination. The additional energy from the electric field increases the number of electron-ion pairs further and this is the signal pulse of interest.

Suppose a proportion of the created ions return from their excited state by internal recombination with the emission of radiation. This fluorescent radiation created takes most of the energy and the ejected electron has a lower momentum and consequently creates a weaker pulse. This weak pulse is often termed the escape peak and corresponds to a low energy peak in the detector response function. Statistically this will be a peak of a fixed ratio with respect to that created by the Auger process. Leakage currents associated with these very high voltages are inevitable and will create a further signal, but just as with the escape peak, it will create a different pulse strength from that of the wanted signal. This signal will be fairly continuous and hopefully well below the pulse strength of a photon interaction with gas amplification. The detector response function is shown in figure 3.7.1(b).

The most important aspect of any detector is the relationship between the signal and the incoming photon flux. The good energy resolution, figure 3.7.1(b) helps enormously in obtaining a stable and proportional response. Clearly though we should isolate the leakage current and the escape peak from the signal current and this is achieved with a pulse height analyser. This analyser works on an anti-coincidence signal from comparing the pulse strength (voltage) with that of two adjustable settings V_1 and V_2. If the signal in the external circuit is greater than V_1 but less than V_2 then it will be accepted. The isolated signal should be proportional to the X-ray flux. The latter can be checked with an instrument configuration (that is insensitive to small focus movements) and varying the tube current or by inserting a series of identical X-ray absorbers placed in the beam. The signal should be proportional to the tube current or to the number of absorbers. For optimum performance this pulse height discrimination should be set very carefully to isolate the signal peak to enhance the dynamic range by removing the noise contribution and maintain proportionality.

Suppose we have the situation where this proportionality is lost at high X-ray fluxes. This can arise when the electrons and ions cannot reach the electrodes before another ionising photon arrives. The local electric field has therefore not recovered and the gas amplification is less effective and photons are missed. Of course this will happen on a statistical basis at lower levels than for a series of photons arriving at regular intervals. Because of the high fields the pulse width is very narrow in time and the loss in proportionality does not occur until >100,000 photons arrive per second. The loss in proportionality is predictable and can be simply measured by increasing the tube current to give count rates above these values. A proportional detector can be used reliably up to about 750,000 photons arriving per second when this correction is included. This correction is often included in the detector control software. The proportional detector is therefore reliable over ~0.2 to 1,000,000 photons arriving per second, before attenuators are required.

3.7.2. *The scintillation detector*

The principle of this detector is again reliant on an ionisation process, except this time in a solid. These detectors are composed of a phosphor, usually NaI with ~1% Tl (at least always >0.1% Tl), followed by a photomultiplier tube to obtain reasonable pulse strengths with minimal noise introduction. When an incoming photon ejects an electron to the conduction band of a matrix atom a positively charge hole is formed that drifts to a Tl impurity causing ionisation. Therefore each event creates an electron in the conduction band and an ionised impurity, i.e. Tl^+. The recombination of the electron and the Tl^+ hole will create characteristic fluorescent radiation that has a lower energy than the surrounding matrix. Because the fluorescent radiation has a lower energy than that of the matrix there is very little absorption and can therefore travel large distances, hence the fluorescence decay is determined by the NaI matrix. The whole process takes about 10^{-7} s and the subsequent processes are generally not rate limiting. The possibility of alternative entrapment of the electrons will lead to losses but this is optimised by the correct choice of phosphor and activated impurity.

The phosphor is optically coupled to a photocathode and this in turn is coupled to a photomultiplier. The fluorescent radiation produces electron emission in the photocathode (Cs_3Sb) and the dynodes of the multiplier create electron multiplication to give a measurable pulse that relates to the incoming signal. The variation in the voltages between all dynodes and imperfections in the optical coupling does create some electron loss and light loss respectively and therefore the pulse strength is much more variable than the gas amplification process, consequently it has much poorer energy resolution.

The number of fluorescent photons created is related to the incoming photon energy as the electron energy transitions are created as a result of photoelectric absorption and by the Compton process, Chapter 2, figure 2.1.1, Compton (1923). These detectors have very good capture efficiency ~100% for typical X-ray photons, but are bulky in comparison, have a poorer energy resolution and generally higher residual noise levels. At one time the scintillation counters were superior to proportional counters at high count rates and X-ray energies, but with improvement the proportional counter out-performs the scintillation counter in dynamic response due to optimisation of the pulse shape from improved design and improved absorption of incoming photons. For high-energy X-rays, Mo *K* radiation and above, the scintillation counter is often preferred because of the capture efficiency compared to the proportional counter.

3.7.3. *The solid state detector:*

The solid-state detector, figure 3.7.2a, is less complicated in understanding the basic physics, although the problems in manufacture can be more challenging than those above. An incoming photon produces an electron-hole pair in a semiconductor in the presence of an electric field. The electrons and holes drift to the respective electrodes and are captured. The read-out chips are reliant on CMOS technology and careful interconnects. To remove the leakage currents and yet achieve sufficient electron-hole capture speeds is a complex design criterion. The advantages however are that the photon capture efficiency is ~100%, for energies up to ~20keV (depending on the stopping power of the semiconductor).

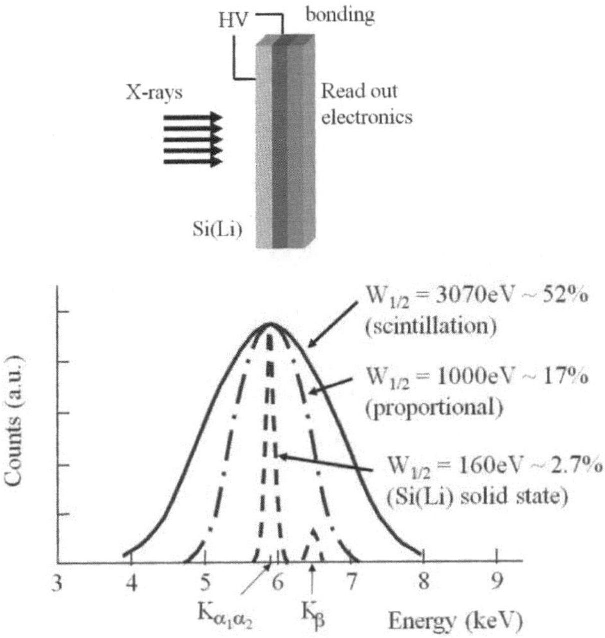

Figure 3.7.2. (a) The geometry of a solid state detector with a front electrode; for a 1D or a 2D detector each line or pixel has to have its own individual bond and electronics. (b) A comparison of the energy discrimination for the solid state, proportional and scintillation detectors, all as 0D detectors; for 1D and 2D solid-state detectors the energy resolution is compromised in present designs is comparable to that for proportional counters. These plots are only relevant to photon counting detectors.

The energy resolution can be very high, ~120eV, although as an area sensitive detector this is more difficult to achieve and the level is more typically 10% ($=\Delta E/E$). To maintain the high field across the semiconductor the resistivity has to be very high and therefore any ionised impurity should be compensated. A common approach is to use Li compensated p-type Si. Li is a fast interstitial diffuser that neutralises the unsatisfied Si bonds associated with the p-type dopant atoms (e.g. B) that are common in Si. From figure 2.1.1 we can see that the major capture process in Si is mainly due to the photoelectric effect and to a far lesser extent the Compton process, i.e. the energy is dissipated by transference to the ejected electron momentum. The amplification of this signal must take place very close to the detector to prevent any disruption of the very weak pulse. The energy resolution for a modern (not position sensitive) solid state detector is shown in figure 3.7.2(b) and is compared with those for a proportional and scintillation detector.

3.7.4. *Position sensitive detectors*

Position sensitive detectors provide a way of collecting data more rapidly, however in general some of the advantages of counting performance can be lost. If the individual regions are small enough they can show variations in the scattering at the micron level. This latter kind of area detector can be X-ray film (Ilford L4 nuclear emulsion plates), which can have a developed grain sizes of less than a micron. The application of fine emulsion film for topography will be discussed through examples in Chapter 5. The pixel size of phosphor linked to a glass capillary for an optical link to CCDs or MOS type storage arrays is decreasing (presently ~ 5 microns) and can be used for low-resolution topographic imaging.

Film has the disadvantages of the development process and CCD storage systems have the disadvantage of high noise levels and slow read out times (several seconds) for dynamic experiments. The dynamic range of these systems is rather limited $\sim 10^2 / 10^3$ whereas image plates (effectively an electronic version of film) can record intensity ranges over 10^5. An image plate requires reading the X-ray generated colour centres with a laser (promoting relaxation from these metastable states to create measurable light) and is quite a major set-up. However earlier problems with leakage and loss of proportionality with time have been improved, but the user must be very careful of artefacts from high-count rate saturation. The pixel size for an image plate is about 200 microns and read out time is a few minutes. This is an area of constant improvement, however some of the more familiar technologies are more suitable for large dynamic ranges. The high noise level in these detectors arises from the lack of discrimination in the counting, leading to an additive noise level that has to be subtracted to achieve the true counts.

The intensity measurement is therefore very inferior to the proportional and solid-state detectors.

In order to obtain good position sensitivity the final detection position has to be close to the initial point of ionisation, this favours the proportional and solid-state detectors. These are true photon counters and do not integrate the noise. To obtain the position sensitivity with solid state detectors or phosphor-based detectors the size of the electrode or optical coupling (glass fibre) defines the spatial resolution. The proportional counter based position sensitive detector works on a different principle. Since the gas amplification is very localised yet the signal is received by an extended wire electrode, it is the time for the charge to be swept along the wire or metal electrode that determines the position of the event. The ratio of the electron pulse transit times to both ends of the electrode defines the position for a one-dimensional detector. For a two dimensional detector a mesh of wires is required, and these run along orthogonal directions: the shared charge will then give both co-ordinates. The energy discrimination will be compromised and the accurate separation of the wires and their dimensions may influence the resolution.

Clearly the transit time limits the total measurable count rate, since the total time from photon capture to an external measurable pulse is rather long. The total count rate limits can therefore be very low ~20,000 photons per second before corrections are required and finally saturation. These flux values are easily achieved with a Bragg peak and therefore considerable care is required if any reliance is needed on the measured intensities. Also the longer the electrode the more severe the problems become. Careful design and attention to perfect uniformity of the wire to maintain good energy resolution are also serious considerations.

For the solid state and phosphor-based detectors the whole basis is quite different and just relates to the scale, since each electrode has its own circuitry. Any detector that becomes too large will have difficulties in that it has to match the geometry of the scattering experiment and even so calibration and checking the response from X-rays with different trajectories should be carried out. Ideally they should be curved about the sample position and the sample position should be very small; but this is rarely possible or practical. The attraction of the individual solid state detectors is the small pulse width and fast read out times, thus overcoming many of the disadvantages inherent with detectors reliant on transit time or localized data storage. A modern solid state detector is given in figure 3.7.3, where the detector is Si for the creation of electron-hole pairs and each 'pixel' is defined by the bonding and read-out chip that can be very sophisticated. The read-out chip is then bonded to the circuitry before information is output. The example given in the figure is of a tiled detector with the circuitry arranged for further expansion. Each pixel is equivalent to an individual solid-state detector described earlier.

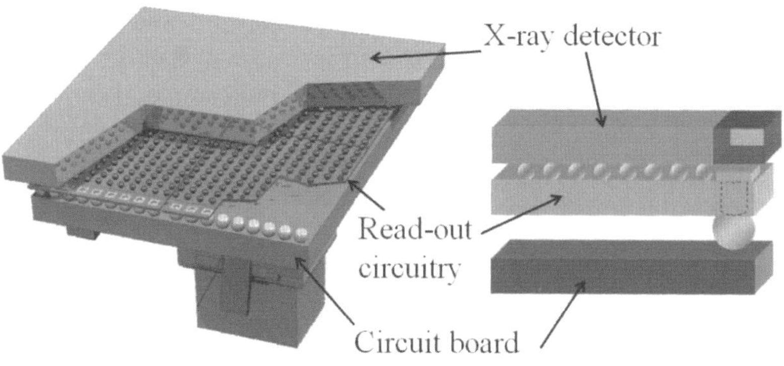

Figure 3.7.3. The position sensitive solid-state detector is shown with a layout indicating how the circuitry can be arranged to create large detector areas.

It is important to realise at this stage the usefulness of position sensitive detectors is confined to experiments that require the resolution defined by the detector. The main applications are in scattering from weak single crystals or polycrystalline materials, polymers and samples with limited crystallinity. As the acceptable count rate increases then the application areas will increase. However state-of-the-art solid-state position sensitive detectors have potential to accommodate impressive count rates, comparable to a proportional counter per pixel, which can have dimensions close to 50 μm.

In general though, the phosphor-based sensors linked to a glass capillary and CCD can have impressively small pixel dimensions, but it is very difficult to eliminate light generated from the phosphor spreading across glass-fibres and this should be considered in any application with these detectors. Similarly it is important to recognize the situation when a photon arrives somewhere close to the boundary of several pixels. In this case the charge will be shared, which will result in the expected energy to be lower, and therefore the pulse-height discrimination levels have to be carefully set to represent this case. This is area of significant technological advances and ideas in solid-state detectors, for example bringing in coincidence counting so that the distributed charge can be associated with a single photon. This is a real advantage of solid-state detectors with sophisticated electronics, to recover a true representation of the photon interaction with the detector.

3.8. Some general conclusions on components

Although the incident and scattered beam components are considered separately, it is clear that this is not easy to consider them in isolation. It is important to know what has happened to the X-rays before they reach a component. This results from the fact that the whole experiment is integrated; the source, the incident beam components, the sample, the scattered beam components and the detector response. This became evident in describing the slit systems, because they try and select angles of beams by position rather than be angle, which clearly depends on their nature before arriving at the slit.

If the component captures by angle, as in single crystal scattering, then the beams are often large to maintain intensity. The combination of all these options of slits and crystals, combined with ray-tracing will show how there are many options in designing an experiment and extracting meaningful information. It was shown in a simple combination of a slit and crystal how the wavelength can be selected for example.

In Chapter 4 we will consider how these components can be combined to create instrument options, and in Chapter 5 we will bring together the instruments, the theory to indicate how we can extract information about the sample. It is also recognized that new technologies, e.g. detector design, will create more options and some of these developments show great promise.

Bibliography

Arndt, U W (1990) J Appl. Cryst. **23** 161

Arndt, U W, Duncumb, P, Long, J V P, Pina, L and Inneman, A (1998) J Appl. Cryst. **31** 733

Bartels, W J (1983) J Vac. Sci. Technol. B **1** 338

Compton, A H (1923) Physical Review **21** (5), 483–502

DuMond, J W M (1937) Phys. Rev. **52** 872

Fewster, P F and Andrew, N L (1999) in *Defect and Microstructure Analysis by Diffraction.* IUCr monographs on Crystallography, Editors; R L Snyder et al, Oxford Univ. Press.

Johannsson, T (1933) *Z Phys.* **82**, 507–528

Kharrak, C P and Cox, D E (1977) J Appl. Cryst. 10 405–411

Komakhov, M A and Komarov, F F (1990) Physics Reports **191** 289

Kirkpatrick, P and Baez, A V (1948) J Opt. Soc. Am., **38** 766–774

Lang, A R (1956) *Rev Sci Instr* **27**,(1) 17–27

Snigerev, A, Kohn, V, Snigreva, I, Souvorov, A and Lengeler, B (1998) Applied Optics **37** 653–662

Wolter, H (1952) Annalen der Physik **10** 94–114

CHAPTER 4

INSTRUMENTS FOR MEASURING SCATTERING PATTERNS

This chapter brings together the components discussed in chapter 3 to create several useful diffractometers with an indication of their overall performance and applicability to materials analysis. The instrument function, or the capture volume in diffraction space, is discussed since this will influence the information that can be obtained about the sample. The diffractometer options presented will range from ultra-high resolution methods often used in semiconductor analysis through to powder diffractometry, as well as some more recent designs based on combining reciprocal and real space attributes. This chapter should give an indication of the limits and usefulness of the various diffractometers for specific analyses.

4.1. Some general points

In chapter 3 we considered various options for conditioning the incident and scattered beams; now we will consider how all this is put together with the sample. The sample can have a very strong influence on the instrument performance, and as we saw from the last chapter there is a plethora of component options. It is important to realise though, we should not assume that a powder diffractometer can only be used for powders, or 'high-resolution' diffractometers, can only be used for studying nearly perfect crystals. To help untangle these options we shall refer to Table 1.2.1 and 1.3.1, i.e. 'structural properties' and 'material types' from chapter 1. It is the structural properties that we wish to know, and it is the material type in combination with these that will determine the most suitable instrument configuration. An attempt to give some guidance is given in Table 4.1.1.

Table 4.1.1. Matching the structural properties of interest and the material type to the most appropriate instrument and section in brackets.

	Nearly perfect epitaxial	Textured epitaxial	Texture polycrystalline	Nearly perfect polycrystalline	Amorphous	molecular
Shape	DC TC MC BS St	TC MC BS CC	FP DP SS	FP CP SS	MC GA	MC
composition	DC TC MC BS St FP	TC MC BS CC	FP DP	FP CP	-	PH
orientation	TC MC	MC	DS PPP	PPP	-	PH
Distortion	TC MC CA	MC	FP CP	FP CP	-	MC
homogeneity	BS St	BS St	-	-	-	-
Interfaces	MC BS	MC BS	PP DP	DP	-	-
Density	MC	MC	MC	MC	MC	-

DC – Double-crystal (4.3.1); TC – Triple-crystal (4.3.2); MC - Multi-crystal (4.3.3); BS – Beam Selection (4.3.4); CC –Double Channel-Cut (4.3.5); St – Static (4.3.6) ; CA – Channel-Cut/area detector (4.3.7); GA – GISAXS/area detector (4.3.8); FP - Focusing powder (4.4.1); SP – slit and parallel plate (4.4.2); CP - Compact powder (4.4.3); SS – single slit for reflectometry (4.5.1); PPP – pinhole parallel-plate (4.5.2); DP - Double parallel plate (4.5.3); DS - Double slit (for texture analysis, not considered).

It is clear from the table that this only helps to reduce the options, or suggests a suitable choice of instrument, but it is a start. This chapter is organized such that the simpler instruments for each two broad types (grey and white cells in table 4.1.1) are considered first, in terms of conceptual understanding, and this is then built upon to describe some of the more complex configurations. The 'high-resolution' instruments (the grey cells), which rely on dynamical theory for their analysis, are considered initially, i.e. composed predominantly of crystals (active components) and this is followed by 'low-resolution' instruments (the white cells), which are largely based on slit systems (passive components). The high-resolution instruments are considered in this order; double-crystal diffractometer, triple-crystal diffractometer, multiple-crystal diffractometer, beam-selection diffractometer, the static diffractometer, which is followed by instrumentation for rapid reciprocal space mapping and instrumentation for GISAXS. The pros and cons of each diffractometer are discussed, which relates to the analysis possibilities in table 4.1.1. The low-resolution instrumentation will be considered in order of popularity, i.e. focusing powder diffractometry, slit and parallel plate diffractometer, the compact diffractometer, which is a high-resolution instrument for small samples. This is followed by instrumentation for low-resolution reflectometry and in-plane scattering.

Before the instruments are discussed the basics of the instrument function will be explained, since this is fundamental to all experiments. We have touched on this briefly in section 2.9 on the scattering from powders (see for example figure 2.9.11), because it is very difficult to consider all aspects of scattering theory and instrumentation in isolation.

4.2. Basics of the resolution function

From the geometry of the dispersion surface, figure 2.3.5 or the basic relation $\boldsymbol{k_H}=\boldsymbol{k_0}+\boldsymbol{S}$, equation 2.2.16, we can draw the condition for scattering, figure 4.2.1. We have restricted the case to reflection from crystal wafers, although if we were to include all cases of reflection and transmission that are accessible then this would be represented by a full sphere. The reciprocal lattice for the sample of interest is represented by dots and this exists in reality as a three dimensional array. A reciprocal lattice point is at a distance $1/d_{hkl}$ from the origin O, where d_{hkl} is the interplanar spacing corresponding to a plane of index *hkl*, section 2.2 (equation 2.2.11). Figure 4.2.1 illustrates the reciprocal lattice for a perfect single crystal. From this figure we can relate our reciprocal space co-ordinates of our scattering vector; $\boldsymbol{q}_x$ and $\boldsymbol{q}_z$ or $\boldsymbol{s}_x$ and $\boldsymbol{s}_z$ to the incident angle ω and scattering angle 2θ

$$s_x = \frac{q_x}{2\pi} = \frac{1}{\lambda}\{\cos\omega - \cos(2\theta - \omega)\}$$
$$s_z = \frac{q_z}{2\pi} = \frac{1}{\lambda}\{\sin\omega + \sin(2\theta - \omega)\} \qquad 4.2.1$$

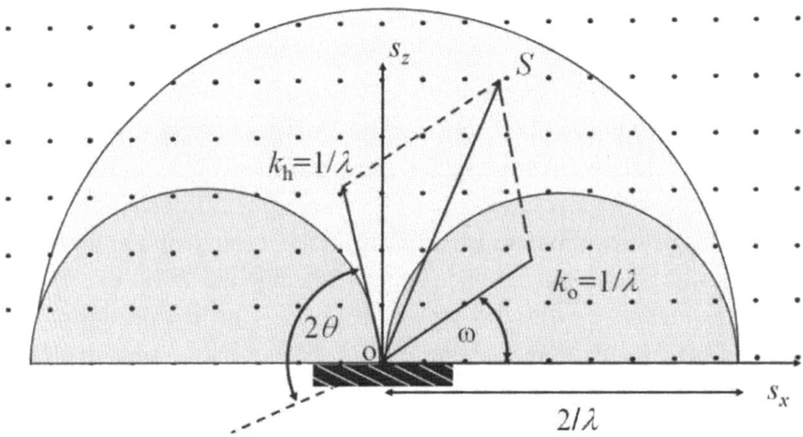

Figure 4.2.1. The region of reciprocal space that can be captured in reflection geometry (pale grey), and the region that is inaccessible, (dark grey), except in special circumstances. The incident and scattered beam wave-vectors indicate the angles of the diffractometer to capture the intensity associated with the scattering vector, for the case when s_z, S, k_o and k_H are all in the same plane.

The reciprocal space co-ordinates $\boldsymbol{s}_x$ and $\boldsymbol{s}_z$ are on the same scale as wave-vectors k_i and the scattering vector $\boldsymbol{S}$ whereas the vector $\boldsymbol{Q}$ is equivalent to $2\pi\boldsymbol{k}$. Throughout this book we have tended to concentrate on using $\boldsymbol{S}$ and $\boldsymbol{k}$ because this gives a reciprocal lattice constructed as the inverse of the interplanar spacings, rather than including the factor of 2π. However, as in the latter part of chapter 2, the factor 2π can simplify some of the exponents. By using '*s*' as in figure 4.2.1, we can simply see that if we change the wavelength the region of capture is changed.

4.2.1. *The shape of the reciprocal lattice point*

The shape of the reciprocal lattice point contains information about the internal arrangement of the atoms in the structure and the distribution and size of the regions having this interplanar spacing. We can therefore consider each point to include contributions from internal strains, angular misorientations, finite size effects and their distributions. In fact all the structural details with varying degrees of each contribution. Immediately we see the reciprocal lattice point has a three-dimensional form and contains a considerable amount of information.

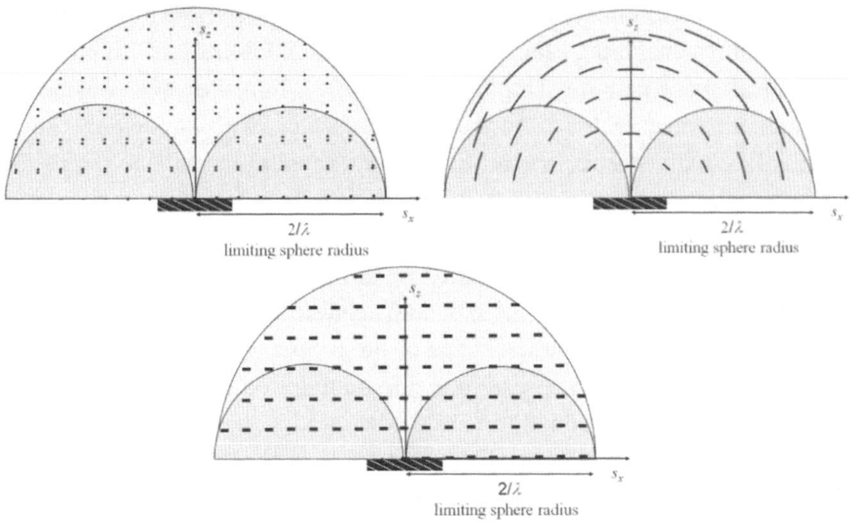

Figure 4.2.2. (a) The modification of reciprocal space for a composite structure of layer and substrate where the lattice parameter in the plane of the interface is the same. (b) An image of reciprocal space for a sample of several crystal orientations and (c) the case for a structure with a limited lateral dimension.

For determining the average structure, as in conventional molecular structure determination, the concern is to capture the total intensity value from each reciprocal lattice point. However our main concern here is to probe any deviation from the average structure, whether this is in the form of a composite layer structure of similar materials or details of the microstructure. These all manifest themselves as intensity variations within the reciprocal lattice spot. Hence our requirements for the reciprocal space resolution are very different from that used for determining the average structure.

For a composite structure, e.g. a semiconductor multilayer, we can superimpose several reciprocal lattice meshes associated with each distinct layer. This provides a good way to visualise the region of reciprocal space to investigate. A few examples are given in figure 4.2.2.

4.2.2. *The shape of the reciprocal lattice probe*

In chapter 3 we discussed the divergence from various X-ray collimators, and in this section it will be explained how this, in combination with the components analyzing the scattered beam, defines the region of reciprocal space that we sample. From figure 4.2.3a, we have considered an incident wave vector $\boldsymbol{k_0}$ on the sample surface, with an incident angle ω that has a divergence spread δω and an angular acceptance range for the detecting system, $\delta(2\theta)$. The finite angular divergence and acceptance create a finite size probe on our reciprocal space mesh. The X-ray wavelength also has a finite range and this is represented by a change in magnitude of the incident and scattered beam vectors, i.e. $1/\lambda$.

To understand the detailed shape of the probe, we shall refer to figure 4.2.3a. By introducing a spread in incident angles on our surface, without moving the detector, we will create a new incident beam vector, e.g. $\boldsymbol{k_0}'$, and a vector set given by the fine dotted line in the figure. It is obvious from this that a spread in the incident angle will produce a streak centred on the reciprocal lattice point and along a direction inclined to the scattering vector, ***S***, by the angle θ. The shape of this streak will mimic that of the divergence. A similar analysis can be done with the spread in acceptance angles by the detector, $\delta(2\theta)$, giving rise to a streak also inclined at θ to the scattering vector. The incident beam and scattered beam vectors have a magnitude $1/\lambda$ and the sphere radius is $2/\lambda$; hence any wavelength dispersion will expand or contract the sphere radius, figure 4.2.1 for a fixed diffraction vector length ***S*** (e.g. ***S*** = 1/***d***). Since our data is collected and plotted including the spread of wavelengths, the effect is to give a smearing of the diffraction vector, represented as $\Delta\lambda$ in the figure. A simulation of the instrument function for a high-resolution diffractometer is given later in figure 4.3.7b.

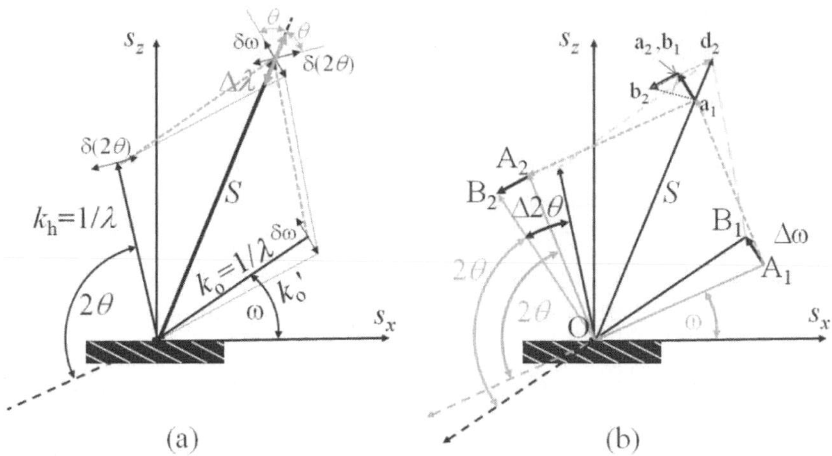

Figure 4.2.3. (a) The influence of the incident beam divergence, analyser acceptance and wavelength spread on the region of capture of reciprocal space. The axial divergence is normal to the plane of the figure. (b) The shape of the probe relationship with data collection; the details are given in the text.

What has been described so far is the probe, so it is pertinent at this stage to discuss what actually happens during data collection when we rotate these two axes, ω and 2θ. We shall refer to figure 4.2.3b. Consider the incident beam A_1O and the detector system set to capture the scattering at 2θ, given by OA_2. Suppose that the incident beam is changed by $\Delta\omega$ to the new incident beam direction B_1O, such that for a stationary detector position (with respect to the beam associated with A_1O) results in a different 2θ, i.e. $B_1OA_2 = (2\theta + \Delta\omega)$. Therefore the probe will move from a_1 to b_1, following our previous argument, however the detector is still at 2θ and therefore the detector is really at B_2. This can be envisaged by assuming the sample is rocked to change the incident angle. This in effect takes point a_2 to b_2, and since $a_1a_2 = a_2b_2 = \Delta\omega$, this rotation in ω will follow the trajectory of a_1b_2 which is normal to the scattering vector ***S***, as we would expect. To sample the intensity along any direction in reciprocal space we then rotate 2θ. For example if we wish to probe along the scattering vector, a_1 to d_2 having arrived at b_2, the probe has to be moved from b_2 to d_2, which from geometry will give $\Delta(2\theta) = 2\Delta\omega$. If we wish to scan normal to the surface of the sample then from geometry $\Delta(2\theta) = \Delta\omega + \{1 + \sin\omega/\sin(2\theta - \omega)\}$, etc.

As mentioned above the reciprocal lattice points do have a finite dimension in all directions, i.e. out of the plane of the figure as well as in the plane. Similarly there is a divergence (axial divergence) normal to the divergence in the scattering plane and therefore this captured region is a projection. The measured intensity is assigned to the midpoint of the angle readout of the axes. Suppose now we wish to map the intensity distribution around a reciprocal lattice point then our probe size should be smaller or at least comparable with any intensity variations formed by the structural details of interest.

The majority of experiments are conducted with reasonable to good resolution in the scattering plane but poor resolution in the axial plane. This can lead to the wrong interpretation; the outline of what is happening will now be discussed with reference to the idea of three-dimension reciprocal space data collection so that a general feeling of the complete probe we are using can be understood. Probing the intensity with three-dimensional reciprocal space mapping has been presented for slit-based systems, i.e. powder diffractometers, with a very small perfect single crystal of silicon and very small slits, Fewster and Andrew (1999); however this only reveals a greater understanding, whereas when this is used with a multiple-crystal diffractometer (MC in table 4.1.1) there is useful information obtained, so the following description refers to the latter, Fewster and Andrew (1995), Fewster (1996).

The reciprocal lattice points are three-dimensional, and therefore the axial divergence will integrate the intensity along this direction. Awareness of this third dimension is important since the conventional two-dimensional reciprocal space map is a projection onto the diffractometer plane. If the axial divergence is of the order of 0.5^0, and this represents an arc of capture (i.e. the projection varies according to its distance from the reference plane, which is our 2-dimensional reciprocal space map) then any intensity collected along this arc will be assigned to the wrong angle in the diffractometer plane. This is noticeable in high-resolution measurements, especially when studying mosaic samples for example. An example will be given in the next chapter, but for now we will briefly describe the technique of three-dimensional reciprocal space mapping. Restricting the axial divergence with crystals is impractical since this leads to a large reduction in intensity, so slits are used, although the eventual probe can still leads to a significant projection. However the view of reciprocal space can be changed dramatically.

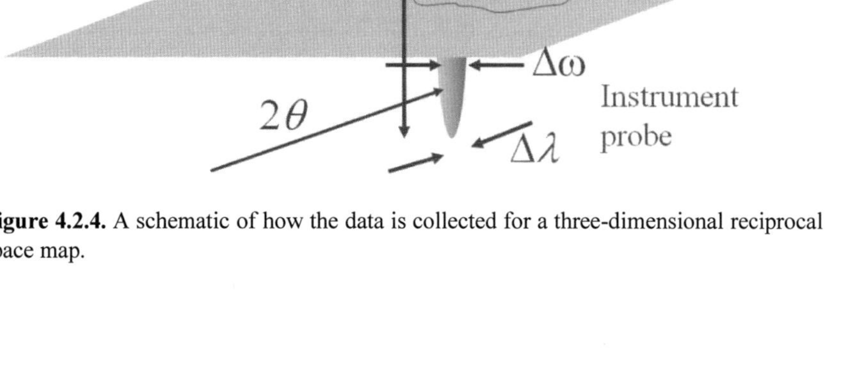

Figure 4.2.4. A schematic of how the data is collected for a three-dimensional reciprocal space map.

Since the divergence in the scattering plane is often defined by crystal optics we only need to restrict the divergence in the axial direction. A 1 mm slit at the monochromator exit and a 2 mm slit in front of the detector is reasonable, using a point focus X-ray source, with the instrument described in section 4.3.3. This will prevent scattering out of the plane of the diffractometer greater than about 0.09^0 from reaching the detector. We cannot actually define the divergence so simply, since as has been stressed throughout this book the whole process from source to detector has to be considered. Suppose the region of the sample that we are analysing (some feature that could be part of a mosaic block for example) is very small, say a few microns then this will define the divergence incident on that region. The actual axial divergence can therefore be exceedingly small for imperfect samples and this is how a significant amount of detail is observable. Although the detail is distributed over a large χ value, this just leads to an effective magnification of the features of interest and this is clear from figure 4.2.4.

To understand the data collection method we should imagine the centre of the reciprocal space probe to be restricted to the plane of the diffractometer but can be moved around in this plane. The height of the probe is $\Delta\xi$ and the reciprocal lattice point to be tilted out of the diffractometer plane by χ, which has a spread in this direction of $\Delta\chi$, will then reveal information on the shape and size of the reciprocal lattice point.

By collecting a series of reciprocal space maps, that are off-set in χ by a small amount ($\sim 0.2^0$) from each other, we will create a three-dimensional array of intensity that represents the distribution of scattering around the reciprocal lattice point. Of course the resolution is different in different directions, but it will create a far-improved data set for interpretation.

So now that we have an understanding of the shape of the reciprocal, or diffraction, space probe, we can begin to understand how the aberrations of the data-collection method influence the sample information we require.

4.3. Diffractometers relevant to nearly perfect crystals

If we know our sample is nearly perfect then it will have large extended regions of perfect material, and the diffraction peaks will be very narrow. This is why high-resolution diffractometers are required to extract detailed information. As we have discussed in chapter 2, section 2.3.2 on dynamical theory, an incident beam will create a well defined scattered beam. It is only imperfect samples that will produce significant scattering over a range of incident angles. Therefore a perfect sample only requires a very well collimated incident beam, and if the sample is rotated the intensity response will correspond to what we would expect from dynamical theory, i.e. $I(\omega)$ as in equation 2.3.41.

We shall now consider specific instrument configurations along with their advantages and disadvantages.

4.3.1. *The Double-Crystal Diffractometer*

One of simplest high-resolution diffractometers is the double-crystal diffractometer, which consists of a collimating crystal and the sample, figure 4.3.1a. The scattering angle from the collimating crystal should match that from the sample, so that the beam emerges from the sample parallel to the incident beam on the collimating crystal, Compton (1917). With this arrangement all the wavelengths will be scattered at the same angular rotation ω: they may take different paths but all wavelengths will scatter together. The advantage of this configuration is that the wavelength dispersion, figure 3.2.1, will not confuse the profile and the only contribution to the broadening will be the intrinsic scattering width defined by dynamical theory. The intensity will obviously be rather high.

As the sample is rotated further from the matching Bragg angle condition, the wavelength contributes more and more to the broadening. This geometrical arrangement is really limited to near perfect crystals that are not bent and match the crystal incident beam conditioner. If the sample is bent then the incident angle on the surface becomes a function of position, and since we are expecting intrinsic scattering widths of a few seconds of arc the profile degrades very quickly. This can be compensated for with a slit placed between the collimator and sample crystals, or with a knife-edge close to the incidence point on the sample crystal. Both these methods limit the illuminated area. There is always considerable extraneous scatter with this configuration and so the slit and knife edge is beneficial in isolating the detector to only capture scattering from the sample.

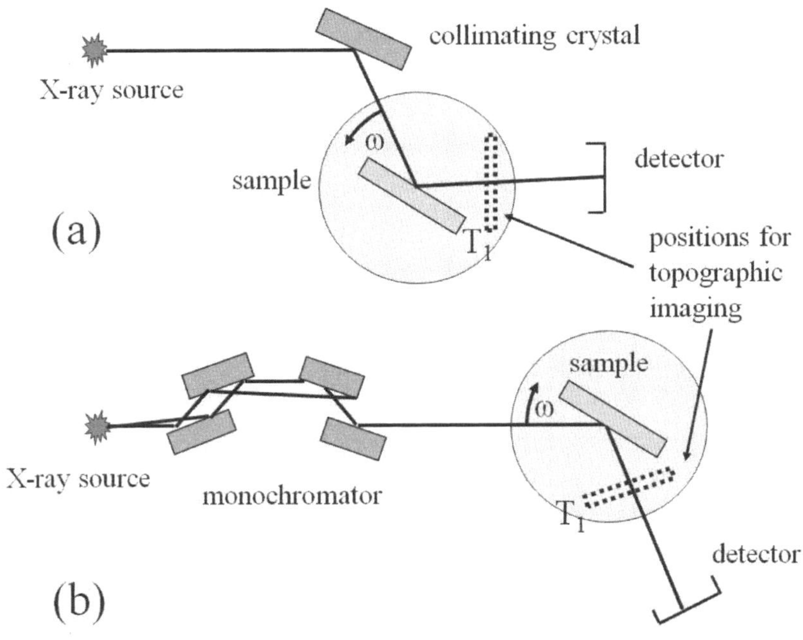

Figure 4.3.1. (a) The geometry of the double-crystal diffractometer, after Compton (1917). (b) The geometry of the diffractometer used by Bartels (1983), to overcome the problems of changing the collimating crystal of the double-crystal diffractometer. The possible locations for topographic imaging are also given.

If the scattering angles of the collimating crystal and sample crystal do not match, and the slit and or knife-edge cannot restrict the wavelength spread to sufficiently small values (using the principle shown in figure 3.4.2) then the collimating crystal should be changed. Because of the simplicity of the instrument replacing or realigning the first crystal to match the scattering angles of the collimating crystal and sample can be fairly rapid, Fewster (1985).

To overcome the frequent exchanging of the first crystal the 2-crystal 4-reflection monochromator can be used, section 3.4.4, after Bartels (1983), figure 4.3.1b. The principle of data collection is similar to that above, however it does not remove the broadening due sample curvature.

The alignment procedure for the sample crystal is applicable to all high-resolution diffractometer analyses. This is discussed here because the alignment is often done in 'double-crystal' mode even for multiple-crystal diffractometers.

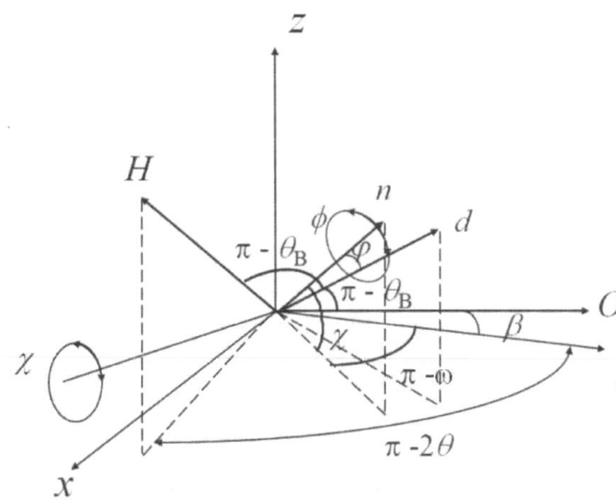

Figure 4.3.2. The angles to be considered for deriving the errors associated with sample and diffractometer misalignments. *0* is the incident beam, *n* is along the ϕ axis (usually the surface plane normal), *d* is the scattering plane normal and *H* is the scattered beam.

4.3.1.1. Alignment of high-resolution diffractometers

The critical consideration is to ensure that the scattering plane normal is perpendicular to the sample rotation axis, ω and in the same plane as the incident beam. The scattered beam should then occur in the plane normal to the sample rotation axis (the diffractometer plane) and the angles measured should be true angles and not projected angles, figure 4.3.2. If this condition is not satisfied then the measured projected angle will be different from the true angle, but also the way in which the reciprocal lattice point interacts with the instrument probe will broaden the profile. For even moderately precise measurements some alignment is necessary, for quality measurements it is imperative.

In the case of most high-resolution diffractometers we have two methods for bringing the scattering plane normal into the diffractometer plane depending on the scattering plane of interest. If the scattering plane is approximately perpendicular to the ϕ axis, then rotating ϕ has little effect but rotating χ is very effective. Generally the latter axis is the refining axis and the former will give an approximate setting for planes inclined to the surface. As a general rule for detailed analyses it is worth setting the azimuthal direction with a ϕ rotation to align another reflection that defines this direction (e.g. using the *444* reflection of a (*001*) orientated sample to set the <*110*> along the incident beam direction for $\omega = 0$). From figure 4.3.2 we can determine the error in the ω angle for various errors in the tilt angle, χ

$$\Delta\omega \sim \frac{\chi^2}{2}\tan\theta\cos^2\varphi \qquad 4.3.1$$

where φ is the angle that the scattering plane makes with the ϕ axis in the plane of the diffractometer. This varies as the angular separation increases. Hence for an error of 1^0 in the tilt angle the angular mismatch is in error by 1 s of arc in 300 s for the *004* reflection from GaAs, Fewster (1985).

4.3.1.2. Applications of the double crystal diffractometer

Since the instrument has two axes it is sometimes referred to as a double axis diffractometer, although during data collection only the sample rocking axis, ω, is rotated. The wide-open detector is placed to receive the scattered X-rays from the sample as it is rocked, sweeping through the diffraction condition. The scattering profile from a perfect crystal can be simulated relatively easily and to a good approximation the instrument function is defined by the profile of the collimating crystal; correlation of these two profiles should match the measured profile. The collimating crystal is stationary and each wavelength will be collimated to within its own intrinsic scattering profile (the beam on the collimating crystal must be from a finite size source, so effectively each illuminated part of the crystal receives a convergent beam). The angle of convergence must be larger than the intrinsic scattering width so that it is equivalent to rocking the collimator crystal at each angular movement of the sample. This mimics the process of correlation.

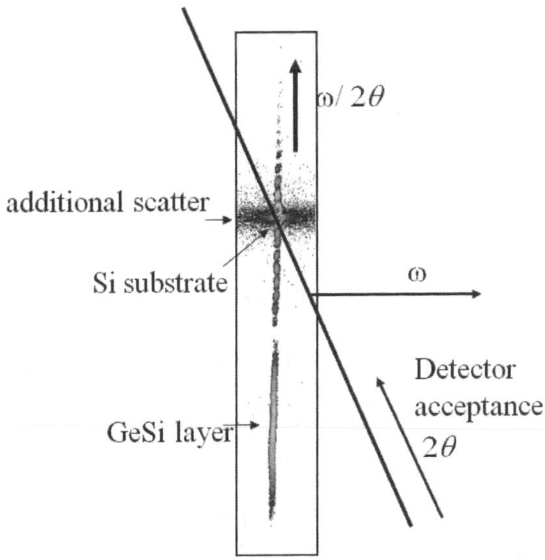

Figure 4.3.3. The instrument function for the double-crystal diffractometer (collimating crystal – sample – open detector window), shows how a nearly perfect crystal will give a very useful profile. The additional scatter is most likely due to defects leading to diffuse scatter.

For a precise measure of the intensity we must be aware of the response across the detector window and the size of the window and how the detector can modify the measured signal. In figure 4.3.3, the detector is shown as the extended $\Delta 2\theta$ capture region (b_2d_2 in figure 4.2.3) and each part of the scattered intensity from the sample will arrive at a different position along $\Delta 2\theta$ if only the ω is scanned. This is usually only a problem with proportional detectors, where the X-rays could enter the window close or far away from the central wire, especially if the wire is normal to the scattering plane. These effects can be easily calibrated, although a well made detector should give a good even response. Alternatively if the detector can be rotated as the sample is rotated, then this can ensure the same area of the detector window is used.

For routine analysis of near perfect semiconductors a double-crystal diffractometer can collect data in a few minutes or even seconds. The intensity can be very high and therefore the possibilities of significant scatter reaching the detector can reduce the dynamic range. An example of this type of analysis will be given in section 5.4.3.1.

The double crystal diffractometer is limited to fairly perfect crystals and this can be a serious limitation if the sample is bent (creating a range of allowable incident angles) and there is no discrimination of the scattered beam. To overcome some of these problems the triple-crystal diffractometer was developed.

4.3.2. *The Triple-Crystal diffractometer*

The triple-crystal or triple-axis diffractometer is an extension of the double crystal diffractometer and can be explained with the simple ray-tracing argument given for the double-crystal diffractometer. The analyser crystal matches the collimating first crystal and will now only pass those scattered X-rays from the sample satisfying the Bragg condition of the analyser. Since all the axes are independent, scanning through the Bragg condition from the sample is very complicated, the sample and analyser crystal need to be rotated, Iida and Kohra (1979). This instrument led very quickly to undertaking reciprocal space maps by setting the sample rotation and scanning the analyser crystal axis, resetting the sample and scanning again, etc., until a full two dimensional distribution of intensity is recorded. With computer control both axes could be moved simultaneously. The complexity of the instrument movements and the lack of versatility, made it very complicated to change to different scattering planes (the whole instrument needed to be rebuilt). This meant that it only existed in a few laboratories around the world, while the double-crystal diffractometer flourished.

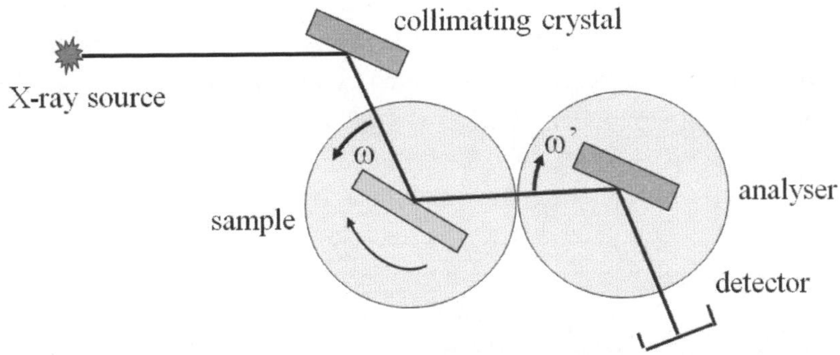

Figure 4.3.4. The geometry of the triple-crystal diffractometer used by Iida and Kohra (1979).

4.3.2.1. Applications of the triple-crystal diffractometer

The instrument can be used as a precision lattice parameter comparator. This requires specialised techniques that have been reviewed by Fewster (1999) and previously by Hart (1981). The problems of resetting the axes and realigning the whole instrument for each region of reciprocal space were overcome with the multiple-crystal diffractometer.

4.3.3. *The Multiple-Crystal diffractometer*

The versatile monochromator described in section 3.4.4 largely overcomes the problems of the double- and triple-crystal diffractometer as regards to changing the first crystal, however a sample that is bent or imperfect will still give diffraction profiles that are difficult to interpret. Basically we wish to create an instrument with an incident beam well-defined in direction and wavelength spread and an analysing system that is also well defined so that no realignment is necessary. This would give the user freedom to analyse any set of scattering planes without changing the instrument. The instrument that satisfies these criteria is given in figure 4.3.5, Fewster (1989). The monochromator controls the scattering plane divergence and the wavelength dispersion giving a well-defined incident beam. The analyser crystal is placed on an axis common with the sample rotation and therefore always points at the sample. The analyser crystal only passes scattered X-rays that are coming from the sample in the specific direction defined by its rotation about the common axis.

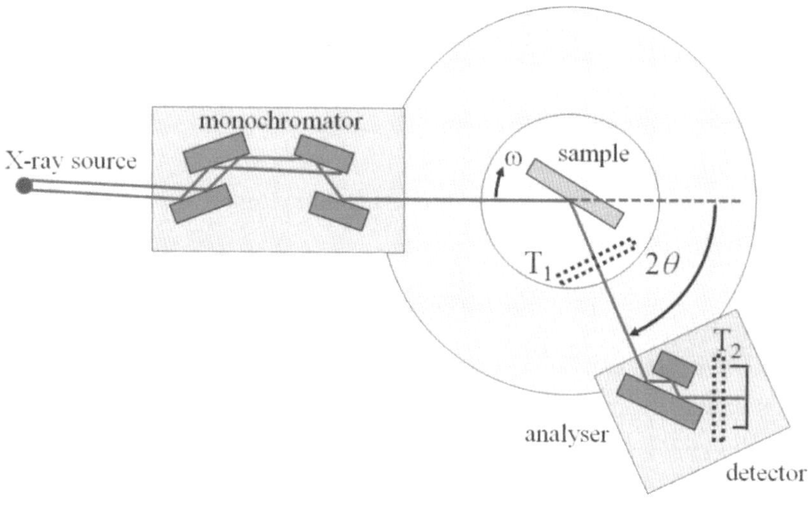

Figure 4.3.5. The multiple-crystal diffractometer is illustrated, Fewster (1989) with the main rotation axes and the optional positions for topographic imaging. To boost the intensity an exchangeable X-ray mirror can be inserted before the monochromator.

The monochromator has been described in detail in section 3.4.4 and it was clear that the wavelength band-pass and divergence is a very complicated function. The analyser crystal is far less complicated but is limited to three internal reflections for several good reasons, section 3.4.3. Firstly from the earlier arguments extending the number of reflections creates little return because of the imperfections present in the most perfect crystals. Secondly the advantage of offsetting the detector (this occurs with odd numbers of reflections) reduces the chance of the directly scattered beam from reaching the detector. This gives a very good compromise to maintain high intensity and low residual background. From calculations of ideal crystals the contribution from the analyser and the monochromator is exceedingly small and unobservable over a dynamic range of 10^9. In practice the practical limit can be checked by scanning the analyser and detector arm through the incident beam direction, figure 4.3.6. Because of the very high intensities this must be done with absorbers in the beam path.

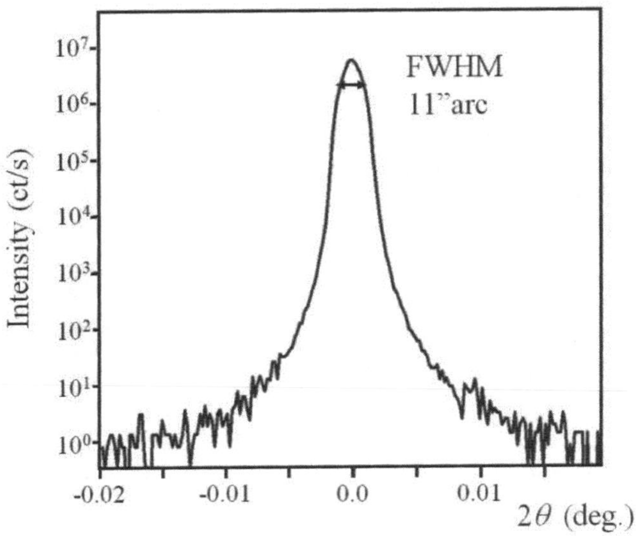

Figure 4.3.6. The variation in intensity as the analyser and detector are scanned through the incident beam direction. The need for absorbers reduces the quality of the profile so it is not optimal.

From the arguments so far we should consider the angular acceptance of the analyser in conjunction with the divergence and wavelength dispersion of the monochromator. For the sake of a conceptual understanding we shall consider the divergence passing through the monochromator to be half that of the intrinsic full-width at half-maximum intensity of the first reflection in the monochromator. If we assume the sample has a delta function response, then rocking the crystal will give a spread in scattered beam directions with an angle equal to that of the incident beam divergence. Therefore our analyser would appear to need to have an acceptance half that of the intrinsic scattering width of the first reflection from the monochromator. However at this point we cannot take this approach any further, and have to refer back to figure 4.2.3a and also the simulation in figure 4.3.7. This is the point when reciprocal space is necessary to understand the details of the instrument response, compared to the 'real space' description that can be applied to the double-crystal diffractometer (tracing the wavelength paths) and typical powder diffractometers that can be described in geometric terms.

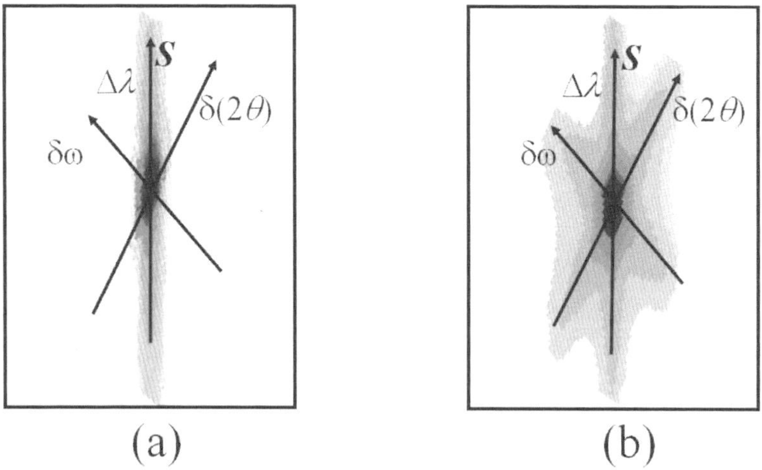

Figure 4.3.7. The simulated reciprocal space map for a perfect Si sample (*004* reflection) with the multiple-crystal diffractometer (a) and a triple-crystal diffractometer (b). The surface truncation rod, or dynamical streak is along the scattering vector direction because the simulation is for scattering planes parallel to the surface. If the scattering is from inclined planes then the dynamical streak is inclined to this scattering vector by the inclination angle.

The instrument function or probe is the overlap of the angular divergence of the monochromator, the wavelength dispersion and the angular acceptance of the analyser. We will neglect the axial divergence for this argument, since the basics were discussed earlier and the analyser does little additional discrimination of the axial divergence. From our discussion on the theory, Chapter 2, it is clear that scattering from a sample with a flat surface will create a profile that can be fully captured by scanning our probe normal to the surface, along s_z. If our sample is perfect then this profile is a line of no width within the assumptions of the dynamical theory. However to capture this information we must have an X-ray source that can only exist with a finite wavelength distribution and divergence. Therefore if we superimpose our probe on this profile we can see that the smearing effect from collecting the data is strongly influenced by the angle of the smearing of the various components, wavelength, divergence and acceptance. The angular spread of these components is a function of the angles ω, 2θ and λ, figure 4.3.7.

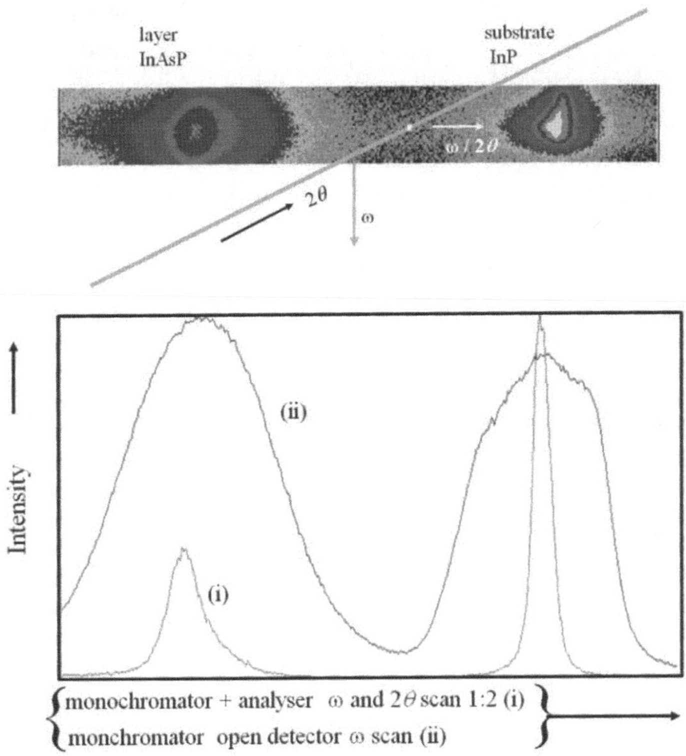

Figure 4.3.8. (a) The interaction of the capture region when used to probe a highly imperfect semiconductor structure; indicating the size of the scattered beam acceptance for the open detector and that from the multiple-crystal geometry. (b) The difference between the double-crystal (open detector) scan and that from the multiple-crystal geometry.

Consider an imperfect sample that gives a distribution of scattering in reciprocal space as in figure 4.3.8a: if the structure was perfect it would have a very small width when scanning in ω, or normal to the dynamical streak, figure 4.3.7a. Now the probe is the region defined by the spread in the dimension Δω and the angular acceptance of the detector $\Delta 2\theta$, figures 4.2.3 and 4.3.7. If the detector is wide open then the whole line along 2θ will represent the detector acceptance. The data collection with this geometry is carried out by "rocking" the reciprocal lattice points in ω through the probe to obtain a "rocking curve."

It is now possible to see the difference in the "rocking curve" with an open detector for an imperfect sample (scanning in ω) compared with a scan combining 2θ and ω along $\boldsymbol{s}_z$ with the multiple-crystal probe (illustrated in figure 4.3.7a). The influence of bend and other imperfections can be made orthogonal to this scan and so the strain can be isolated and modeled, figure 4.3.8b. Alternatively the whole map of figure 4.3.8a can be projected onto the $\boldsymbol{s}_z$ and modeled, thus removing all the influence of tilt and bend but capturing all the intensity from the probed region of the sample.

Let us consider the details of the probe for mapping the intensity in reciprocal space. If the sample is perfect then the addition of an analyser has little or no effect, since the probe may be larger than the intersection of the monochromator and the sample profile. Hence we can see that when the instrument function is too large it is just not used to the full and we have to be aware of this in our modeling. For perfect samples, the analyser can have a large acceptance range and will result in a simple rocking curve similar to the data collection method of the double-crystal diffractometer. From this we can collect all the data we need to model the data and extract information. If the detector window is smaller than the 2θ range of the scattered beams, then we can also scan the detector, to ensure all the intensity is captured. This two-axis scan also has the advantage of ensuring that the same region of the detector collects the data at each point for reflections with planes parallel to the surface. For large rocking curve scans scanning the detector (2θ) and rocking angle (ω) along ω/2θ where ω = θ, i.e. along s_z, figure 4.3.3, will also help to prevent missing data from the limited size of the detector window.

4.3.3.1. Alignment of multiple-crystal diffractometers

As in any high precision measurement the alignment of the scattering planes to coincide with the diffractometer plane is crucial to prevent the measurement of projected angles. This was partly covered in section 4.3.1.1, but with the addition of an analyser we do have more possibilities. Generally the alignment and setting of the sample on the diffractometer should not be too different from that suggested for the double-crystal diffractometer, however this only brings the scattering plane normal into the plane of the diffractometer (preferably by the minimum ω method) and optimises the ϕ rotation. This is obtained with an open detector since the scattering does alter the scattering angle and it is the intensity we are measuring. When the analyser crystal is substituted for the open detector then we know the sample is scattering and therefore we need to scan the detector / analyser axis (2θ). The maximum intensity gives us the setting for this 2θ axis and then we are in a position to undertake a scan or obtain a reciprocal space map.

To put reliance on the measurement of the various angles we need to determine the dependence of these on the various diffractometer angles. The details of the geometry will not be repeated here but the errors in the measured values of ω and 2θ due to sample misalignments and the incident beam not being parallel to the diffractometer plane are given by

$$\omega = \sin^{-1}\left\{\frac{\sin\theta - \cos\Delta\chi\sin(\chi-\Delta\chi)\sin\beta}{\cos(\sin^{-1}(\cos\Delta\chi\sin(\chi-\Delta\chi)))\cos\beta}\right\}$$

$$-\cos^{-1}\left\{\frac{\cos^2\Delta\chi\cos^2(\chi-\Delta\chi)+\cos(\sin^{-1}[\cos\Delta\chi\sin(\chi-\Delta\chi)])-1-\cos^2\Delta\chi+2\cos\Delta\chi\cos\varphi}{2\cos\Delta\chi\cos(\chi-\Delta\chi)\cos(\sin^{-1}[\cos\Delta\chi\sin(\chi-\Delta\chi)])}\right\}$$

$$2\theta = \cos^{-1}\left\{\frac{2\sin\theta_B\cos\Delta\chi\sin(\chi-\Delta\chi)\sin\beta+\cos 2\theta_B}{\cos(\sin^{-1}(2\sin\theta_B\cos\Delta\chi\sin(\chi-\Delta\chi)))\cos\beta}\right\}$$

4.3.2.

and

$$\Delta\chi = \sin^{-1}(2\sin(\varphi/2)\sin\phi \qquad 4.3.3$$

φ is the angle between the diffracting plane and the sample mount normal, which is tilted at an angle χ with respect to the plane of the diffractometer, figure 4.3.2. The angle ϕ is the rotation about the normal to the plane of the sample mount to bring the diffracting plane into the plane of the diffractometer. ϕ is zero when the projection of φ on the plane of the diffractometer is at a maximum. If the tilt is small and φ is sufficiently large then the tilt can be aligned by rotation of ϕ, equation 4.3.3. $\Delta\chi$ is the angle of the scattering plane with respect to the surface normal to the plane of the diffractometer. The variation of the angles ω, 2θ with χ are given in figures 4.3.9a, b and c.

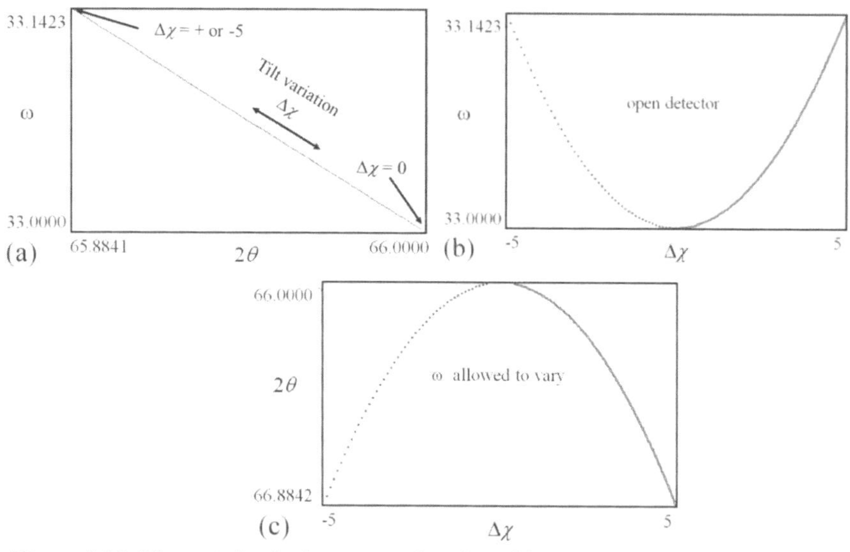

Figure 4.3.9. The variation in the measured peak positions as the tilt error is altered. The correctly aligned position for all three examples (a), (b) and (c) are when $\omega = 33^0$ and $2\theta = 66^0$.

These derivations clearly relate to the condition when the reciprocal lattice point and the various divergences are negligible. These are not serious omissions but an extended reciprocal lattice point and an extended probe in the axial direction can cause problems in alignment.

For example the diffraction plane normals of each mosaic block of a crystal or those of a layer and a substrate may not be able to be brought into the diffractometer plane at the same time. This projection effect and the subsequent errors can be significant, equations 4.3.2 and 4.3.3. To overcome this, another data collection method should be employed for the most precise measurements, i.e. 3-dimensional reciprocal space mapping discussed in section 4.2.2 and with examples in chapter 5.

At this stage we have largely given a theoretical approach to alignment of the angles; the setting of the sample in the beam and practical examples will be given in chapter 5.

4.3.3.2. Applications of the multiple-crystal diffractometer

The real benefit of the multiple-crystal diffractometer is in the study of imperfect samples, i.e. real samples. Imperfect materials will have regions that can cause diffraction broadening, they may be orientated with respect to each other, the sample could be bent, there could be many defects that create diffuse scattering, etc. The importance of the probe size for these studies now becomes rather crucial. The profile of imperfect materials will be broadened considerably and the range over which the incident beam will create scattering will be larger than the probe dimensions, e.g. figure 4.3.8a. Clearly if the analyser acceptance is large then the intensity assigned to the position (ω, 2θ), the centre of the probe, will be the sum of all the contributions. These contributions can include bend, mosaic spread, finite size effects and strain with little chance of isolating these effects.

The smearing effects associated with an analyser having a large angular acceptance will therefore confuse the contributions of the various sample properties that we wish to determine. The ideal combination could be for the analyser acceptance to match that of the incident beam divergence. As discussed in section 3.4.4 the divergence of the monochromator is a very complex relationship of wavelength, and this will have an impact on the instrument function because of the various wavelength contributions that will influence the $\delta\omega$ as well as the $\Delta\lambda$ profile, figure 4.2.3a. Also from our theoretical analysis we know that the strongly scattering planes have broad profiles, so depending on the reflection being analysed the influence of instrument broadening effects, e.g. wavelength dispersion, the inclination of the monochromator and analyser streaks, which depend on ω and 2θ, must all be considered to assess the appropriate configuration for the information required. So there are numerous possibilities that will depend on the material to be analysed and the information wanted. We can have an acceptance twice that of the divergence to maintain good intensity, increase the divergence with asymmetrically cut first crystals or reduce the acceptance with asymmetrically cut crystals. Without modeling the whole system including a typical sample, estimating the most beneficial combination is only guesswork.

A diffraction space map is obtained by scanning ω and 2θ as described in section 4.2.2. This gives a radial sector of reciprocal space, figure 4.3.10a, which can be converted to form a reciprocal space map. Alternative maps can also be collected along any direction by using something other than the 1:2 step ratio by offsetting in $\Delta\omega$ and $\Delta(2\theta)$ dependent on the particular region of reciprocal space of interest. Any of these maps can then be converted to reciprocal space using the relationships given in equation 4.2.1. The data can also be collected directly in reciprocal space (e.g. along s_z and s_x), which can have advantages in interpretation. The angles that the diffractometer needs to be moved to, in terms of the reciprocal space co-ordinates, are given by:

$$2\theta = 2\sin^{-1}\left(\frac{\lambda\sqrt{\left\{s_z^2 + s_x^2\right\}}}{2}\right)$$

$$\omega = \tan^{-1}\left\{\frac{s_x}{s_z}\right\} + \theta \qquad 4.3.4$$

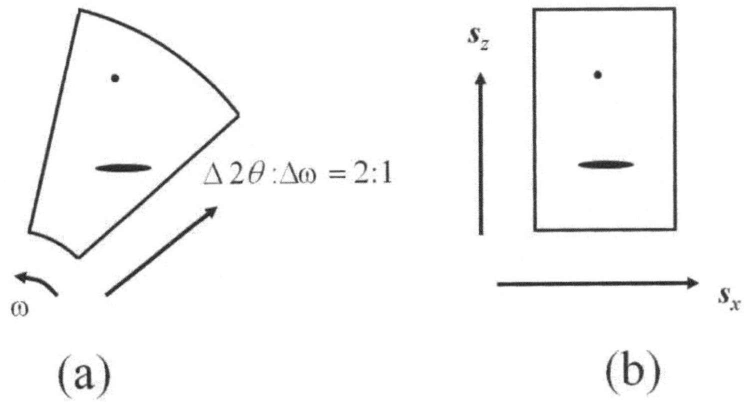

Figure 4.3.10. Two data collection strategies for reciprocal space maps; (a) collecting a segment of reciprocal space by defining angular movements and (b) collecting the data directly in reciprocal space.

The reciprocal space map can therefore be obtained in any parallelogram shape in reciprocal space. We have taken the reciprocal space co-ordinates with respect to the surface and this can be useful for wafers, however this may not be very convenient for samples with irregular surfaces or structures that do not have a simple direction parallel to the surface. In these cases it may be more convenient to define s_z parallel to a crystallographic direction, in which case a reflection along the defined s_z should be used to define the ω direction such that $\omega = 0.5(2\theta)$, and similarly the $\chi = 0$ should be set at this optimum scattering condition. To search for reflections without the surface reference can be achieved by finding two independent reflections, assuming an approximate unit cell dimension, from which a third can be predicted and found. This makes it possible to create an orientation matrix from which an estimation of where all the other reflections of interest can be found, Hamilton (1974).

This reciprocal space probe has many applications and will be covered in greater detail with examples in chapter 5.

4.3.4. *The Beam-Selection Diffractometer*

The double-crystal diffractometer by its very nature will result in very high intensity; however any sample imperfections will smear the information content. The multiple-crystal diffractometer will isolate all the components of the scattering, but the intensity will suffer. Adding an X-ray mirror will increase the intensity ~10 fold, but the sampled area increases along with the axial divergence. Neither can easily isolate a region on the sample for detailed analysis. The Beam-Selection diffractometer can be a very useful compromise, Fewster (2004).

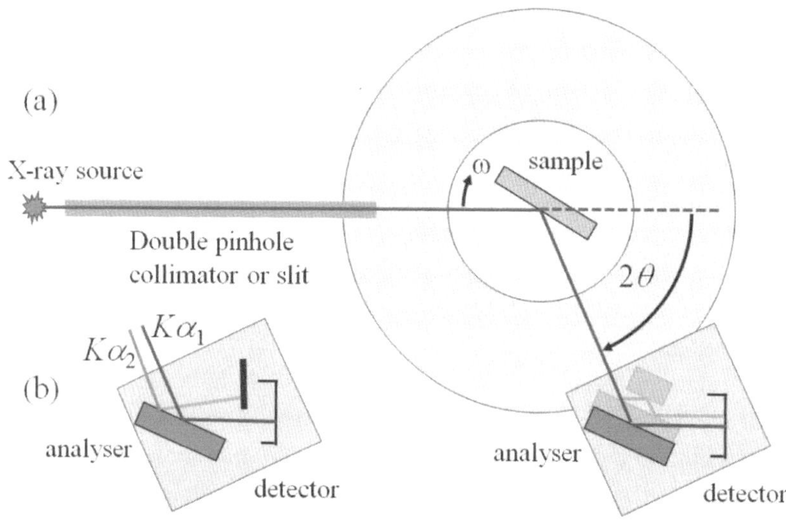

Figure 4.3.11. The Beam Selection Diffractometer, Fewster (2004), is shown with some of the options. The basic instrument comprises of an analyser that is Bragg angle matched (as close as possible) to the reflection from the sample (a). If the sample and analyser are a long way from matched then the spatial separation of the major wavelength components can be used to eliminate a contribution (b).

Although the appearance may be similar to the double-crystal diffractometer, figure 4.3.11a, it is conceptually different and in fact more like the multiple-crystal diffractometer where the monochromator is substituted for a slit or pinhole or an X-ray mirror. If the analyser crystal has the same scattering angle as the sample then as we have seen for the double-crystal diffractometer, all the wavelengths scatter at the same time but take different trajectories. This condition creates the highest intensity and the narrowest peaks if the sample is nearly perfect. If the sample scattering angle does not match that of the analyser, the analyser can be changed rather simply or the different trajectories of the different characteristic wavelengths can be blocked with an absorber, figure 4.3.11b. Placing the absorber in the correct position can be achieved by scanning across the direct beam, figure 4.3.12, that will produce a profile containing both $K\alpha_1$ and $K\alpha_2$; the spatial separation of the characteristic wavelengths will depend on the distance of the X-ray source (and its size) from the absorber and the analyser reflection (see figure 3.4.2), and typically this will be ~1 mm. The profile with the absorber is also given in figure 4.3.12.

Although the wavelengths are spatially separated, the angular separation is very small when the sample is inserted, figure 4.3.11a. Therefore the actual broadening from the wavelength dispersion is very small (depending on the degree of match of the scattering angles). The profile in figure 4.3.12 does not represent the resolution. The separation of two wavelengths λ_1 and λ_2 in angular space for scattering angles $2\theta_1$ and $2\theta_2$ from the sample and analyser is given by:

$$2\theta_{\alpha 1} - 2\theta_{\alpha 2} = \left|\sin^{-1}\left(\frac{\lambda_{\alpha 1}}{2d_i}\right) - \sin^{-1}\left(\frac{\lambda_{\alpha 1}}{2d_j}\right)\right| - \left|\sin^{-1}\left(\frac{\lambda_{\alpha 2}}{2d_i}\right) - \sin^{-1}\left(\frac{\lambda_{\alpha 2}}{2d_j}\right)\right| \quad 4.3.5$$

For example using a *004* Ge analyser with a GaAs and Si sample these separations are 0.7 and 20.5 seconds of arc respectively, although a Si analyser would be more appropriate with the latter. The spatial separation with this configuration and diffractometer radius of 320 mm, is 2.3 mm; therefore the $K\alpha_2$ contribution is easily removed with an absorber. Since the FWHM of the $K\alpha_1$ profile is known, section 2.2, the spreading of the peak from the wavelength for these examples are 0.2

and 5.9 seconds of arc. If we consider the combination of the *002* reflection from GaN and an analyser of *220* Ge, then the $K\alpha_1$-$K\alpha_2$ separation is 53 arc seconds and the $K\alpha_2$ can be removed if an absorber can be placed within 0.7 mm, i.e. half the spatial separation of these two contributions. The broadening due to this added wavelength dispersion is 15 seconds of arc, which has little effect on evaluating thickness values in GaN. However the broadening of the profile can be blurred by the overlap of the incident beam divergence with the spread of the scattering perpendicular to the diffraction vector; this can be reduced by limiting the pinhole or slit size, figure 4.3.11. This can be visualized with reference to figure 4.2.3a, by assuming that $\delta\omega$ is extended and $\delta 2\theta$ is very small for this instrument, so that a small contribution to the spread in the scattering perpendicular to $\boldsymbol{S}$ will be included in the capture during a scan parallel to $\boldsymbol{S}$.

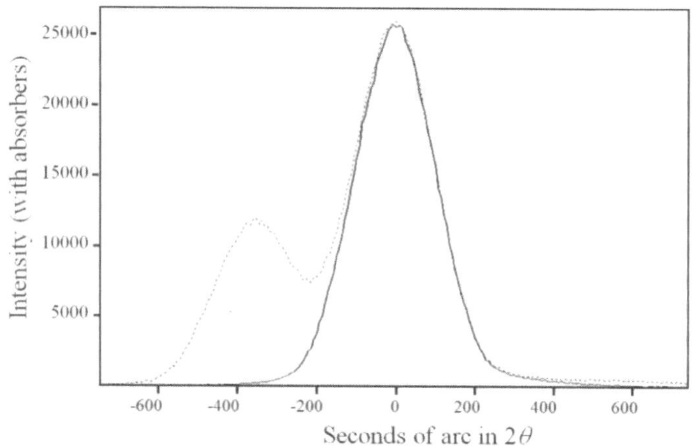

Figure 4.3.12. The scan through the direct beam ($2\theta = 0$) is shown for the Beam Selection Diffractometer without (dotted line) and with (solid line) a beam absorber to isolate the Cu $K\alpha_1$ contribution.

4.3.4.1. Alignment of the Beam Selection Diffractometer

One of the significant advantages of this diffractometer, and the reason for its name, is that the analyser will select only those beam trajectories that lie close to the plane of the diffractometer. Clearly the beam will be axially divergent as well as in the scattering plane, however if the analyser crystal is aligned to only accept beams in the plane of the diffractometer then the exact sample alignment is largely irrelevant, figure 4.3.13b. This becomes obvious when compared to the sensitivity to tilt alignment of the double-crystal diffractometer, figure 4.3.13a.

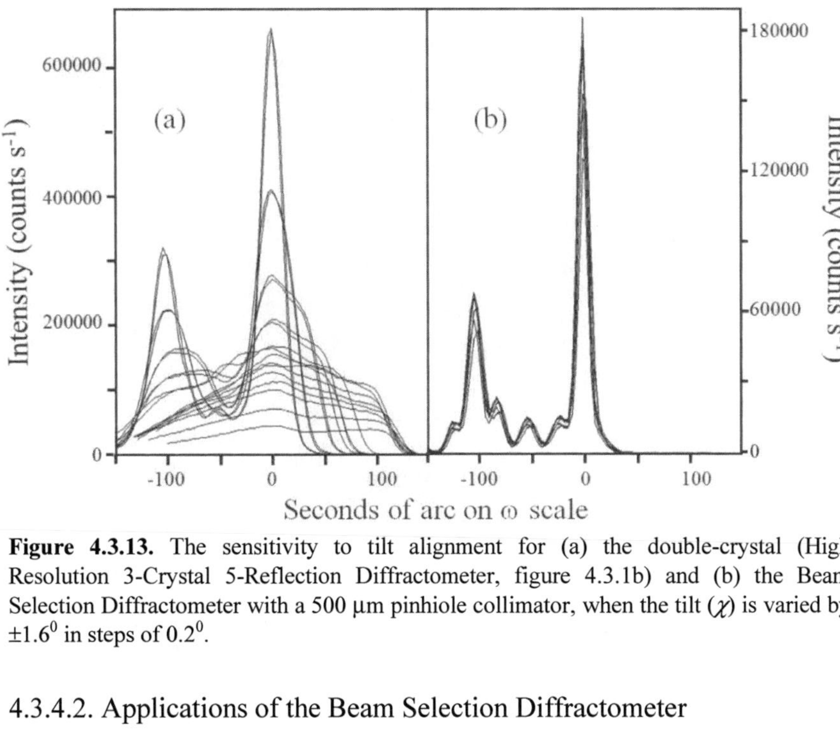

Figure 4.3.13. The sensitivity to tilt alignment for (a) the double-crystal (High Resolution 3-Crystal 5-Reflection Diffractometer, figure 4.3.1b) and (b) the Beam Selection Diffractometer with a 500 μm pinhiole collimator, when the tilt (χ) is varied by $\pm 1.6^0$ in steps of 0.2^0.

4.3.4.2. Applications of the Beam Selection Diffractometer

The two main advantages of this high-resolution diffractometer are the very high intensity and the control of the sampled area. These make it suitable for mapping the scattering profiles across a wafer with dimensions down to ~50 μm. Because the angular divergence can be restricted with a slit combined with a line focus, figure 4.3.11, or an X-ray mirror, figure 3.4.7, the instrument capture volume can be very small and can be used to map the intensity in diffraction space. Figure 4.3.14 illustrates an example of a diffraction space map collected in 5.6 h on an area 1 mm × 200 μm. With the very high intensities the instrument capture artifacts are always more noticeable, especially on a logarithmic scale, and this can be seen in this diffraction space map at the peak by the inclined elongation of the scattering. If the area of the sample is increased and an X-ray mirror is used then this map can be collected in ~4 min, and with improved resolution; such that the incident beam divergence is reduced from ~0.137^0 down to ~0.04^0.

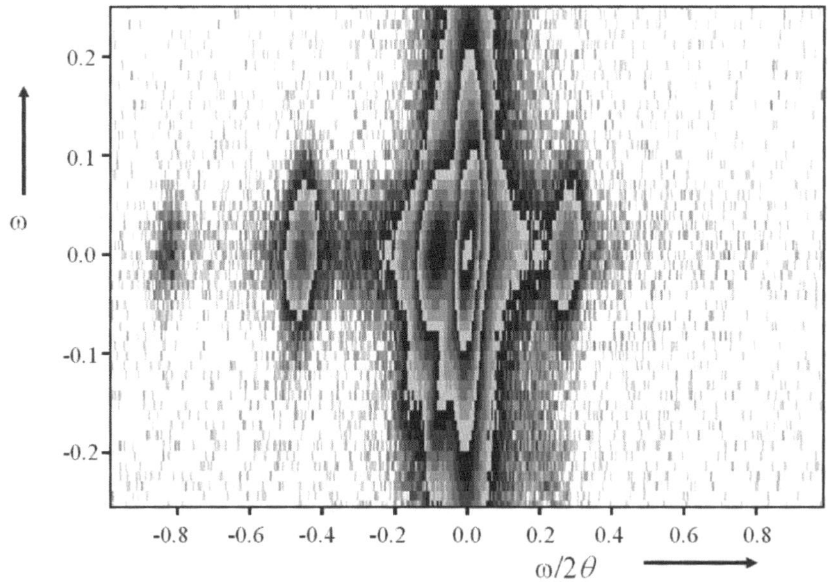

Figure 4.3.14. This figure illustrates a diffraction space map obtained from an InGaN/GaN multi-layer structure with the Beam Selection diffractometer on a small area approximately 0.2 mm × 1 mm. The satellites, average layer peak and substrate peak are all clearly seen. The large extent of the scattering in ω is indicative of the defects in the structure.

This very high intensity and high resolution can be useful for extracting information about interfaces as well as composition and thickness parameters, either in a superlattice or a single layer. Examples of these will be given in sections 5.4.5.5.4.and 5.4.3.6. This geometry is also useful for mapping the inhomogeneity of parameters across a sample.

4.3.5. *The Double Channel-Cut Diffractometer*

This instrument is similar to a triple-crystal and multiple-crystal diffractometers and is a compromise to boost intensity but still maintain reasonable resolution. The geometry is given in figure 4.3.15a, where the combination of the X-ray mirror and the incident beam channel-cut crystal does improve the wavelength dispersion that is sufficient to remove the $K\alpha_2$ component. The channel-cut crystal will lead to a loss of intensity, because the mirror has a divergence of ~0.04^0 and the crystal has an acceptance of ~0.0034^0 (the intrinsic diffraction width for *220* Ge from Cu $K\alpha$). This acceptance is sufficient to isolate the Cu $K\alpha_1$ wavelength component. This leaves some dispersion associated with the width of this component, which equates to ~0.017^0 ($\Delta\lambda/\lambda$=0.0007, θ_{220}=22.63^0 inserted into equation 3.4.2).

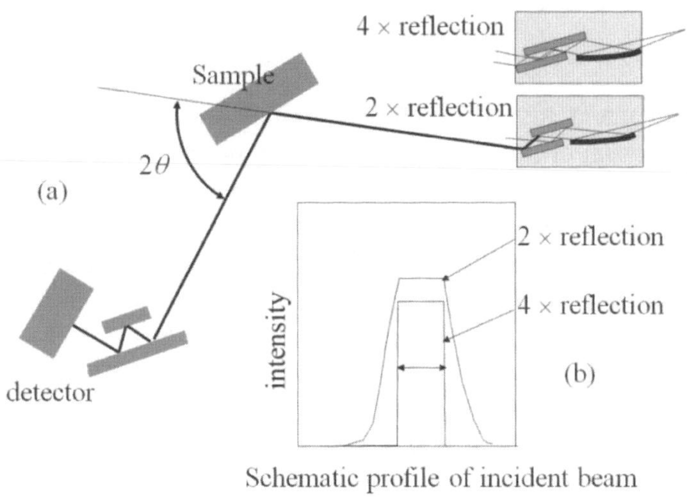

Figure 4.3.15. (a) The Double Channel-Cut Diffractometer can be compared with the multiple-crystal and triple-crystal diffractometers. The combination of the mirror and channel-cut can suppress the Cu $K\alpha_2$ contribution to give a reasonable resolution and good intensity. (b) gives the profiles for two options of the channel-cut reflections.

This does not mean that the experiment can never improve upon this profile with this configuration; we have to consider the sample influence, whether it can accept all this divergence, and whether the wavelength broadening is reduced in the manner described by equation 4.3.5. Similarly when an analyser is included, that will also influence the profile width. Another important consideration is the chosen scattering planes from the sample and the direction in which the information lies, e.g. normal or parallel to the surface plane. As we are beginning to see, the complexity is growing and the whole sample and instrument cannot be considered in isolation, i.e. there is no simple instrument function that will describe any instrument. However we will try and give an indication of the performance of the various optical configurations in section 4.3.9.

4.3.5.1. Alignment of the Double Channel-Cut Diffractometer

Because the incident beam is well defined, the sample has to be aligned carefully to the incident beam. The method is therefore identical to the double-crystal diffractometer, given in section 4.3.1.1.

4.3.5.2. Applications of the Double Channel-Cut Diffractometer

This instrument just offers another option, more intensity than the multiple-crystal diffractometer but a compromise on the resolution. The comparison with the Beam-Selection diffractometer is interesting, in that it would appear at first the Double Channel-Cut diffractometer offers higher resolution, with some loss of intensity; however this requires clarification. These latter two instruments can be considered identical in one of the configurations (X-ray mirror and Ge *220* channel-cut analyser), except for the incident beam channel-cut. Suppose that the sample reflection being measured is close to the Ge *220* scattering angle $2\theta = 45.26^0$ then the intensity loss of including this incident beam channel cut is ~12× ($0.04^0/0.0034^0$, referring to the divergence of the mirror and that of the channel-cut), and in both cases the resolution is the same and optimized. As we move a long way from this optimal condition the divergence of the incident beam channel-cut limits the resolution to 0.017^0, less than half that of the mirror alone of 0.04^0. Therefore the resolution is higher than the Beam-Selection diffractometer in this situation but the intensity is ~5× weaker. If the analyser of the

Beam-Selection diffractometer is changed to better match the reflection being studied, then the resolution of the Double Channel-Cut diffractometer can be poorer and its intensity considerably weaker. Clearly there is no simple answer, and the nature of the information required must decide the configuration.

4.3.6. *The Static Diffractometer*

All the instruments described so far require scanning, which will always add to the data collection time and require precision goniometers. The Static Diffractometer requires no scanning, the alignment is simple and fast and the data collection can be seconds. As with any instrument offering such possibilities it becomes more restrictive, however it can be a very good approach for routine analysis.

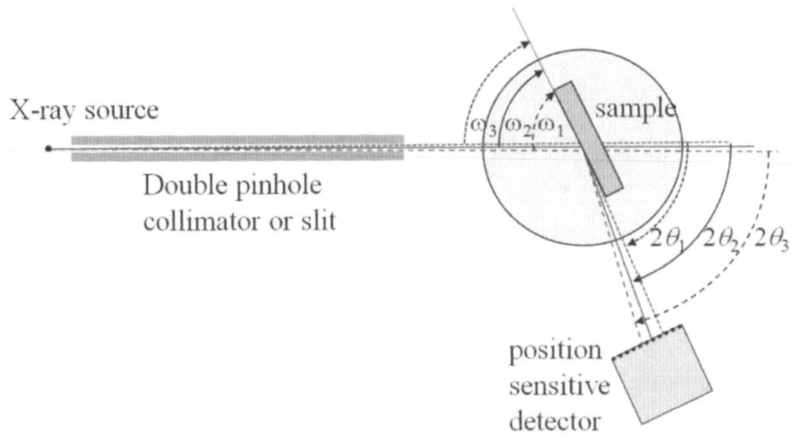

Figure 4.3.16. The geometry of the Static Diffractometer, Fewster (2005), relies on the small divergence of the incident beam to give a wide spread of scattered beams when a small grazing exit angle is used. This allows the whole profile to be captured on a position sensitive detector with no scanning.

The principle relies on an understanding of dynamical theory, and the combination of diffraction space and 'real' space resolution, Fewster (2005). The components relevant to this instrument are simply, an X-ray source, a slit and position sensitive detector. The sample does the rest, figure 4.3.16. Consider equation 2.4.6, reproduced here

$$\beta_H \approx \left\{2\sin\theta_B \cos\varphi(1+\chi_0)^{1/2} - (1+\chi_0)^{1/2}\gamma_0 - \gamma_H\right\}\gamma_H$$

that relates the deviation parameter β_H, to the incident angle ω. Suppose we select a reflection where 2θ - ω tends to small values, then γ_H is small such that a small change in ω will have a large impact on β_H. This grazing exit condition will result in a very compressed 'rocking curve', which can be observed in figure 2.4.5, i.e. a small movement in ω will result in a rapid movement over the profile. The dispersion surface diagram for the grazing exit condition has been given in figure 2.4.4b and is reproduced in figure 4.3.17. Suppose that the original incident beam direction is k_{01} that produces a scattered wave k_{H1} and probes a region in diffraction space defined by S_1; then an alternative wave with k_{02} will create a scattered wave associated with S_2 (equation 2.4.6 given above), which is given by k_{H2}. The first thing to notice is that the small change in incident beam direction produces a large deviation in the scattered beam direction. We can also describe this in terms of conservation of energy; the incident beam has a certain width and divergence, however on scattering the width is compressed and therefore the divergence must be increased to compensate.

Therefore the small divergence of the incident beam can be such that the scattered beam divergence can be large enough to be captured with a position sensitive detector. The compression of the scattered beam origin results from the 'end on' view experienced by the detector, giving the impression that all the scattering comes from the same position on the sample. The combination of the divergent scattered beam and the apparent central point of the scattering give the diffractometer high-resolution performance. The only requirement is find a suitable set of scattering planes inclined to the surface to create this grazing exit condition. If there is no suitable set of planes then another wavelength can be used.

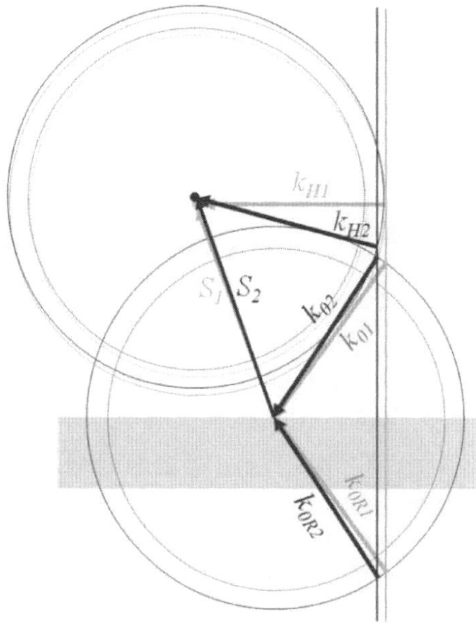

Figure 4.3.17. The dispersion surface construction for the grazing exit condition is shown, illustrating how a small difference in incident beam direction can lead to a large separation of scattering angles.

Figure 4.3.18a illustrates a diffraction space map obtained with the Static Diffractometer, where it is clear to see the substrate peak associated with the Cu $K\alpha_1$ and Cu $K\alpha_2$ characteristic lines in figure 4.3.18b and the layer peak with thickness fringes, figure 4.3.18c. The profile given in figure 4.3.18c is that normal to the surface, whereas figure 4.3.18d is the profile captured by the detector that contains a significant proportion of the information. This example is for a 50 nm $Si_{0.2}Ge_{0.8}$ layer on a *001* Si substrate capped with 500 nm of Si using the *113* reflection, where $(2\theta - \omega) = 2.79^0, 2.82^0$ for $Si_{0.2}Ge_{0.8}$ and Si respectively. This accounts for the difference between figure 4.3.18c and d. The latter profile is collected in 1 second, and in this case the detector contained 128 strips 70 μm wide at a distance of 320 mm from the sample. The information contained is sufficient to extract the $Si_{0.2}Ge_{0.8}$ layer thickness and composition, and the thick cap layer thickness has been obtained with a detector with narrower strips.

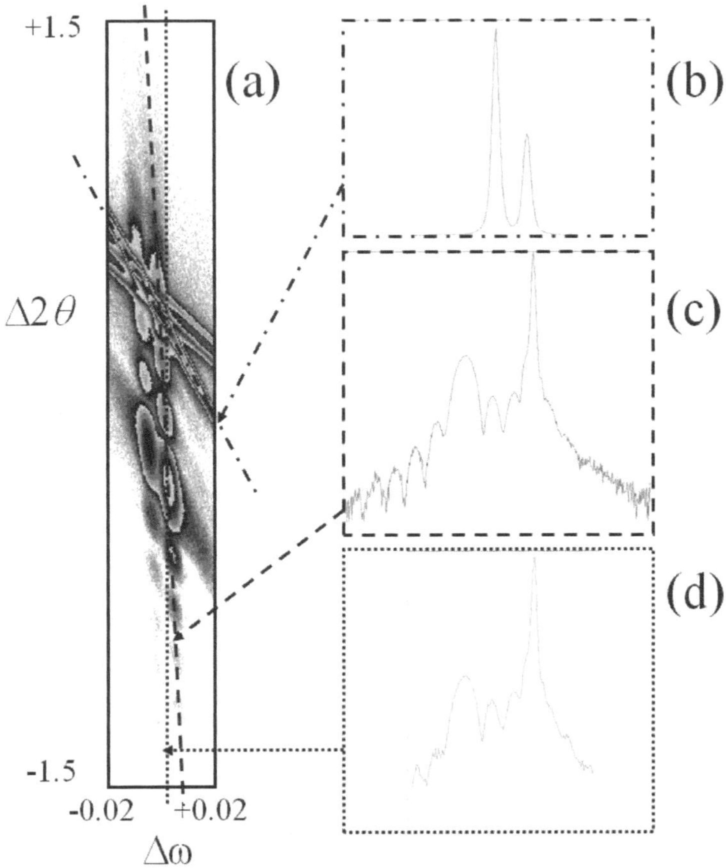

Figure 4.3.18. (a) is a diffraction space map (obtained with the Static Diffractometer) for a SiGe/Si structure. The two nearly vertical rows of features correspond to the Cu $K\alpha_1$ and Cu $K\alpha_2$ wavelength contributions, which can be shown by the extracted line (b). The full structural information is captured along the line given in (c) and compares with the information obtained with the Static Diffractometer in (d). The two parallel streaks evident near the substrate peaks are a tube artifact that can be suppressed.

The positions of the peaks, the potential overlap of Cu $K\alpha_2$ contributions is best resolved through simulation. For GaAs, where $(2\theta - \omega) \sim 1.6^0$ for the *113* reflection, the profile obtained extends further because the alignment is closer to the surface normal. The capture range will also be a function of the detector size and divergence of the incident beam, which is usually set to remove the Cu $K\alpha_2$ contribution. This is illustrated for a 15 nm layer of $In_{0.15}Ga_{0.85}As$ on GaAs and a 0.1 nm cap layer of $Al_{0.25}Ga_{0.75}As$, in figure 4.3.19 showing both the simulated and experimental profiles as the divergence slit is changed. It is seen that the Cu $K\alpha_2$ contribution is removed for slit widths below 175 μm, and for the instrument configuration used this can be fixed (incident beam long fine-focus X-ray source to sample is 320 mm, and slit set at 100 mm from the source, and sample to detector is 320 mm). The axial divergence has little effect on the profile under these conditions, for a fuller explanation the reader is referred to Fewster (2005).

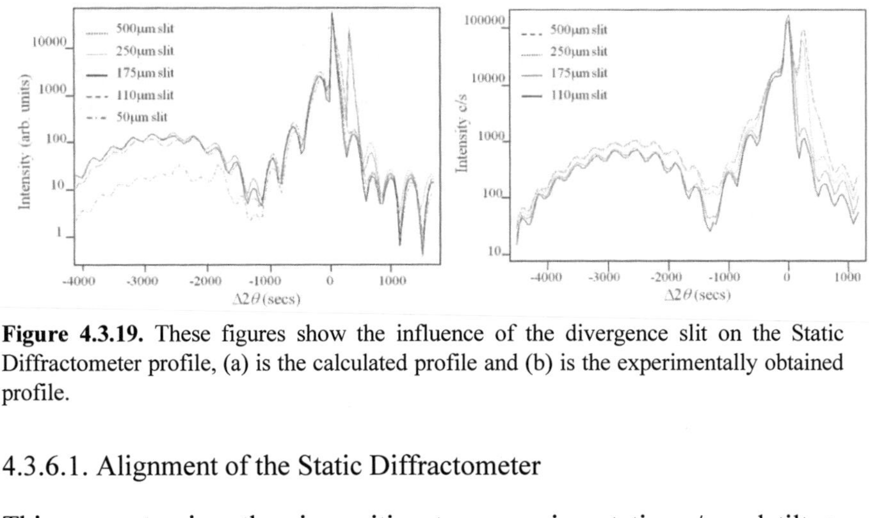

Figure 4.3.19. These figures show the influence of the divergence slit on the Static Diffractometer profile, (a) is the calculated profile and (b) is the experimentally obtained profile.

4.3.6.1. Alignment of the Static Diffractometer

This geometry is rather insensitive to errors in rotation ϕ, and tilt χ, making the setting up very simple. There is no measureable change in the profile if ϕ is kept within $\pm 1^0$ and χ is kept within $\pm 0.8^0$. Since most wafers are orientated within 0.25^0 the tilt requires no alignment in the presence of X-rays, and often the wafer will come with a flat that defines the reference point for rotation. The instrument already has a fixed 2θ for the chosen reflection and therefore the only significant setting angle is ω. The sensitivity of this is best considered with reference to figure 4.3.18. Suppose the detector is set to accept all the intensity along 2θ, and the divergence slit is set to that required to isolate Cu $K\alpha_1$, e.g. 175 μm at 100 mm from a 40 μm source, then a scan in ω will give a profile dominated by Cu $K\alpha$ doublet for the substrate. This very intense and rapid scan will allow the ω setting to match that of the Cu $K\alpha_1$ substrate peak or some intermediate position between the substrate and dominant layer peak. Following this the single instantaneous scan can be obtained or a reciprocal space map as shown in figure 4.3.18a.

An important point here is to emphasise that the divergence created by the 175 μm slit at 100 mm from a 40 μm source is 0.123^0, so it may suggest that the information would be blurred. This is clearly not the case in figure 4.3.18a. The reason why this is not the case is fundamental to how the instrument works and will be briefly explained. Any intensity appearing at a specific 2θ value will require a specific incident angle ω, which as described above is very exacting (i.e. a small movement in ω will probe a different region of the profile). Therefore, if we have a small or distant source there will be only one position on a perfect sample that will produce intensity at a specific 2θ value corresponding to a specific ω. If we accept that we have a finite size source at a finite distance from the sample, then for the example of sample 320 mm from a 40 μm source the effective divergence on the sample is 0.014^0 ($=2\tan^{-1}\{0.04/320\}$). We can now see that resolution can be enhanced by increasing the source to sample dimension or using a smaller focus.

4.3.6.2. Applications of the Static Diffractometer

The rapid data collection with the Static Diffractometer with a reasonable size beam makes it possible to go to small spot sizes and still collect data very rapidly. This has advantages in wafer mapping, and since the profiles are insensitive to alignment data can be collected by just moving the sample on an $x-y$ stage and collecting for a short time at each point. From the discussion in the previous section we established that the probe size is approximately the size of the focus for a specific 2θ contribution; so consider the layer peak in figure 4.3.18a that is displaced from the substrate peak by $\sim 0.0029^0$ in ω, and the spread of incident angles in the 40 μm region is $\sim 0.0143^0$, the centre of the sampled region for the substrate reflection is displaced from that of the layer peak by 8 μm. This is a small effect, which would suggest that with this particular instrument configuration probing at the 50 μm level would mean that the profile can be analysed reliably. Generally the sample is likely to be homogeneous over a larger scale than this.

If we use an X-ray tube with a focus dimension of 15 mm × 0.04 mm, then it is possible to define several separate beams along the 15 mm height to increase the number of profiles collected in parallel.

It is also perfectly feasible to undertake a diffraction space map at each wafer map point. The inclusion of the Cu $K\alpha$ doublet is not really a problem, since as we can see in figure 4.3.18a that the contribution can be separated rather easily. A diffraction space map at the 40 μm level can therefore be obtained to measure the composition, layer thickness and the degree of lattice relaxation present. Examples of this will be given in sections 5.4.3.9 and 5.5.5.3.1.

4.3.7. *Configurations for rapid reciprocal space mapping*

As we can see from figures 4.3.14 and 4.3.18a, reciprocal space maps can be obtained exceedingly quickly with the Beam Selection and Static diffractometers respectively. If the beam size on the sample is a reasonable size then the reciprocal space map in figure 4.3.14 can be obtained in ~4 min, and the area of the map containing all the information in figure 4.3.18a can be captured in about 40 s. The difference between the two methods is that the Beam Selection diffractometer collects the data sequentially and relies on the high intensity, whereas the Static diffractometer collects data along 2θ in parallel, making data collection very fast. For this section we shall refer 'rapid reciprocal space mapping' to data collected partly in parallel.

It must be recognized that these methods have limited resolution, so their application apart from using the Static diffractometer, relies on the divergence of the incident beam and the dimensions of the detector pixels. We can see from figure 4.3.18a, that as the angle (2θ - ω) increases, the capture along 2θ moves away from the information normal to the surface. The data collection has to be in map form, and the greater (2θ - ω) becomes the larger the mapping needs to be, figure 4.3.20, and the benefits diminish. The greatest benefit is always to try and use reflections that come closest to grazing exit, i.e. (2θ - ω) small. To capture the information along s_z, figure 4.3.20, the detector centre pixel needs to be offset in ω from *S* to *A* followed by a series of instantaneous captures along 2θ offsetting in Δω between each. If the data is collected by scanning ω and 2θ, each pixel should have a relationship to ω and 2θ that is less straightforward. As discussed in section 4.2.2 the change in ω will alter the position of 2θ, which can be calculated through a simple geometrical relationship; each step in Δω will change 2θ to 2θ - Δω.

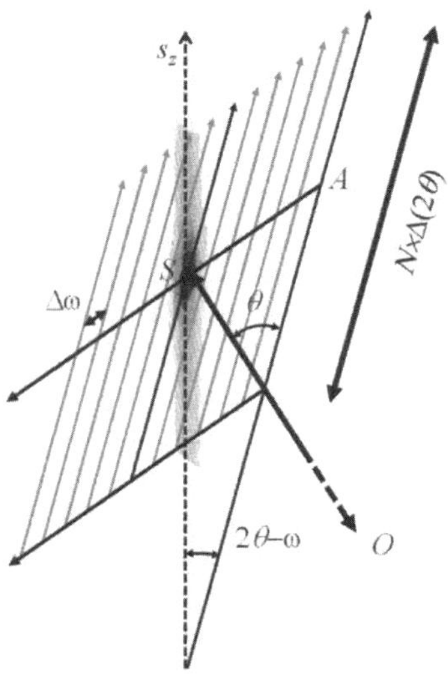

Figure 4.3.20. The capture region is given for rapid diffraction space mapping. *OS* is the diffraction vector, *AS* represents the direction in ω when the sample is rotated and the data is collected simultaneously in 2θ.

The data collected is based on the position of the detection point and this requires that the scattering captured comes from a very small region on the sample; otherwise the 2θ resolution is degraded. The spread in ω of the incident beam will also give a spread in 2θ, from the discussion above. So depending on the required resolution, the incident beam should be narrow with low divergence and to benefit from fast data-collection the intensity should be high and monochromatic. This really limits the possible incident beam optics to the channel-cut combined with a mirror, and unless the exit beam from the sample can be reasonably close to grazing the incident beam should be reduced in width with a slit. It really all comes down to what information is required, how reliable the data should be, etc. Collecting the 2θ angle by position cannot be too reliable without some internal calibration, but relative peak positions are possible, but it must be remembered that the data is collected on a flat detector (in general) and not on an arc, and requires a correction or large radii to limit the effect. This can be resolved through calibration as in section 4.4.3.1, when there are several contributing factors. Examples of rapid reciprocal space mapping will be covered in sections 5.4.3.9 and 5.5.5.3.1.

4.3.8. *Configurations for laboratory-based GISAXS*

GISAXS has had limited appeal outside synchrotron facilities because of the conflicting requirements for high intensity and high-resolution in both s_y and s_z to achieve a reciprocal space map as in figure 2.11.13. The method though is perfectly feasible with standard laboratory X-ray tubes and components. The most direct approach is to achieve high-intensity and a small beam, so the scattering comes from a small projected region, and if this is combined with a large distance to the detector high resolution can be achieved. The intensity sacrifice with this configuration can be a concern in which case a high-brilliance source can be used, or the resolution requirements can be relaxed, depending on the problem. Another approach is to use focusing mirrors, e.g. based on the Kirkpatrick-Baez arrangement, section 3.4.8, or an elliptical mirror with a focal point on the detector, section 3.4.5. We shall also consider an alternative approach that has a different way of resolving the s_y resolution associated with sample size effects.

4.3.8.1. The direct approach to GISAXS instrumentation

A small narrow beam impinging on a sample will give a resolution in s_y limited by the length of the sample × sin2θ. For a double pinhole configuration (figure 3.3.2) with diameters of 0.1 mm, 100 mm apart, the maximum divergence is $\pm 0.03^0$ for the umbra region. This is comparable to the sample size effects which amount to 0.04^0 along s_y at 0.64^0 2θ (0.25 of the width of a 14 mm detector at 320 mm) for a 10 mm × 0.1 mm illuminated area. However quite reasonable data can be achieved with this configuration as discussed in section 4.3.8.6. An example of a compressed beam optic, similar to cutting a crystal with the bottom-half of the crystal in figure 3.4.4b followed by the top-half of the crystal in figure 3.4.4a, is given by Jergel et al (2013) that can give a beam compression of ~15 and ~21 and a beam size of ~0.06 mm and ~0.05 mm when combined with a microfocus tube. Because each crystal in this V-groove will have a different refractive index peak-shift (section 2.3 and 3.4.3), which is especially evident at high compression ratios this has to be compensated for with a SiGe alloy gradient rather than pure Ge. The resolution in s_z is applicable to all these methods and is discussed in section 4.3.8.4.

4.3.8.2. The use of focusing beams in GISAXS instrumentation

An alternative approach is to consider a beam from an elliptical mirror, section 3.4.5, that is focused onto the detector. Each beam trajectory will have a width that is comparable to the width of the X-ray focus and any scattering in a direction 2θ approximately parallel to the surface plane will arrive at the same point on the detector. If we assume that the convergence from the mirror is ~0.04^0, see section 3.4.5, then the combined effect will give an instrumental blur of ~0.046; assuming that the mirror is perfectly elliptical and its focus is precise across the detector, which is unlikely but the principle is clear. If the capture angle is comparable to that given in figure 3.4.7, i.e. 0.8^0, then the maximum converging beams subtend an angle of ~0.2^0, which defines the level of uncertainty for surface orientation information.

The resolution in s_z could be defined by another elliptical mirror (the Kirkpatrick-Baez configuration, section 3.4.8); this defines the angular divergence directly. The spread in the incident angle will have to be constrained with slits to limit this variation and loss in resolution in s_z. The general discussion of the s_z range and resolution is covered in section 4.3.8.4.

4.3.8.3. The monochromator and slit approach to GISAXS

The approach described in this section has been used in our laboratory and can result in a variable range of resolution in s_y. The instrument uses a 2-crystal 4-reflection monochromator, section 3.4.4, combined with a narrow slit and an area detector, figure 4.3.21. The instrument could be made very compact. The scattering angle parallel to the surface plane is selected by the position on the detector and the distance from the slit, and the resolution is defined by the detector pixel size and the slit width and their separation. If the highest resolution is not required then the beam can be manipulated by reducing the cross-slit (closer to a pencil beam on the sample) and the slit on the sample widened.

The monochromator produces a highly parallel monochromatic beam that defines $\boldsymbol{k_0}$ and the slit and detector define $\boldsymbol{k_H}$. From the figure; the intensity associated with point C will be all the scattering that comes from the line AB, then the scattering angle associated with this scatter is OSC. B is the position of the sample edge. Similarly scattering angle OSC_1 comes from the line A_1B_1. This configuration gives the opportunity to change the resolution with the slit size, to reduce the sample size broadening, and therefore reduce the detector to slit distance, OS. The proximity of the slit to the edge of the sample modifies the s_y range accessible. Clearly the scattering intensity will vary due to the geometry above the angle OSC_1, because the line of material accessible for scattering changes from A_1B_1 to AB, and requires a geometrical correction to the measured intensity. For a 10 mm sample or illuminated length with a 0.4 mm wide beam (point focus source) the angle OSC_1 is 1.14^0, which corresponds closely to the angle subtended by a 14 mm solid-state area detector at 320 mm to capture both -2θ and $+2\theta$ when held stationary, i.e. 1.26^0. This also coincides with a typical GISAXS area of capture. The width of the beam from a mirror is ~1.2 mm and the angle OSC_1 increases to 3.4^0. The mirror has the advantage in increasing the intensity, but will then need slits to limit the angular acceptance range in s_z, see section 4.3.8.4. The s_y resolution is defined by the slit and the pixel size on the detector, and if these match, i.e. 55 μm in the example above, will amount to ~2.2×10^{-4} Å^{-1} corresponding to a step of 0.01^0 in 2θ for a slit to detector distance of 320 mm.

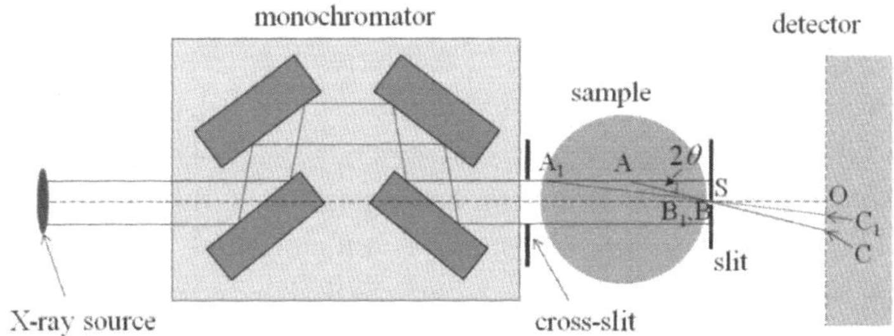

Figure 4.3.21. The GISAXS with this configuration makes use of a large parallel monochromated beam to select a narrow high-resolution beam in s_y defined by the slit and detector, after Fewster (unpublished work). The slit after the sample also reduces the unwanted scatter.

4.3.8.4. The resolution in s_z for GISAXS experiments

Each incident angle will produce a full diffraction space map, specific to that incidence angle. Ideally a very small intense focus at a large distance from the sample is a rather trivial analysis; the map can be calculated at a specific incident angle and that is it, section 2.11.5. The scattering varies with incident angle, so for a large source and small path lengths it becomes more of a problem unless each incident angle within the spread of incidence angles is calculated and combined. Our interest here is to indicate what is possible in the laboratory with standard equipment. Also the data will be collected on a flat detector, which will distort the diffraction space map, and so this has to be considered. Ideally the diffraction space map should be simulated, taking into account all the instrumental effects, although a considerable amount of information can be extracted by measuring peak positions and widths. However it is helpful to have an indication of typical features and the resultant scattering. The method discussed here concentrates on determining the intensity at the detector through simulation, and applying the instrumental aberration, whereas in section 4.3.8.6 some examples are presented to give an indication of what is possible.

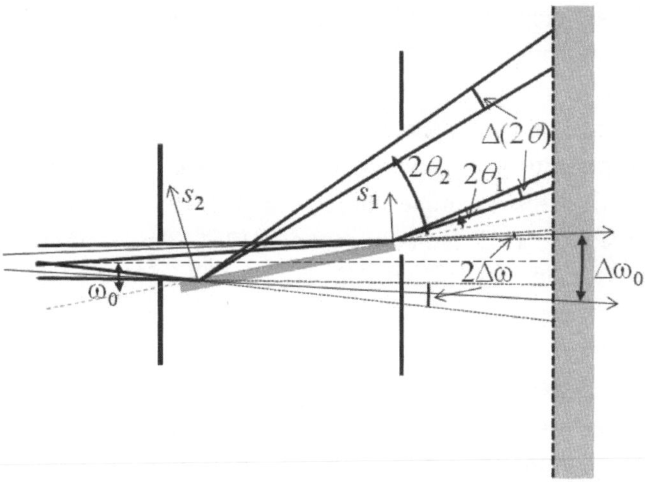

Figure 4.3.22. The beam trajectories in the plane normal to s_y for a GISAXS measurement are given, illustrating the convergent beams that give a spread of $\pm\Delta\omega$; this will relate directly to the incident intensity, at a point that has an average incident angle ω_0. The spread of the average incident values is given by $\Delta\omega_0$.

We shall consider the influence of the data-collection and the instrumental aberrations on the unadulterated simulation of the reciprocal space map. There are a few considerations; the varying projection of the sample surface to the detector with scattering angle, the divergence of the beam on the sample that gives rise to a spread in incidence angles. The scattering at a point on the sample will be a combination of the intensity arriving at that point, the average incidence angle and its spread. The determination of these latter values is based on those ideas discussed in section 3.3.1. Figure 4.3.22 illustrates the convergence of the beams, $\pm\Delta\omega$, at positions on the sample for regions within the X-ray coherence length, and the spread in the average direction of these convergent beams across the sample ω_0. The angular convergence at each position relates to the flux at that position.

The projected area of the sample on the detector will increase with s_z and therefore reduce the resolution. The resolution in s_z is calculated for a 0.4 mm point source 320 mm away from the 10 mm sample. The projection of the sample on to the detector is ~0.07 mm for an incident angle of 0.4^0, rising to ~0.13 mm for an incidence angle of 0.8^0, these equate to a blurring of ~0.012^0 and ~0.023^0 in 2θ respectively, normal to the s_y plane.

The angular spread of the scattering along s_z due to the incident beam will correspond to the convergence of the beams on each feature that scatters, and in this example amounts to 0.0717^0 and the spread in the average incident angle amounts to $\pm0.0125^0$. These combined effects result in a spread in s_z of 2.0×10^{-3} Å^{-1} at 0.4^0 incidence and a fraction more at 0.8^0 incidence. This resolution should be sufficient for most applications.

Increasing the divergence with a larger source size can be achieved with an X-ray mirror as in figure 3.4.7, which typically creates a beam of 15 mm × 1.2 mm, or with the configuration with the elliptical mirror, section 4.3.8.2. But as we can see from the discussion above the source size does start to impact on the resolution in s_z, so it becomes a balance of masking much of the source, the benefits of intensity, the resolution losses and of course the sample being studied. Using a micro-focus tube, section 4.3.8.1, will create a small convergence at each position (a 0.04 mm focus results in $2\Delta\omega$ ~ 0.007^0), however the spread in average incident angles will still be similar to above, $\pm\Delta\omega_0$ ~ $\pm0.0125^0$. This reduces the instrumental smearing effects further.

4.3.8.5. Alignment for GISAXS experiments

These geometries rely on careful alignment, which is slightly different for each. For the direct approach (section 4.3.8.1) that has a very narrow beam parallel to the surface with divergence spread normal to the surface; the sample surface normal should be aligned accurately to this and similarly to the detector pixels. This is where the area detector and specular beam help in defining this setting. The double pinhole arrangement creates significant stray scatter making setting-up more difficult, so careful masking is required. The focusing mirror approach will be more sensitive to sample tilt, because it depends on the illumination of a larger area, so the alignment of the specular beam with the detector pixels and loss of focus will have an additional effect. The monochromator and slit based system (section 4.3.8.3) will require the detector pixels, the slit and the specular profile to be coincident. The sample should be orientated such that the specularly scattered beam passes through the slit, and the length of the profile will be indicative of the closeness to alignment. Because of the dominating intensity of the specular profile, this should be absorbed with a narrow beam stop when the weak diffuse scatter is collected.

4.3.8.6. Applications for GISAXS

GISAXS is equivalent to the more widely used Small Angle Scattering (SAXS) but in reflection mode. It is the added complexity associated with the generally large illuminated area on the sample and the use of reflection mode that makes GISAXS more difficult in the experimental set-up. Scattering at very low angles is only sensitive to long length scales, this makes GISAXS suitable for nano-material shapes including layer structures, patterned surfaces, etc. The higher the resolution the larger the length scales that can be explored. For smaller length scales, which are correlated, we move into the realm of in-plane scattering, section 2.11.2, which can be carried out on more conventional instruments discussed in section 4.5.2 and 4.5.3.

Suppose we have a structure containing a random distribution of 15 nm diameter spheres with a 10% dispersion in sizes, then the scattering will appear like that in figure 4.3.23a (the raw simulations on the left-hand side of the image and that with the instrumental aberrations included on the right-hand side). This simulation is calculated for the monochromator-slit configuration with the dimensions as in examples above, i.e. a 0.4 mm × 0.4 mm focus at a distance of 320 mm from a 10 mm sample that is 320 mm from a 256 × 256 0.055 mm square pixel detector, a 0.1 mm slit is set 1 mm from the sample, figure 4.3.21. The simulation follows the ideas discussed in section 4.3.8.4, where the scattering angle ranges from z-pixel=40 (=0.4^0, the minimum angle with this incidence angle, although normally we would expect some scattering below this from the spread in the incident angle) to z-pixel=256 (=2.52^0). The intensity maximum that occurs at z-pixel=81 (=0.8^0) is the specular scattering effects from the sphere. The circular fringing relates to the size of the spheres, however care must be used in directly interpreting these because of the large dynamical effects. The approximations of a kinematical model will be inadequate (this will give the maximum intensity at z-pixel=0), and that of the semi-kinematical model (where only beams associated with the wave-vectors q_1 and q_2 are included, figure 2.11.3) only differs subtly from the full DWBA model, section 2.11. Although this simulation is presented for the configuration with the monochromator and slit, the other approaches given in sections 4.3.8.1 and 4.3.8.2, will produce similar results except their s_y resolution will be sample size dependent.

The second example is for simulations of cylindrical structures distributed on a surface with lattice type ordering, this is given in figure 4.3.23b. The axes of the cylinders are normal to the surface plane. Again the incident angle is set at 0.4^0 and exactly the same sample size and configuration is used as in the above example. It is interesting to note that the maximum intensity in figure 4.3.23a is the same as that in figure 4.3.23b and corresponds to $2\theta = 0.8^0$.

To give an impression of the type of data that can be obtained in the laboratory with standard equipment, we shall refer to figure 4.3.24. The data in figure 4.3.24a are obtained with a simple double pinhole system, section 4.3.8.1, and the data in figure 4.3.24b is obtained with the monochromator-slit system (section 4.3.8.3). This scattering is from cylindrical voids (axis parallel to the surface) in a SiO_2 matrix that exhibits hexagonal order. Depending on the information of interest, figure 4.3.24a illustrates data that can be captured very rapidly with a double pinhole. Clearly the resolution is degraded and great care is required to mask the inevitable stray scatter. The stray scatter around the masking can be seen along the specular peak; because the beam contains wavelengths that are very penetrating it is difficult to remove. This also illustrates quite well how GISAXS experiments can be performed with standard components, and it really comes down to the sample and information wanted. Data collected using the elliptical mirror on this sample is given in figure 2.11.13.

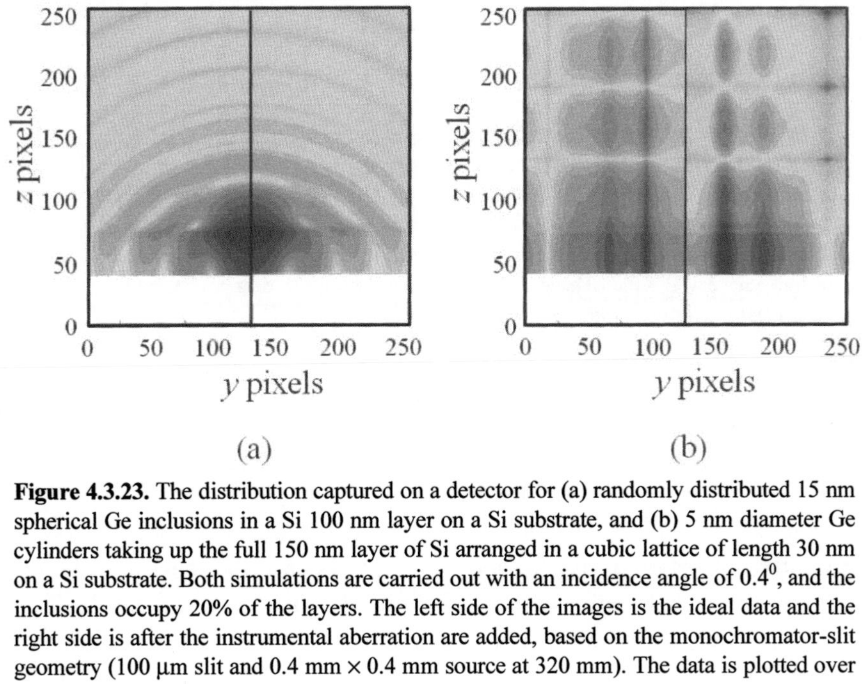

Figure 4.3.23. The distribution captured on a detector for (a) randomly distributed 15 nm spherical Ge inclusions in a Si 100 nm layer on a Si substrate, and (b) 5 nm diameter Ge cylinders taking up the full 150 nm layer of Si arranged in a cubic lattice of length 30 nm on a Si substrate. Both simulations are carried out with an incidence angle of 0.4^0, and the inclusions occupy 20% of the layers. The left side of the images is the ideal data and the right side is after the instrumental aberration are added, based on the monochromator-slit geometry (100 μm slit and 0.4 mm × 0.4 mm source at 320 mm). The data is plotted over ~7 orders.

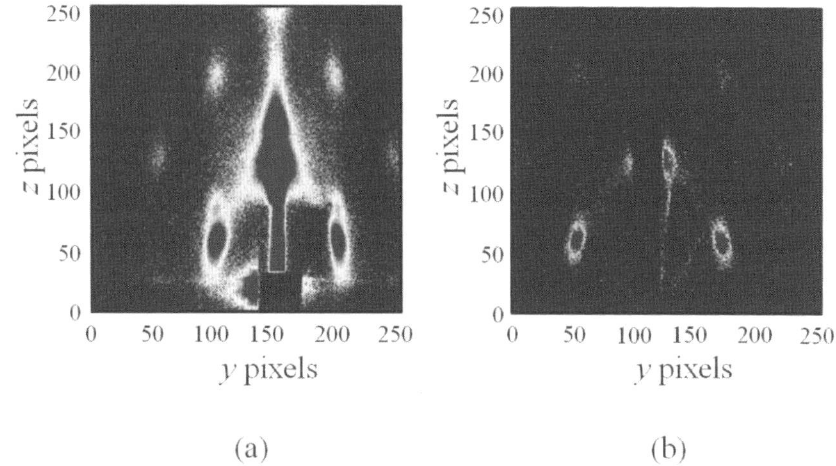

Figure 4.3.24. (a) The distribution of scattering from an array of cylindrical voids in a SiO_2 layer of grains, where the cylinder axes lie parallel to the surface, captured in the equivalent of 5 h with a conventional point focus tube and two 100 μm pinholes,. (b) is captured with the mononchromator-slit system in the equivalent of 6.5 h for a stationary detector, courtesy P Kidd. These can be compared with the data in figure 2.11.13 from the elliptical mirror configuration (section 4.3.8.2).

4.3.9. *A guide to the performance of high-resolution optical configurations*

It should be clear at this stage that there is no simple instrument function, or capture volume, that can be considered in isolation. Nor is there a definitive instrument for a specific experiment, the choice depends entirely on what information is required. The sample can have a significant influence and this becomes more important as we use some of these 'fast' methods. However some knowledge of the sample and the information of interest will influence the instrument choice, e.g. Table 4.1.1, but as we can see from the discussions the sample can define the instrument function. In an attempt to give some general guidance on the resolution of these "high-resolution" instruments, the 'instrument function' as a function of 2θ is given in figure 4.3.25.

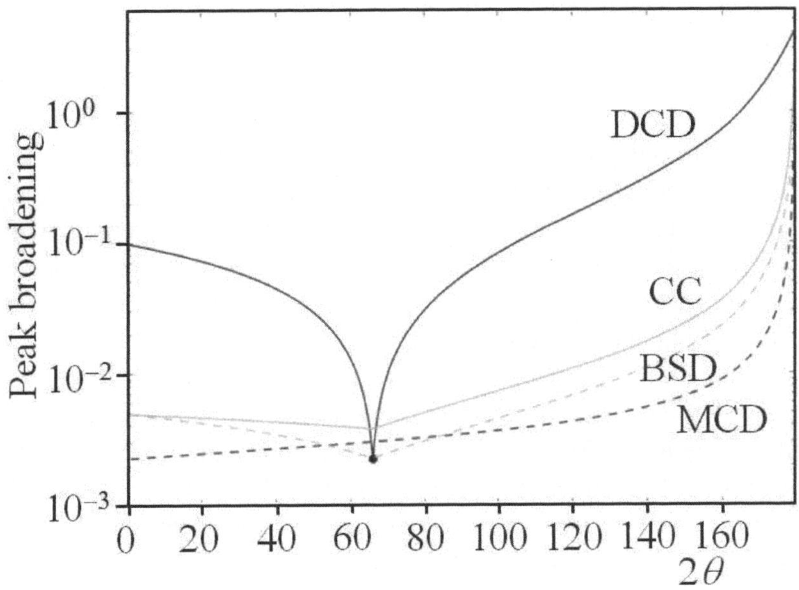

Figure 4.3.25. The variation in the peak width for various instruments, DCD: the double-crystal diffractometer, CC: the double channel-cut diffractometer, BSD: the beam selection diffractometer, MCD: the multiple-crystal diffractometer. The DCD is the only instrument that does not isolate the Cu $K\alpha_1$ characteristic line, the DCD and BSD are optimized for the *004* reflection from Ge(~34^0 θ). The MCD and CC are modeled using *220* Ge reflections.

From figure 4.3.25 it can be seen that for probing the *004* of Si, Ge and GaAs semiconductors the double-crystal diffractometer and the beam-selection diffractometer perform very well, although the latter will out-perform the former if the sample is bent or in any way imperfect. The multiple-crystal diffractometer has a small peak broadening over a large angular range, making it suitable for investigating a large range of reflections. The double channel-cut has the advantage of greater intensity, but as can be seen the peak broadening is significantly greater. These estimates of the peak widths do rely on the sample being perfect, and the use of an analyser as in the beam-selection, multiple-crystal and double channel cut diffractometers does help. But if the minimum peak width for the θ of the reflection being investigated is too large then the broadening due to the imperfections will still contribute, making interpretation more difficult. Hence this general categorization of 'resolution' or minimum peak width must be treated with caution.

Boosting the intensity with any of these configurations with an X-ray mirror has advantages, although it must be recognized that the beam is enlarged considerably and similarly the axial divergence, so again this is a compromise and depends on the size of the sample and information required. Combining any of these methods with topography will be discussed in chapter 5, however limiting the divergence in both directions aids the spatial resolution, so in general a mirror does not help apart from methods developed specifically for this in mind.

The next section will concentrate on diffractometers for polycrystalline materials, but this only reflects their common use and can still be useful for investigating semiconductors. This really depends on the information wanted, for example; the high intensity can be useful for investigating interfaces in superlattice structures, the high angular resolution from high-order reflections can give the alloy composition, etc. Some of these aspects are covered in section 5.4.5.5.1, however the traditional application is covered in the following section.

4.4. Diffractometers relevant to polycrystalline materials

The scattering from polycrystalline aggregates is dominated by contributions remote from the Bragg condition, section 2.9.3. However there is enhancement at the Bragg angle, regardless of the orientation of the crystallites provided they are within the ranges associated with the data collection method. This new description of the scattering of X-rays, Fewster (2014), implies that the number of crystallites required to create a scattering pattern is small and reliable and stable intensity estimates can be achieved rather easily. This has an impact on the diffractometer design as well as the consequent analysis. This section will consider the instruments typical of those designed with the conventional theory in mind, followed by designs that exploit some of the consequences of the new theory.

As it can be seen from section 3.5, it is very difficult to separate the performance of the diffracted beam analysers from the incident beam conditioning without defining a diffractometer. The following section will therefore bring together the designs based on the focusing geometry, which is currently the favoured instrument configuration.

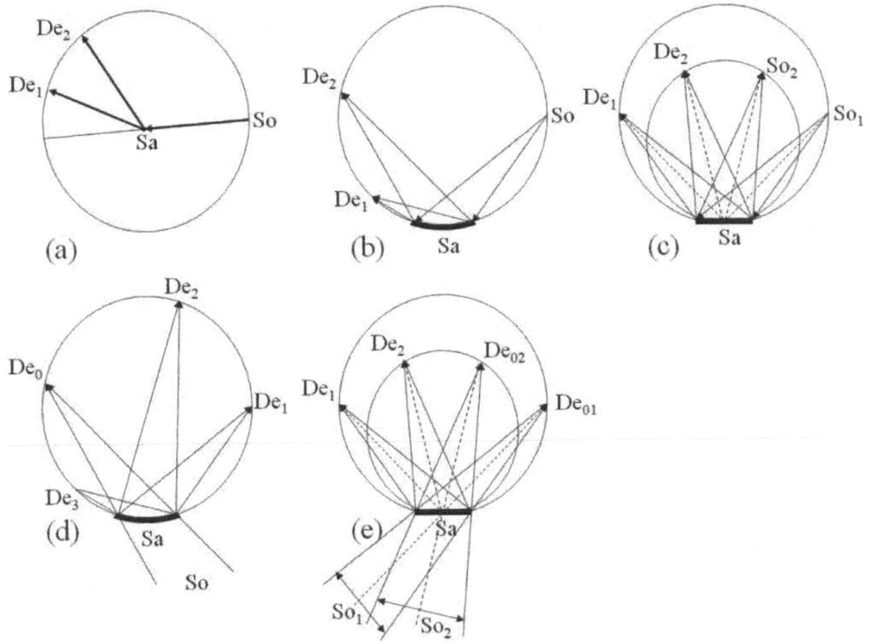

Figure 4.4.1. Some of the configurations used in powder diffractometry (a) the Debye-Scherrer geometry, (b) the Seeman-Bohlin geometry in reflection, (c) Bragg-Brentano geometry in reflection and (d) and (e) Seeman-Bohlin and Bragg-Brentano geometries in transmission mode. So represents the source and the De detector position.

4.4.1. *Diffractometers based on the focusing geometry*

The combination of a small source and narrow slit (or detector) in the diffracted beam that are placed at equal distances from the sample will create the focusing condition. Because the source is very small compared with the incident beam slit the angular divergence is roughly constant over the sample illumination area and therefore the analysing slit basically defines the resolution. However there is little point in reducing the analysing slit below the size of X-ray source, typically 40 μm.

Figure 4.4.1b to 4.4.1e illustrates some focusing diffractometer configurations that can be compared to the simplest instrument configuration, figure 4.4.1a. The configuration in figure 4.4.1a is typical of the Debye-Scherrer diffractometer that has limited resolution because of the divergence of the incident beam, which is defined by pinholes, and the size of the sample, which is usually a powder within a glass capillary. Reducing the pinhole dimensions, making the sample small and increasing the radius all improves the resolution, however the intensity becomes weak. Whereas with the focusing arrangements the sample can be large and the resolution defined by the source size and the receiving slit or detector. The focusing arrangements given in figure 4.4.1b and 4.4.1d rely on the sample being bent to the radius of the focusing circle and have the advantage that the data can be collected in parallel, Seeman (1919) and Bohlin (1920). The configurations in figure 4.4.1c and 4.4.1e do not require the sample to be bent, with the disadvantage that the sample and detector have to be rotated together, Bragg (1921), Brentano (1946).

The flat sample option of figures 4.4.1c (reflection geometry) and 4.4.1e (transmission geometry), is the favoured configuration and has proved very popular. The data collection is sequential, however with solid-state strip detectors a certain level of parallel recording is possible without significant loss in resolution through focusing errors. With these detectors the resolution is defined by the strip dimension, which is the equivalent of the receiving slit.

All the configurations in figure 4.4.1 are discussed without reference to monochromation of the beam and therefore the resolution will be dominated by the wavelength dispersion, especially the $K\alpha_1$ and $K\alpha_2$ separation. In the reflection case, figure 4.4.1b and 4.4.1c a bent single crystal is often used as a monochromator, to isolate the $K\alpha_1$ characteristic line, which also captures some of the divergence of the X-ray source and brings it to a focus at S_0. Any imperfections in the curvature can be corrected with a slit at that position. In the transmission case, figure 4.4.1d and 4.4.1e, the beams are brought together at the detector, and the resolution relies heavily on the quality of the bend of the monochromator. Obviously if the sample is made smaller then the quality of the focusing is less stringent.

4.4.1.1. Enhancements to the focusing powder diffractometers

Two problems with the geometry described above come from unwanted scatter. The angular acceptance of the analysing optics can be improved with the addition of a second set of slits (anti-scatter slits). These can be set to only allow scattering from the direction defined by the illuminated region of the sample and the receiving slit. This improves the signal to noise ratio. These slits can be automatically controlled for maintaining the Bragg-Brentano geometry.

An advantage of the automatically controlled divergence and anti-scatter slits is that they maintain a constant area of illumination on the sample, section 3.5.2. This means that the divergence and the X-ray flux incident on the sample vary with the incident angle. If the intensities are to be related to those from a fixed slit arrangement (typical database values) then a correction factor is required. The slits will also open asymmetrically to maintain the illuminated area central to the diffractometer axis and require precision engineering for this to be achieved. This configuration does enhance the high angle intensities and suppress the low angle intensities, which compensates for the natural fall-off in the intensities with scattering angle. If α and β are the divergence angles for the beam below and above the central beam to the goniometer axis that makes an angle θ to the flat sample, then the number of crystals illuminated at a radius R is given by:

$$N \propto R\left\{\frac{\sin\alpha}{\sin(\theta+\alpha)}+\frac{\sin\beta}{\sin(\theta-\beta)}\right\} \qquad 4.4.1$$

To maintain constant illumination over an area of x will require α and β to be adjusted so that:

$$\alpha = \tan^{-1}\left\{\frac{R\sin\theta}{R\cos\theta - x/2}\right\} - \theta$$
$$\beta = \theta - \tan^{-1}\left\{\frac{R\sin\theta}{R\cos\theta + x/2}\right\} \qquad 4.4.2$$

The incident beam flux available will also vary with x, because it will be proportional to $\alpha+\beta$ whereas the flux associated with the fixed slit will be proportional to the selected divergence. To achieve a comparison this ratio should be taken into account.

Some unwanted scatter could be associated with the sample from fluorescent X-rays created by the incident beam. This is clearly a sample characteristic, some of which can be removed by the detector pulse height analysis, but fluorescent X-rays that are close in energy to the coherent scattered beam will need to be removed. For example, samples containing Fe studied with Cu $K\alpha$ radiation will result in a large background intensity. The only way to reduce this is to increase the energy resolution of the detection system and a very simple way to achieve this is with an analyser crystal, section 3.6.1. For the study of polycrystalline materials graphite with a controlled mosaic spread works very well, although a factor of 3 in intensity can be lost. This graphite diffracted beam monochromator also removes the $K\beta$ radiation. A typical graphite crystal will have a mosaic spread of ~0.5°. Ideally the monochromator crystal should fit on a radius including the analysing slit and detector window, however a flat crystal is acceptable because of the mosaic spread in these crystals.

4.4.1.2. Alignment of diffractometers based on focusing geometry

The alignment of these instruments is quite involved, because the beam from the source is divergent, making the determination of the zero difficult, and also the source, sample and receiving slit should be kept on the focusing circle. The most important reference is to ensure that the divergence slit is accurately set in line with the source and the centre of the diffractometer main axis. The divergence slit should be as narrow as possible and ideally the main axis reference should be a narrow slit. It is assumed that the line focus, main axis and detector strips are all parallel. The rationale in having a slit at the main axis reference is that any asymmetric setting, for example a knife edge, can create specular scatter and give a systematic offset in the beam position if the method relies on locating the half-intensity setting. Because of the very divergent incident beam the $2\theta = 0$ position can only be set approximately.

Probably the most reliable method for setting the sample stage is a mechanical method, by using a straight-edge from the incident beam slit and the $2\theta = 0$ position. These diffractometers benefit from high precision engineering to help with the alignment and maintain reproducibility.

4.4.1.3. Applications of diffractometers based on focusing geometry

The obvious application is in the study of powder samples, e.g. for identification of structural phases and their proportions. The identification relies on the comparison of the peak positions, and to a lesser extent on the intensities, with database information. The proportions of the phases is obtained by comparison of their relative intensities, which must be referenced to the calculated values, i.e. there are no absolute intensity values in the databases. The calculated values in terms of $|F_{hkl}|^2/V$, where V is the volume of the unit cell, will give the intensity per unit volume. The intensity associated with this phase will be the intensity per unit volume multiplied by its weight fraction divided by its density. The absorption path for each *hkl* reflection, i.e. each 2θ, is the same for these scanning focusing geometries (Bragg-Brentano configuration figure 4.4.1c and 4.4.1e) making the calculation of the weight fraction of the phases in a mixture rather straightforward. There are numerous texts describing various methodologies, from internal standards to create calibration curves to fitting based on modeling the full profile.

Several general crystallographic methods from structure determination, estimation of texture, stress, etc., have been performed on powder samples. There is a very extensive literature on these methods, however take note of the comments in the Epilogue and section 2.9.

This geometry can also be used for studying semiconductor multi-layer structures, especially if the kinematical approximation is valid. The diffraction profiles are dominated by wavelength dispersion (the sample intrinsic scattering range will be considerably narrower than this), therefore the alloy composition can be obtained as well as a fairly complete analysis of periodic multi-layers. It can also be used for low-resolution reciprocal space mapping. When this method is combined with position sensitive detectors (as in section 4.3.7) this data can be collected very rapidly. Examples of these are given in section 5.4.5.5.1 and section 5.7.2.1 respectively. This slit based system can also be used for reflectometry, which is discussed in section 4.5.1, and compared with other configurations that use passive components.

4.4.2. *Double slit incident and parallel plate collimator scattered beam diffractometers*

The discussion above touched on a few problems associated with single slits and controlling the view of the illuminated area, with the incident beam and the scattered beam slits. This is a limitation for studying scattering planes inclined to the sample surface, from anything other than highly perfect crystals. This can be partially overcome with the geometry illustrated in figure 4.4.2. The parallel plate collimator acts as a double slit assembly but extends over an area of the sample that can be very large. Hence for low angles of incidence the whole scattered beam can be captured, for example. The resolution of the incident beam optics is really quite easy to modify, however the manufacturing tolerance (width of blades and separations) make for less easily controlled scattered beam optics. The resolution of this combination is therefore rather poor but is much more even than the single slit geometry described for studying scattering planes inclined to the surface.

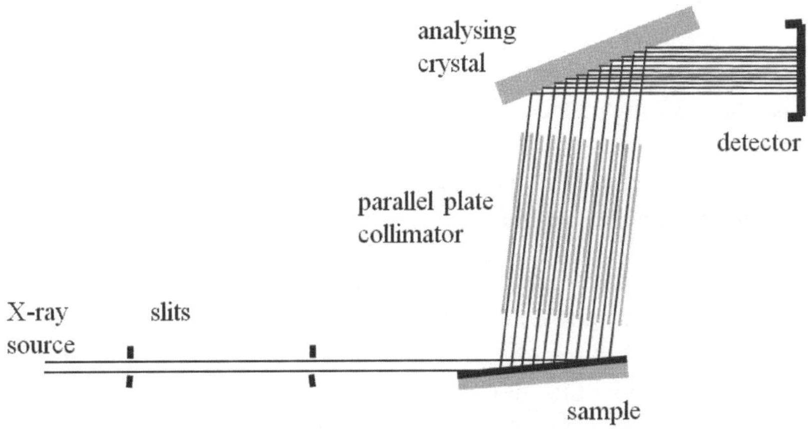

Figure 4.4.2. The use of a parallel plate collimator for capturing the scattering from large regions on a sample and still maintaining a reasonable resolution, after Parrish, Hart and Huang (1986).

This geometry does give enhanced signals from very thin polycrystalline layers by using incident angles of ~1.5° and just scanning the detector axis, 2θ. The advantage of low incidence angles is that the substrate contribution to the scattering is often reduced, although the crystallites could have preferred orientation that could alter the intensity distribution. This geometry is generally applicable to finding the positions of peaks and not the detailed shape. The alignment requirement for this combination is not very critical, because the peak positions are defined by the direction of the scattered beam, rather than the relative positions of the sample and analysing slits.

4.4.2.1. Enhanced double-slit incident and parallel-plate collimator scattered beam diffractometers

As with the arguments of unwanted scattering described above, the addition of an analysing crystal after the parallel-plate collimator can improve the signal to noise ratio and remove unwanted characteristic wavelengths, e.g. Cu $K\beta$. However because the better angular acceptance of this collimator compared with a single slit, the improvement is less marked. A graphite or lithium fluoride crystal placed after the parallel plate collimator can define the resolution, unless the divergence is smaller than the angular spread of the crystal mosaic blocks. If that is the case then the advantages depend very strongly on the application.

4.4.2.2. Alignment of the double-slit and parallel plate collimator scattered beam diffractometers

A significant advantage of this configuration is that the parallel plate collimator in the main detects an angle, and the resolution depends on the length of the collimator and plate separation as well as the beam divergence from the double-slit. If the poorer resolution of this instrument is acceptable then this can have advantages. The sample position is less critical compared with the focusing geometries, and makes it suitable for analyzing large components rather than powders.

4.4.2.3. The applications of the double-slit parallel plate collimator scattered beam diffractometers

This instrument was promoted as a thin-film configuration because it can be used at low incidence angles and the spread of the beam on the surface is not problematic, because the scattering is captured by angle. However as has been shown in the discussion on penetration depths, figure 2.10.2 the incident angle should be not more than a few degrees even for a 1 μm layer. If the polycrystalline layer has a rough surface then the reflectivity is lower and the penetration increases, refer to figure 2.10.3. So this assumption must be treated with care. However if the drop in resolution is acceptable compared with the focusing geometry, there are advantages in investigating the scattering from large objects because the sample height is less critical, and only the 2θ arm needs to be scanned. This geometry is also used for reflectometry measurements and this is discussed in section 4.5.1.

4.4.3. *Compact high-resolution powder diffractometers*

It was established in section 2.9.3 that the scattering contributions at the Bragg angle do not depend on any crystallite satisfying the Bragg condition. The consequence of this is that each crystallite can contribute to a large number of Bragg angles at any orientation. This suggests that far fewer crystallites are required to build a powder diffraction pattern than indicated by conventional theory, and therefore this should not be a restriction. This makes it possible to limit the X-ray beam on the sample and to probe the sample with a 'parallel' beam. By having the beam defining the centre of the scattering the resolution is not degraded by having a large sample.

The major difficulties associated with the focusing geometries described above are; complications in creating a monochromatic beam, especially in transmission, and the requirement to bend the sample to be able to collect the data in parallel. The following descriptions by-passes these difficulties and are based on the work of Fewster and Trout (2013). These bring together aspects of dynamical theory and the theory of Fewster (2014), section 2.9, to understand why and how it works.

4.4.3.1. The compact monochromatic transmission diffractometer

The first requirement is to create a narrow beam that is monochromatic, yet has a divergence greater than the scattering width of a typical crystallite at the Bragg angle. From the description of the 'static' diffractometer, section 4.3.6 and dynamical theory for scattering when the exit beam is close to grazing exit, section 2.4.3.1. The scattering from a single perfect crystal will result in a significant spread in 2θ for a very small movement in the incident angle ω. This suggests that the exit beam divergence can be increased by moving closer to grazing exit. The significant advantage of grazing exit is that the exit beam width is very small, since this is the projected area on the sample at a very small angle. The other significant advantage is that this geometry will produce a separation of the $K\alpha_1$ and $K\alpha_2$ wavelength components spatially and in angle, which makes it trivial to remove the $K\alpha_2$ component with a slit. This can be achieved if the beam on the crystal has significant divergence. Alternatively if a well collimated beam is used, e.g. with an X-ray mirror, then $K\alpha_1$ can be selected by the rotation of the crystal. With the appropriate inclined planes from a single crystal this can create a monochromatic beam with a dimension from a few microns up to tens of microns. The geometry of the monochromator is given in figure 4.4.3a, and the calculated beam size at the sample is given in figure 4.4.3b(i).

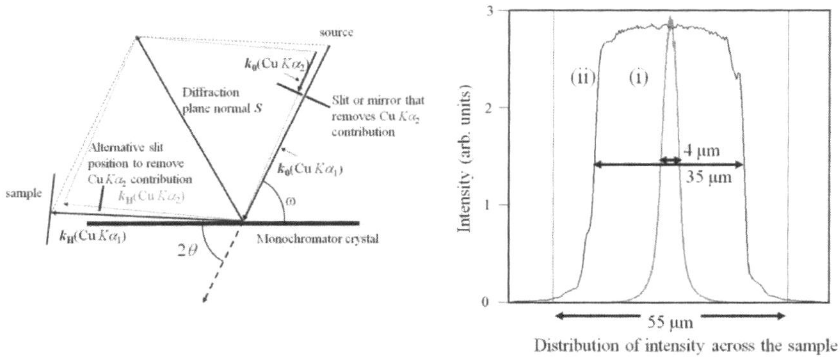

Figure 4.4.3. The figures illustrates the configuration of the monochromator crystal used in the compact diffractometer (a) and how it isolates the Cu $K\alpha_1$ beam with by limiting the divergence on it with a slit (this can be achieved with a slit or with an X-ray mirror). (b) shows the calculated profile of the resultant beam at the sample position for (i) without the mirror and (ii) with the X-ray mirror.

This geometry can be used with or without an X-ray mirror. The divergence of the incident beam on the monochromator crystal when using the mirror will be sufficiently small for the Cu $K\alpha_1$ component to be isolated by adjusting ω; e.g. for the *113* reflection of GaAs and Cu $K\alpha$. Without a mirror, the Cu $K\alpha_1$ component can be isolated with a slit before or after the monochromator crystal. When a mirror is used the intensity is increased by 10× and the beam is increased in size by 10×, as in figure 4.4.3b(ii), but in this case it is still less than the pixel size of the detector used in the following description.

The layout of the diffractometer, with and without the mirror, is given in figure 4.4.4. It is shown with an array of position-sensitive solid-state strip detectors, section 3.7.4, although a single detector is adequate. The scattered beams from the crystallites will be similar in divergence and size to the incident beam apart from the addition of spread due to diffraction broadening. The incident beam divergence on the sample for the configuration given here is ~0.01^0, whereas the intrinsic scattering width of a crystallite is ~0.0015^0 for a 5 μm diameter crystal. The size of the beam is comparable to the size of the crystallites. So this monochromator is well suited for powder diffraction. The sample in this case is captured on a strip of self-adhesive tape, and mounted on a holder that rocks or rotates. The vertical dimension of the beam is ~14 mm, matching the X-ray line source, monochromator and detector. A Soller slit is placed in front of the detector to limit the axial divergence.

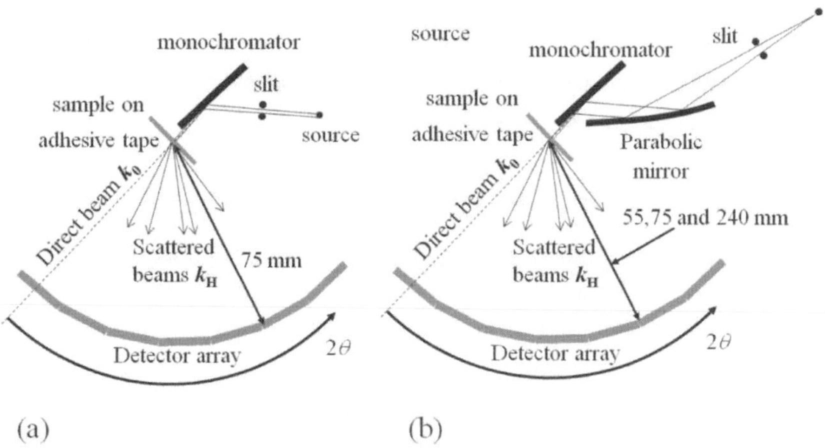

Figure 4.4.4. The geometry of the compact diffractometer (after Fewster and Trout, 2013) is shown for two modes, (a) without and (b) with an X-ray mirror. The data collection can be captured with a single movable detector as in the examples of this section, but could be collected in parallel with many detectors.

The strip detector used in the instrument described has a width of 55 μm and defines part of the resolution of the instrument. Also because the instrument can be modeled with the aberrations, the data can be corrected to give very precise and accurate peak positions. For example the sample mount that defines the sample position and where the incident beam impinges on the sample, may not be above the detector centre of rotation, similarly the detector is flat. All these factors can be corrected through calibration to give very reliable peak positions.

4.4.3.1. Alignment of the compact powder diffractometer

It is assumed that the line focus, monochromator tilt and detector strips are all aligned. The monochromator is rather insensitive to the source position since it will simply extract a suitable beam direction that scatters strongly. The $2\theta = 0$ position can be set with the direct beam, although this is very stable because the monochromator crystal defines the beam trajectories from the source that satisfy the $2\theta_{Bragg}$, and any movement x of the scattering point along the crystal will only displace the beam by $x\sin(2\theta_{Bragg}-\omega)$. The sample will be one or two crystallites thick, so the displacement error is small and just depends on the positioning of the tape, which is very reproducible.

4.4.3.2. Applications of the compact powder diffractometer

This instruments is used with a stationary detector (256 × 55 μm strips), at a radius of 75 mm, which will capture scattering over 10.75^0 in 2θ with a step size of 0.042^0. At 240 mm radius it will capture over a range of 3.36^0 in 2θ with a step size of 0.013^0. The data collection with a stationary detector can be observed during collection to decide when sufficient information exists in the pattern for analysis. Two examples are given in figure 4.4.5.

The profile in figure 4.4.5a was captured using the configuration that includes an X-ray mirror and with a radius of 240 mm; the sample is LaB_6 composed of 2-5μm crystallites. The data is collected close to the known diffraction peak positions and indicates the peak widths for this standard reference material with this configuration. The accurate position of these peaks and their broadening compared with that expected can give micro-structural information, for example distortions through the variations in d, i.e. the atomic plain spacings, and the approximate crystallite size through equation 2.7.10.

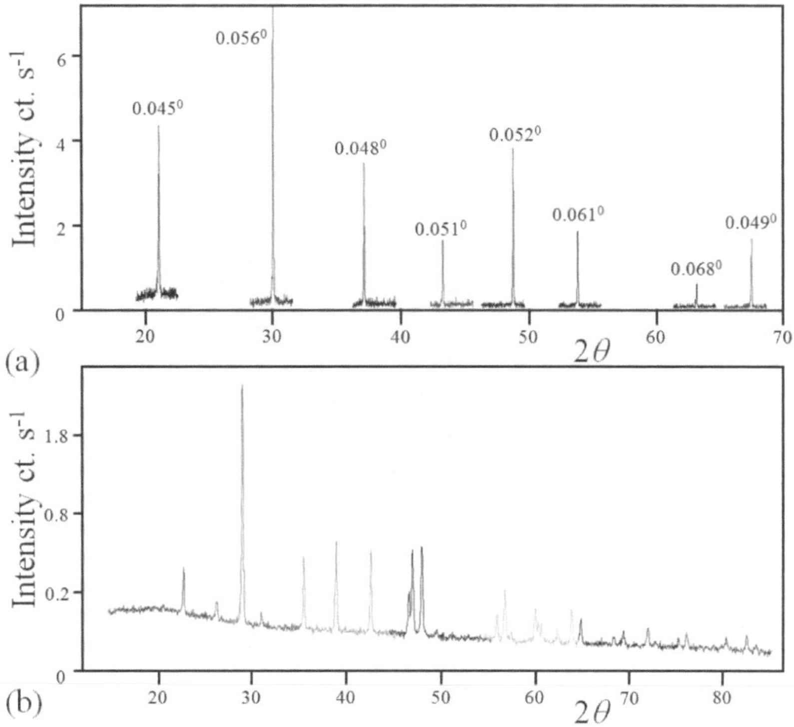

Figure 4.4.5. Two examples of the data collected with the compact diffractometer, (a) for a sample of LaB_6 with the configuration given in 4.4.4b with a radius of 240 mm, and (b) is the profile for a geological sample of several phases captured with the configuration in figure 4.4.4a and a radius of 75mm.

Figure 4.4.5b illustrates a profile obtained for the configuration without a mirror and with a radius of 75 mm given figure 4.4.4a. The sample in this case is a geological sample, mainly chalk, which is plotted on a square-root scale to enhance the weaker peaks. The positions of the peaks and their magnitude make it possible to determine several phases and their proportions. So despite the sample being very small, phase proportions down to below 1% can be detected and analysed. The advantage of high resolution, i.e. narrow profiles and monochromatic radiation, makes it much simpler to isolate peaks. For example those close to 60^0 2θ, would appear as a broad asymmetric peak at lower resolutions, yet this conveys interesting information.

4.4.4. *Some general comments on powder diffractometers*

The powder diffractometer configurations discussed above are far from exhaustive; even the multiple-crystal diffractometer can collect powder diffractograms. This high-resolution geometry has been used to obtain reciprocal space maps of the Debye-Scherrer rings in both 2-dimensions and 3-dimensions (Fewster and Andrew, 1999). This though, is a pretty specialized application but does indicate that defining an instrument as a powder diffractometer or a high-resolution diffractometer only suitable for perfect crystals can be very misleading. Of course not using one of the application labeled instruments may make interpretation more of a challenge, but the information will be there.

4.5. Slit based diffractometers for grazing incidence studies

Reflectometry (section 2.10), in-plane scattering (sections 2.11.2 and 2.11.4) and GISAXS (section 2.11.5) can be undertaken on most instruments provided the incident beam and diffracted beam dimensions are made small. In the latter case (GISAXS) this has been discussed in section 4.3.8.1, so the concentration will be on the instrumentation for reflectometry and in-plane scattering using slit based systems. High–resolution instrumentation for reflectometry and in-plane scattering is discussed with examples in chapter 5. The conversion of the standard Bragg-Brentano geometry and the double-slit parallel plate scattered beam diffractometer are covered in the next section.

In-plane scattering requires a different approach. Reflectometry is only concerned with the scattering in one plane, i.e. the plane of the source, sample surface normal and the detector, whereas in-plane scattering also requires good resolution approximately parallel to the surface. One of the perceived advantages of in-plane scattering is the sensitivity to depth penetration. This is quite difficult to achieve for the reasons illustrated in figure 2.11.9, however if the structure is a multi-layer then the reflectivity at each interface can contribute to creating a standing-wave. This standing-wave associated with the modulation of the electric-field, is very sensitive to the incident angle and therefore depth information can be extracted through modeling, section 2.11.4. The instrumentation for obtaining this data is also discussed.

4.5.1. *The single slit based systems for reflectometry*

Because of the low angle of incidence the projection of the sample length will be small, so if this is combined with a small source the divergence spread will be small and create a high resolution probe. The reflectometry profile is achieved by rotating the sample in ω from close to 0^0 up to several degrees. If the sample is bent with a curvature that exceeds the incident divergence then the features in the profile will be smeared. So the use of slit-based diffractometers for reflectometry will require some thought, which is discussed in the following sections. Making use of the high-resolution of the multiple-crystal diffractometer, section 4.3.3, will give a very good profile on an absolute scale for accurate density measurements, whereas these slit-based systems have the advantage of very high intensity. The effects of wavelength dispersion are less of a problem because this is a $\tan\theta$ effect and obviously θ is very small, equation 3.4.2.

The focusing configurations discussed in section 4.4.1 can be used for reflectometery, however because the $2\theta = 0$ position cannot be set precisely this would also have to be a variable in any fitting, making it difficult to obtain a precise density determination. Suppose that the incident beam can be narrowed down to give a divergence of 0.03125^0 ($1/32^0$) then the optimum receiving slits can be derived for the various geometries. For the specular scan $2\omega = 2\theta$ precisely, the angular acceptance varies very slowly with angle and all the intensity is collected with a 0.7 mm slit at the same distance, 320 mm, as the source from the sample centre.

For scans to investigate the diffuse scattering the 'instrument function' is not so simple. As we scan away from the condition for which

$2\omega = 2\theta$ then the resolution is changed. For $2\omega > 2\theta$ the divergence slit is too large and excess residual scattering can reach the detector. For $2\omega < 2\theta$ this analysing slit does not pass the whole of the expected scattered beam because the illuminated area has enlarged. The acceptance slit can be narrowed below the optimum without significantly changing the profile shape, except to reduce the diffuse scattering, but it will reduce the signal strength and modify the dynamic range to a small extent because of the varying instrument probe at small angles. To expand on the latter comment we have to recognise with this geometry the scattering angle is related to the position on the sample where the scattering occurs. A very large sample for example may create a similar angular distribution of specular beams as a small sample, but the beams are very well separated spatially and this defines the requirements for the slit. An alternative approach is to reduce the incident beam slit to a very small value and place a double slit at the acceptance end or just use a very small sample!

These approaches are methods for defining the area on the sample and there have been many ways to achieve this. Perhaps the simplest way of defining the beam on the sample is a knife-edge placed close to the sample. The illuminated area of the sample that can be seen by the scattered beam is then defined geometrically, figure 4.5.1. The intensity modification from the use of a knife-edge is approximately given by

$$I \propto 2\mathrm{d} \cos \omega \qquad 4.5.1$$

where d is the distance of the knife-edge to the surface. L is the length of the sample and assumes that the incident angle $\omega > \tan^{-1}(2\mathrm{d}/L)$, below this the intensity will go over the edge of the sample. Clearly if the knife-edge is close to the sample the intensity falls and the intensity is roughly constant over small values of ω. The intensity available for scattering falls by 0.4% out to $\omega = 5^0$ and is independent of d, unless the beam is allowed to pass over the end of the sample. To prevent this, the knife-edge should be brought closer to the surface.

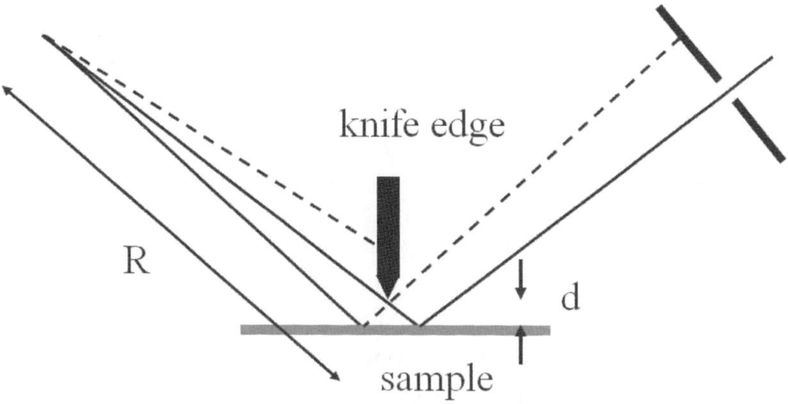

Figure 4.5.1. The principle behind defining the region on a sample with a knife-edge.

From figure 4.5.1, the effect of a knife-edge is that not only does it restrict the sampled area but reduces the divergence, which will vary with the incident angle. The incident intensity will be proportional to the incident beam divergence. This divergent beam combined with a sample with negligible diffuse scattering can produce good results. The limited area can be used to remove the effects of curvature and the detector slit should be set to select out the specular beam from the intensity distribution. If the acceptance slit is too wide then any diffuse scattering will contribute to the signal. It is better to use a narrow slit to collect reflectometry data and scanning with ω (=θ) and 2θ coupled together.

A similar result can be achieved with automatic divergence slits, but it must be remembered that although the area on the sample is maintained, making it less sensitive to curvature, the increase in divergence with angle will still require an acceptance slit as for the case with a knife-edge. Basically there are numerous options for obtaining a reflectometry profile, however divergence should be considered carefully if for example, the rate of decline of the intensity is to be interpreted in the presence of diffuse scattering. A combination of the approach using high-resolution optics, section 4.3.3, and high intensity slit systems discussed in this section can be a good compromise for material analysis.

All these methods have their merit and depend on the nature of the problem, the precision required and the quality of the sample.

4.5.2. *Double-pinhole and parallel plate collimator for in-plane scattering*

For general powder diffractometry the geometry given in figure 4.4.2 is used with a line focus and slits (i.e. considerable axial divergence is acceptable). Substituting the slits for a double-pinhole collimator or beam pipe capillary and using a point focus source will constrain the incident beam divergence sufficiently for in-plane scattering, figure 4.5.2. The procedure for setting up such an experiment is as follows.

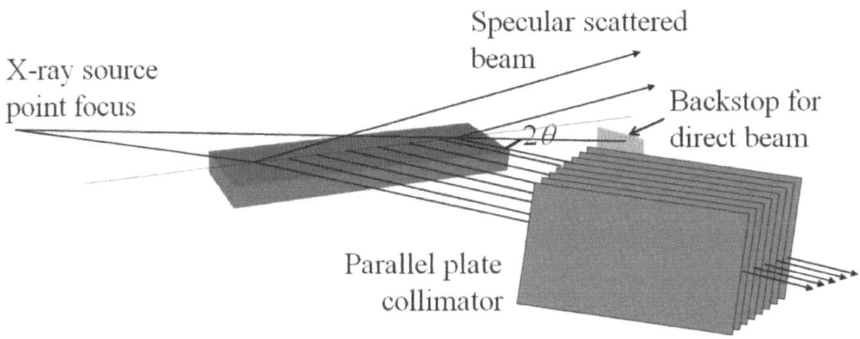

Figure 4.5.2. The application of the parallel plate collimator for analysing scattering from planes normal to the surface plane.

The choice of scattering plane, approximately normal to the surface should be decided first, so that the scattering plane normal approximately bisects the angle between the incident beam and the expected scattering beam direction. We will see later that a large tolerance is acceptable. The sample with its surface parallel to the normal scattering plane is moved to cut the incident beam in half. The sample should be rocked in α_1 figure 2.11.8, to ensure the sample surface is parallel to the incident beam direction, to an approximation this occurs when the maximum intensity is measured at the detector. Because some of the incident beam is specularly reflected the α_i angle will be biased making the sample height obtained to be over-estimated. This is a problem with this geometry that can be corrected for with a narrow horizontal slit at the detector to remove the specular component. Within the precision of the method these are small effects, so can be ignored.

The large detector window can accommodate scattering from the sample up to several degrees out of the normal plane of scattering. A beam-stop should now be placed in front of the detector to just isolate and remove the incident beam passing over the sample, figure 4.5.2. The sample can now be scanned about the axis α_i, and a reflectometry curve obtained. The resolution of this depends on the collimation, but the angle of inclination of the beam to the surface gives an indication of the depth from which the scattering is obtained, figure 2.11.9. The finite divergence of this incident beam gives a distribution of depths from which the scattering originates. However these angles are always small and generally just above the critical angle for the sample, where there is measurable specular and in-plane scattering. The X-ray beam incident above the critical angle will penetrate the sample and will scatter at angles above the critical angle. A reflection can be found in the normal way with an open detector (or parallel plate analysing slits if the scattered beam direction is known).

Suppose now we have found a reflection then it will be clear that the incident beam will be distributed over a large length of the sample, because of the low angle of incidence. The analyser acceptance should therefore be small in angle for resolution but accept the scattering from a large area. This configuration is ideal for the parallel plate collimator. If the initial incident angle in setting up the experiment is kept small, then there will be a small change in its value as the sample is rotated about the main goniometer axis, ω, and not about the surface normal, ϕ.

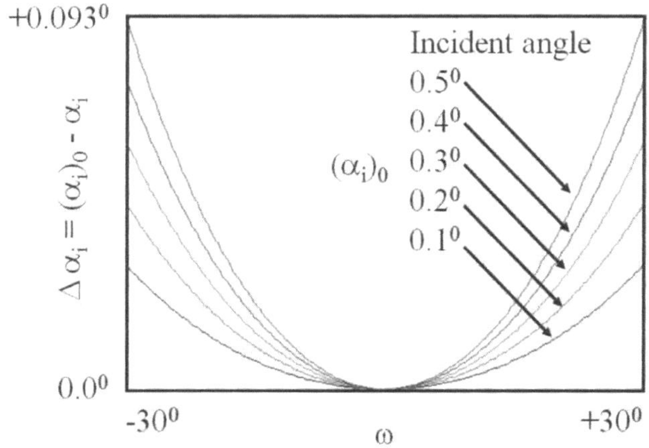

Figure 4.5.3. This shows the change in the incident angle with respect to the surface when the sample is not rotated about the surface normal. In effect the original α_i value $(\alpha_i)_0$ can be modified whilst rotating in ω.

This has been calculated for a large rotation in ω from the optimum position and for several initial incidence angles; the changes to the incident angle introduced are shown in figure 4.5.3. It is clear that for small incidence angles large angular ranges are perfectly acceptable and as the initial incident angle is increased the error is still quite acceptable when the beam divergence and uncertainties about sample flatness are considered. This change in incidence angle leads to a change in penetration depth, which is most sensitive at very small incidence angles. Therefore creating an extra axis to precisely rotate about the surface normal is unnecessary for this type of application and resolution.

This geometry can easily be used for examining peak positions for determining approximate in-plane lattice parameters or relative orientations between layers and substrates. An example of this application is given in section 5.6.3.3.

4.5.3. *Double parallel-plate diffractometer for in-plane scattering*

As we could see from the last section, structural information as a function of depth should be accessible, but this depends strongly on how well the incident beam divergence is defined, figure 2.11.9. If the structure is periodic or a multi-layer then this depth sensitivity can be enhanced, section 2.11.4, and the geometry used for that work is described here. The depth resolution relies on a low angle of incidence and the X-ray line focus being parallel to the sample surface. The scattering plane divergence and acceptance is limited by parallel plate collimators, figure 4.5.4. The incident beam divergence normal to the sample surface is determined simply by geometry from the source size (typically 40 μm) and a 40 μm slit 155 mm from the source to give a divergence <0.03^0, section 3.3.1. The combination of two parallel plate collimators will define the scattering angle resolution, section 3.3.3. A suitable combination is 0.18^0 for the incident beam divergence and 0.09^0 for the scattered beam acceptance. This will keep the instrumental broadening function to <0.1^0.

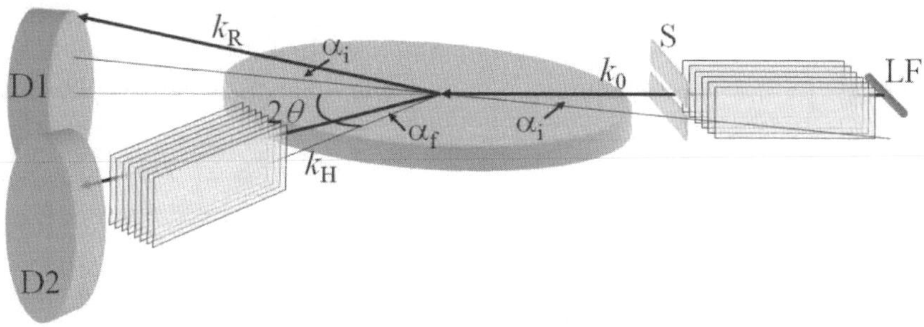

Figure 4.5.4. The configuration for in-plane scattering that was used for enhancing the depth sensitivity in a multi-layer structure (after Fewster, Andrew, Holý and Barmak (2005)).

In this configuration the detector is open and the intensity is integrated over scattering angle. From the discussion in section 2.11.1 and figure 2.11.3, this can be considered in reverse, such that a position sensitive detector can be used and a fixed incident beam. The angular acceptance on the parallel-plate collimators depends on the resolution required in 2θ. In this particular experiment, periodically-deposited polycrystalline metal layers, were studied and the sample was kept stationary at α_i for each 2θ scan. The incidence angle was changed and the scan repeated to build up a diffraction space map.

4.5.3.1. The alignment of the double-parallel plate collimator diffractometer

The instrument itself requires the slit and the focus to be aligned with the axis α_i, so that the centre of the scattering does not traverse too far during rotation. At low incidence angles this can be considerable, so to maintain a reliable intensity the beam foot-print on the sample should be within the capture dimensions of the parallel-plate collimator. The umbra from this focus and slit combination is ~5.7 mm for $\alpha_i = 0.4^0$; at 0.2^0 this dimension is doubled and close to the width of the parallel-plate collimator housing, therefore to achieve reliable intensity and to ensure that approximately the same area sampled is maintained, the beam should go through the axis as closely as is possible. At higher incident angles the beam extension shrinks and is less of a problem. Depending on the material being analysed the critical angle is ~0.2^0 to 0.4^0 when meaningful intensity starts to emerge. The sample rotation axis should be parallel to the X-ray source. An example of an analysis with this geometry is given in section 5.7.6.

4.6. General Conclusions

It should be clear from the discussions in this chapter that there are many possibilities for altering the instrument resolution to suit the problem to be solved. The discussion concerning the combination of components is far from complete, but an indication of the complexity and arguments for some arrangements have been given, which should give an indication of the possibilities. Very simply the laboratory-based instrument is now becoming a very versatile tool for the analysis of most materials with intensity to saturate a detector and a resolution comparable to the most perfect materials available. The following chapter will discuss various examples and therefore indicate the performance and the type of analyses possible with some of these combinations.

Optimising the experiment to achieve the required data is always an important consideration and one aspect that is relevant to the discussion above is the size of the instrument function or probe. To obtain a complete set of data whether it is in the form of a reciprocal space map or a scattering profile the angular step for each data point should just touch each other or overlap. This will ensure no unexpected variations to occur between data points. This is a particular problem with step scans when the diffractometer stops and measures the intensity at each position over a fixed time interval. This can be resolved more successfully by continuously scanning and counting at the same time, the accumulated intensity is then associated with the angular mid-position of the count time interval. Hence a given count time determines the speed of the scan axis rotation. This is the most appropriate data collection procedure for diffractometers with encoders on the final axes. Stepper motors with encoders mounted on the motor axis are more suited to the step scan mode of data collection. Having encoders on the final axis is the ideal situation to determine the angular positions of the axes, since it removes errors associated with slack gears and linkages. These are very important considerations for precision angular measurements. The two data collection strategies have been compared and no difference is observed provided that the step size does not exceed the probe dimensions.

Bibliography

Bartels, W J (1983) J Vac. Sci. Technol. B **1** 338
Bohlin, H (1920) Ann. Phys. {Leipzig} **61** 421–439
Bragg, W H (1921) Proc. Phys. Soc. Lond. **33** 222–224
Brentano, J C M (1946) J Appl. Phys. **17** 420–434
Compton, A H (1917) Phys. Rev. **10** 95
Fewster, P F (1985) J Appl. Cryst. **18** 334
Fewster, P F (1989) J Appl. Cryst. **22** 64
Fewster, P F (1996) Critical Reviews in Solid State and Materials Sciences, **22** pp. 69–110.
Fewster, P F (1999) J Mat. Sci.: Materials in Electronics **10** 175
Fewster, P F (2004) J Appl. Cryst. **37** 565–574
Fewster, P F (2005) J Appl. Cryst. **38** 62–68
Fewster, P F (2014) Acta Cryst A **70** 257–282
Fewster, P F and Andrew, N L (1999) Reciprocal space mapping and ultra-high resolution diffraction of polycrystalline materials: *Defect and Microstructure Analysis by Diffraction*, Ed: Snyder, Fiala and Bunge, IUCr series 10: pp 346–364. Oxford University Press.
Fewster, P F and Andrew, N L (1995) J Phys. D: Appl. Phys. **28** A97–103

Fewster, P F and Trout, D R D (2013) J Appl. Cryst. **46** 1626–1639.
Fewster, P F, Andrew, N L, Holý, V and Barmak, K (2005) Phys. Rev. B72 174105.
Hamilton, W C (1974) International Tables for Crystallography **IV** 280
Hart, M (1981) J Cryst. Growth **55** 409
Iida, A and Kohra, K (1979) Phys. Stat. Solidi A **51** 533
Jergel, M, Siffalovic, P, Végsö, K, Majková, E, Korytár, D, Záprazný, Z, Perlich, J, Ziberi, B, Cornejo, M and Vagovic, P (2013) J Appl. Cryst. **46** 1544–1550
Parrish, W, Hart, M and Huang, T C (1986) J Appl. Cryst. **19** 92–200
Seemann, H (1919) Ann. Phys. [Leipzig] **59** 455–464

CHAPTER 5

A PRACTICAL GUIDE TO THE ESTIMATION OF STRUCTURAL PARAMETERS AND INTERPRETATION OF SCATTERING PATTERNS

This chapter brings together the analysis of materials using X-ray scattering, it cannot and is not fully comprehensive, but hopefully some of the approaches and ideas can be translated onto specific solutions methods for the reader. To make it as general as possible it will cover the basic analyses of bulk (typically substrate) materials, leading through to characterising multi-layers that are assumed to be perfect, including periodic structures, to increasingly imperfect materials (mosaic structures), and finally to polycrystalline materials. The emphasis is on the measurement of composition, thickness, interface structure, strain and strain relaxation as well as mapping the homogeneity across a wafer. The material structure can determine the most appropriate instrumental configuration and this will be discussed as well as the precision and accuracy one might expect.

5.1. General considerations:

So far we have built an understanding of the structural properties of our sample, the scattering theories and how the instrument collects the data. In this chapter some examples of typical analyses will be given. This will not be comprehensive but should give examples of the methods that have been developed. Firstly we should reiterate some of the discussion so far.

The sample is defined in terms of several structural properties, i.e. shape, composition, crystallinity, etc., and these properties determine the appropriate experimental arrangement as well as the parameter of interest. It is also very important to have an understanding of the interaction between these parameters, when trying to extract a single parameter value, since other parameters from the one of interest may have a significant influence on the scattering. The type of experiment is therefore greatly influenced by the knowledge that we already have concerning the sample.

The scattering theory to choose will depend on the material properties and the precision required. In chapter 2 we covered a range of scattering theories, all of which have assumptions and must therefore only be used beyond these boundaries with care. However it is clear that we do have theoretical methods to cover from the most imperfect to the most perfect crystal samples. The depth of information that can be obtained is quite considerable.

In chapter 4 we discussed the instrumentation for analysing samples. It is crucial to recognise that we cannot isolate the influences of the sample and instrument without very good knowledge of one or the other. Basically if we wish to know detailed structural information about the sample then we require a very well defined and predictable X-ray probe. If we know considerable detail about the sample then we can use cruder experimental techniques. It is therefore important to emphasise that the instrument should be versatile so that it can be adapted to the problem of interest. We cannot simply define an instrument function and a sample response and form a correlation of the two influences. The reason for this is that the sample may not make full use of the X-ray probe instrument function. This can lead to profile widths being narrower than the instrument function, for example.

The conclusion of all this is that the diffraction experiment should be considered as a whole; X-ray source, incident beam conditioning, scattering from the sample and detection with scattered beam conditioning. Of course as we understand more about the sample problem to be tackled, the influence of other structural parameters and the performance of the appropriate instrument configuration we can simplify the analysis considerably. The following examples presented use a range of techniques all having their merits, some for speed, some for precision, etc., depending on the level of assumptions that are made. There are further examples on a whole range of material types discussed in Fewster (1996a).

5.2. General Principles:

From our understanding of reciprocal space we can see that we have a large range of scattering to analyse in three dimensions. Depending on the nature of the sample, some of this will be inaccessible. To limit any confusion I have used 'reciprocal space map' loosely; strictly many of the maps presented are plotted in angular coordinates and are therefore 'diffraction space maps', the differences are explained in section 4.2. The choice of wavelength and diffractometer geometry will also limit the accessibility, but in general the information will be over defined and much of the analyses can be carried out using a few reflections. As discussed in chapter 1, semiconductors that are manufactured will have details of their structure known but not to the precision necessary, it is therefore only the fine detail of exact composition, thickness and perfection that is often required. In this case the analysis could be restricted to the region close to a single reflection. If the structure is composed of several layers and some critical thickness is reached, so that the normally assumed strained state has relaxed, then we require more reflections to analyse this strain state to obtain compositions, etc. If the relaxation has progressed to the stage of breaking the sample into small mosaic blocks then these details require reciprocal space mapping to analyse their shape and so on.

Clearly with the analytical tools in X-ray methods we can produce an excellent understanding of our sample, by combining reciprocal space mapping, topography and simulation. For very complex structures often the number of parameters can be too large for a complete 'all in one analysis' and the dominant parameters need to be determined first. From this we can carry out further experiments to obtain a complete picture of our sample, by using several reflections (some are more sensitive to different properties). It is important to establish uniqueness as in any analysis and in general X-ray experiments use a very small fraction of the total available data and hence there are plenty of opportunities to quickly test the proposed model. Generally all the parameters are correlated in the scattering pattern and therefore the full simulation of the sample with the most exacting theory is one of the most rigorous tests we can conduct.

As mentioned above X-rays are very sensitive to deviations from perfection, which is a significant advantage, however this can make interpretation more difficult. It is the purpose of this chapter to help with this through examples. It should also become clear that the resolution of the probing instrument should be carefully matched to the problem and therefore there will be significant referral to the sections in Chapter 4. Once the understanding becomes more complete then there are many assumptions that we can make and the analysis can be exceedingly fast.

We shall start with examples that illustrate the use of direct interpretation and how this can also lead to errors. Various approaches to some of these measurements have been previously discussed in Fewster (1996a), it is more the purpose here to concentrate on the preferred method (simplest or most accurate, etc.), and describe the chosen experimental conditions.

5.3. Analysis of bulk semiconductor materials:

Bulk semiconductors are primarily used for substrate material or ion implantation and their quality is often the basis for good epitaxial growth. The important parameters associated with bulk material will therefore relate to the general "quality", the chemical composition, the absolute lattice parameter throughout the sample (hence the state of strain), the orientation of the crystallographic planes with respect to the surface and the surface quality.

5.3.1. *Orientation:*

Orientation is fundamental to all subsequent experimental techniques since this defines the region for more detailed analyses. Generally a wafer ready for epitaxy is orientated and specified by the supplier and therefore this is often assumed. However details of the directions associated with [*110*] and [*1-10*] for a [*001*] surface of GaAs, for example are rarely presented.

5.3.1.1. Surface orientation – the Laue method:

The Laue method is perhaps the simplest experiment in X-ray scattering. It consists of simply placing the sample in an X-ray beam and collecting the scattered beams on a photographic plate. The geometry of the back-reflection Laue method and a convenient precision camera is given in figure 5.3.1.

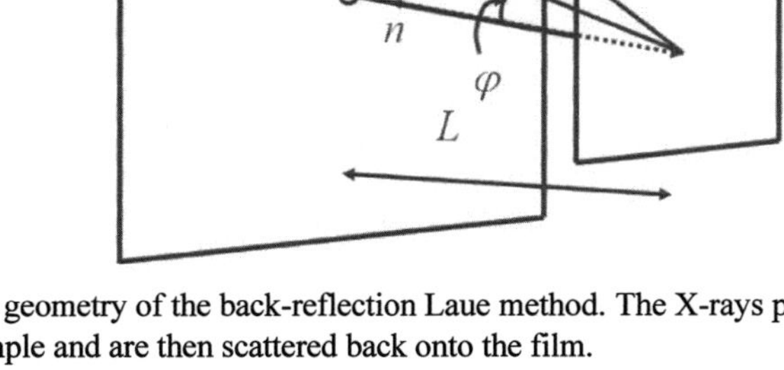

Figure 5.3.1. The geometry of the back-reflection Laue method. The X-rays pass through the film to the sample and are then scattered back onto the film.

Ideally the mechanical design should be such that removing and replacing of the sample holder maintains a precision for obtaining misorientations to within 0.05°. A typical scattering pattern is given in figure 5.3.2, for a GaAs wafer taken with a crystal to film distance of 3 cm. The symmetry of the pattern is at first sight clearly 4-fold and any distortion of this pattern such that the centre of symmetry is displaced from the centre of the pattern will relate to the misalignment of the crystallographic planes with respect to the surface. The symmetry of the pattern relates to the symmetry of the surface normal. Methods of interpretation for general orientations have been given by (Cullity, 1978, Fewster, 1984, Laugier and Filhol, 1983). These methods work well for high symmetry space groups and become progressively difficult with lower symmetries.

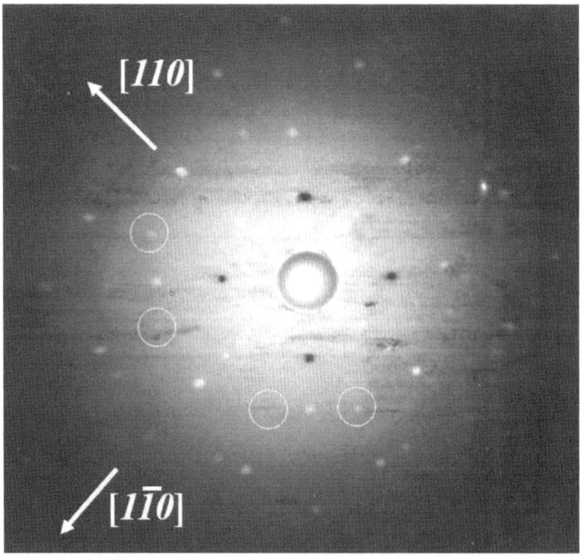

Figure 5.3.2. The Laue back-reflection image from a (*001*) GaAs sample. The symmetry of the pattern is close to 4-fold however the small intensity differences (circled, strong pairs indicate the [*110*] direction) show up the 4-fold inversion axis that has 2-fold symmetry. The black spots are a camera guide for the film centre.

The pattern in figure 5.3.2 is taken with Polaroid film with an exposure of 5 minutes using a 0.8 kW X-ray Cu source. The preferred

X-ray source anode for this work is tungsten since it gives a broad region of white radiation. The exciting voltage across the X-ray tube is limited to 20 kV to reduce the fluorescent yield that increases with higher energies, and the tube current is increased to boost the flux from the source. The principle of the technique is illustrated in figure 5.3.1. The full spectral distribution of the source is used and for a stationary sample each crystal plane orientation with respect to the surface plane, φ, and inter-planar spacing, d, will strongly scatter if it satisfies the following condition based on the Bragg equation

$$\varphi = \frac{\pi}{2} - \sin^{-1}\left\{\frac{\lambda}{2d}\right\} \qquad 5.3.1$$

Therefore with a large spread in wavelengths each inter-planar spacing accessible will scatter in a direction that is defined by the sample itself. However the intensity of the scattering is very complex and each maximum can have contributions from many wavelengths. We can determine the misorientation of anyone of these crystal planes from the surface normal by determining the length l in figure 5.3.1, and from simple geometry

$$\varphi = \frac{1}{2}\tan^{-1}\left\{\frac{l}{L}\right\} \qquad 5.3.2$$

Suppose now that we want to measure the misorientation of the surface with respect to the [*001*] direction in figure 5.3.2, then we have to use symmetry-related maxima to define the centre of the pattern. Just using the intersection of straight lines between these maxima is adequate for small displacements where the central maximum is hidden, although strictly they lie on hyperbolae.

The above approach will achieve accuracies of the order of 0.3° with care. To improve this accuracy considerably we can move the sample further away from the film (increase L, equation 5.3.2.), and also account for any geometrical alignment errors in the camera, Fewster (1984). Suppose that we take two photographs of the Laue pattern, the second after rotating the sample through 180°, then the pattern should rotate about its surface normal that should coincide with the camera centre if it is perfectly aligned. Take the measured centres of the two patterns as having co-ordinates (x_{C1}, y_{C1}) and (x_{C2}, y_{C2}) then the true length l is simply

$$l = \left[\left\{ \frac{(x_{C1} + x_{C2})}{2} \right\}^2 + \left\{ \frac{y_{C1} + y_{C2}}{2} \right\}^2 \right]^{1/2} \qquad 5.3.3$$

where x and y represent the distances horizontally and vertically from the perceived camera centre, C for the two exposures, 1 and 2. This also gives a method of aligning the camera.

The errors associated with this method are discussed in Fewster (1984) and for a 5cm crystal film distance these permit measurements to within 0.1° and at 10 cm it is possible to achieve measurements to within 0.05°.

5.3.1.2. Determining the orientation by diffractometry:

In this section we shall discuss two methods where the choice is determined by the instrumentation available. In the first method we require a well defined incident beam (e.g. 2-crystal 4-reflection monochromator, section 3.4.4, or a double pinhole slit arrangement, figure 3.3.2) and precision ϕ and ω axes (i.e. the ϕ axis should be normal to the sample mounting plate). The second method relies on multiple crystal methods (i.e. the multiple crystal diffractometer, section 4.3.3) and precision axes (both ω and 2θ). The principles of the two methods are quite different and the latter is considerably more precise than the former. Any method relies on defining the surface plane very precisely and relating this in some way to the crystallographic planes.

5.3.1.2.1. Monochromator and open detector method:

This is most suited to obtaining approximate values of misorientation. The sample is mounted so that its surface is accurately parallel to the mounting plate. The precision of this basically defines the precision achievable. The larger the sample wafer the smaller the error, also if the sample is not parallel sided then the sample should be mounted with its surface facing the surface mount.

A suitable reflection should be found with the detector in the wide-open position, figure 5.3.3a, at the Bragg condition $\theta = \theta_B$. The incident angle now corresponds to the Bragg angle (ignoring refractive index corrections) plus the crystal to surface plane angle. By rotating the ϕ axis the intensity will start to decrease unless the misorientation is zero. If the rotation is a full 180^0 then the angle through which ω must be rotated to recover this intensity is twice the angle between the sample surface and the scattering planes, φ.

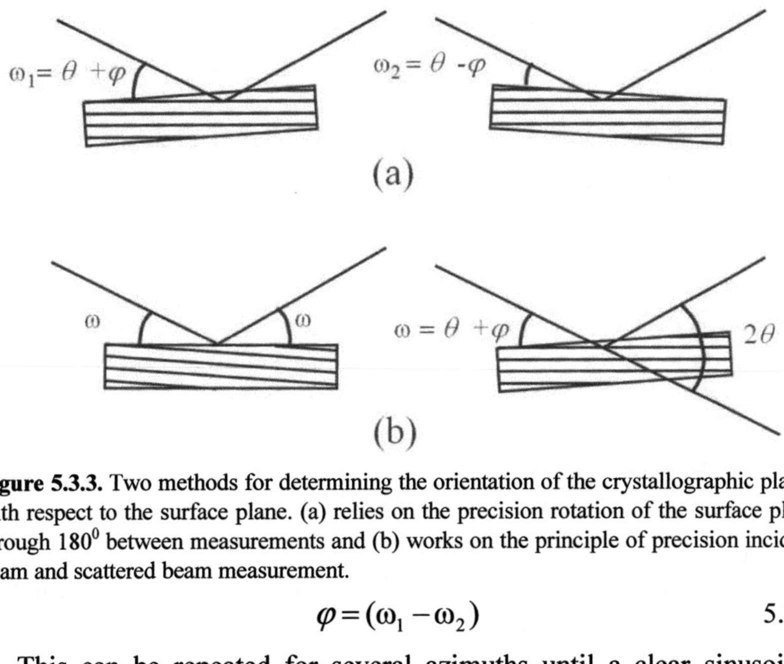

Figure 5.3.3. Two methods for determining the orientation of the crystallographic planes with respect to the surface plane. (a) relies on the precision rotation of the surface plane through 180^0 between measurements and (b) works on the principle of precision incident beam and scattered beam measurement.

$$\varphi = (\omega_1 - \omega_2) \qquad 5.3.4$$

This can be repeated for several azimuths until a clear sinusoidal variation is observed. The maximum in this oscillation is the largest tilt direction and the azimuth at which it occurs. If however the measurements are carried out at two orthogonal azimuths, i.e. rocking curves at 0^0 and 180^0 to obtain φ_0 and at 90^0 and 270^0 to obtain φ_{90}, then the maximum φ is given by

$$\varphi_{\max} = \tan^{-1}\left\{\sqrt{\tan^2 \varphi_0 + \tan^2 \varphi_{90}}\right\} \qquad 5.3.5$$

and this occurs at the azimuth given by

$$\phi_{\max} = \tan^{-1}\left\{\frac{\tan\varphi_{90}}{\tan\varphi_0}\right\} \tag{5.3.6}$$

5.3.1.2.2. Multiple crystal diffractometer method:

The difficulties in referencing the sample surface to sample mounting plate are removed with this approach. From section 2.10 we could see that at low incident angles the incident beam is specularly reflected and therefore the scattering angle, 2θ, is precisely twice the incident beam angle ω. This is achieved by placing the analyser assembly at some angle just above the critical angle and scanning ω until the maximum intensity is found. We now have ω and 2θ on an absolute scale. $2\theta = 0$ is precisely determined from scanning the analyser/detector through the direct beam.

We now find a suitable reflection and rotate the ω and 2θ axes until the scattering condition is maximised. Initially the reflection will be found with no analyser, but with a good precision goniometer moving between the two does not loose this reference. The ω angle can be determined from the peak position with the open detector and once set at this maximum in the intensity the analyser assembly can be invoked and scanned to determine the scattering angle, figure 5.3.3b. If $2\omega = 2\theta$ then the surface and crystallographic planes are parallel to within the accuracy of the instrument resolution, i.e. a few arc seconds. If $2\omega \neq 2\theta$ then the misorientation is given by

$$\varphi = (\omega - \theta) \tag{5.3.7}$$

The maximum tilt is now given by equation 5.3.5 and its azimuthal position by equation 5.3.6. The precision of this method is exceedingly high and only relies on the precision of the main ω and 2θ axes, and judgment of the peak positions that are defined by the intrinsic scattering width. The most precise determination of the peak position is achieved by performing a very small reciprocal space map around the peak with the minimum step size available.

5.3.1.3. Determining polar directions:

From Chapter 1 we could see that the zinc blende structure that is characteristic of GaAs, InP based materials, etc., are not centrosymmetric. This is also evident in the manner in which the bonding at the free surfaces of these materials reconstruct and this will clearly influence the epitaxial growth on these materials. Some of these influences may be minor for [*001*] orientations but for [*111*] orientations this can become very pronounced, where the bonding normal to the surface is either Ga to As or As to Ga for example.

To determine these details we have to consider more subtle influences on the scattering patterns. The measured intensity depends on the structure factor, equation 2.3.9, which is the phase sum of the scattering factors of the atoms, i.e. its value depends on the distribution of atoms in the unit cell. Equation 2.3.9 is reproduced here

$$F_{hkl} = \sum_{j=1}^{N} f_j e^{2\pi i(hx+ky+lz)} \qquad 5.3.8$$

Now if the scattering from each atom involves no absorption (i.e. f_j is assumed real), then the structure factor associated with the planes (*hkl*) and (*–h-k-l*) will be complex conjugates of each other. If we take the simple kinematical description of intensity ($I_{hkl}=F_{hkl}F^*_{hkl}$) the intensities will be identical. However f_j is not wholly real because the presence of absorption. The degree of absorption depends on how close the X-ray frequency matches the resonant frequency of one or more of the sample atoms. The imaginary contribution to the scattering factor (f'') will always retard the phase of the scattered wave by

$$\Phi = \tan^{-1}\left(\frac{f''}{f+f'}\right) \qquad 5.3.9$$

Since the contribution (i.e. phase retardation) is the same for F_{hkl} and F_{-h-k-l} they are no longer complex conjugates and the intensities of reflections from the (*hkl*) and the (*–h-k-l*) planes are no longer equal. We can use this difference to compare X-ray scattering from (*hkl*) and (*–h-k-l*) crystal planes.

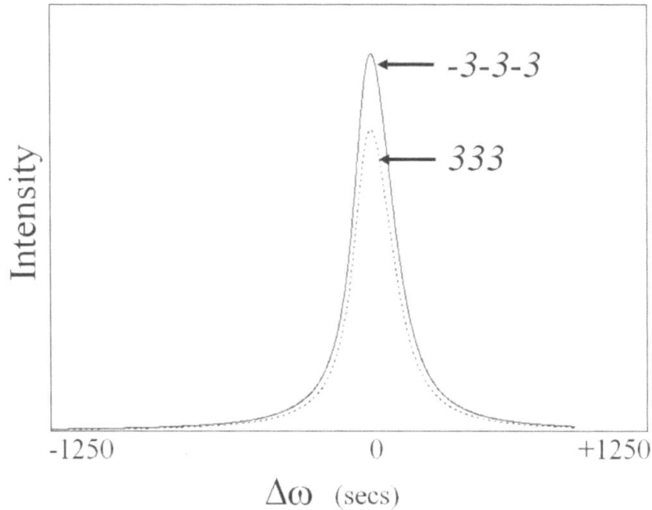

Figure 5.3.4. The difference in the scattering strength from (*333*) and (*–3-3-3*) crystallographic planes from CdTe with Cr $K\alpha$ radiation.

Figure 5.3.4 gives the scattering profiles and the integrated intensities from the *333* and *–3-3-3* reflections of CdTe using Cr $K\alpha$ radiation. The choice of wavelength here is to enhance the values of f' and f'' for both Cd and Te and emphasise the effect. We can see quite clearly that the intensity difference is very significant and easily measurable. Since we may wish to determine the surface to grow on for epitaxy (some material systems produce far better growth on some surfaces than others) we have to remount the wafer. When comparing opposite sides of the same wafer, the incident beam should be monitored and all the scattered intensity measured (in case the sample is mosaic or bent) or an insensitive reflection as an internal reference should be used. A suitable choice in this case is the *222* reflection, which has the same intensity as the *–2-2-2* reflection (Fewster and Whiffin 1983).

Another important polar direction that can influence the over growth of epitaxial layers is that of distinguishing directions in the surface plane, although the surface plane and its reverse are identical. The principal directions, i.e. {*110*} type, can be observed directly from the Laue pattern of figure 5.3.2, however [*110*] and [*1-10*] are not the same in non-centrosymmetric crystals. We indicated earlier that the Laue method is multi-wavelength and therefore we can make use of this by finding a reflection with an appropriate *d*-spacing that selects a wavelength between the Ga and As absorption edges for example, equation 5.3.1. Firstly select a wavelength range that enhances the effect and then find a range of surface to diffraction plane angles, φ, that can be observed (depending on the film size) and determine appropriate d values and hence *hkl* that will show this. For the (*001*) orientated GaAs example of figure 5.3.2 a reliable reflection is the *139* type, that occurs in eight positions in the pattern and scatters with a wavelength of 1.1183Å. The reflection with indices *139, -1-39, 319, -3-19* are all stronger than the –*139, 1-39, 3-19, -3,19* reflections. So once we have established the reflection to use any wafer can be quickly orientated in this way, (Fewster, 1991a; Schiller 1988).

5.3.2. *Revealing the mosaic structure in a bulk sample:*

As we could see from section 2.3.2 the intrinsic scattering profiles are very narrow for nearly perfect materials, and any regions that have different orientations from the average will scatter with different strengths for a fixed incident angle. We can consider a mosaic block as a region of perfect crystal surrounded by low angle boundaries. A mosaic block structure can be formed in a crystal from the convection currents close to the growing liquid-solid interface. By their very nature the blocks will be tilted and if they are very small they could create diffraction broadening effects. Mosaic crystals can therefore create dramatic changes to the scattering.

5.3.2.1. Mosaic samples with large tilts:

The first example we shall consider is a sample of $LiNbO_3$ that has large mosaic crystal blocks (few mm) with significant relative misorientations ($>0.05^0$). The simplest method for analysing samples similar to this is by low-resolution topography, Berg (1931), Barrett (1945). The principle of the method is illustrated in figure 5.3.5a. The divergent X-ray source relies on the different orientation of the sample scattering planes to select the appropriate incident beam direction. Since the scattering angle is the same for each block the scattered beam leaving the sample is rotated by the misorientation angle. Hence each image from each mosaic block is displaced, figure 5.3.5b. The degree of overlap or the size of the gap along with knowledge of the sample to film distance will give the angle of relative tilt. The resolution and nature of the geometry leads to the benefit that a measurement can be made along any direction in the plane of the topograph.

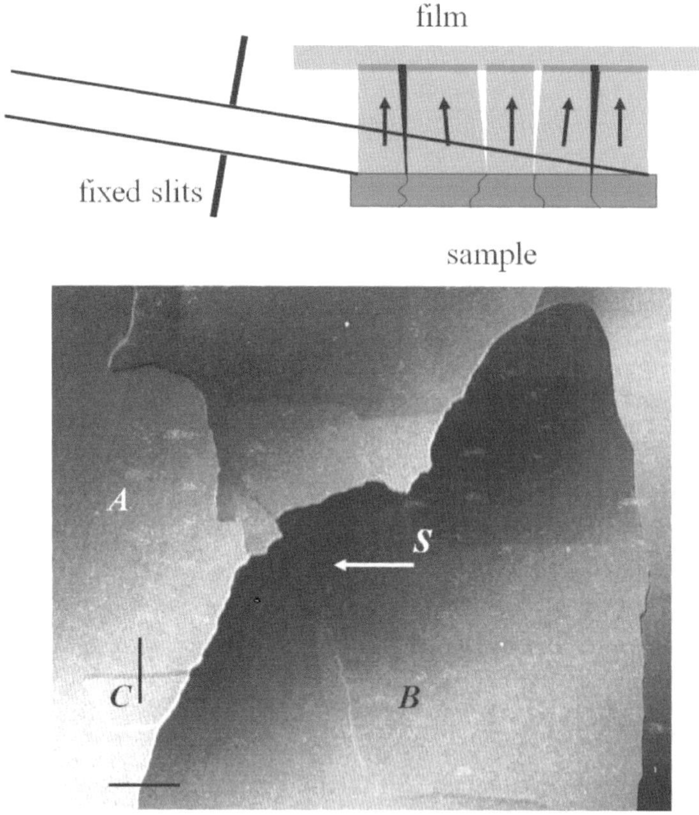

Figure 5.3.5. (a) The principle of the Berg-Barrett topographic method for analysing mosaic crystals. (b) The topograph of a $LiNbO_3$ crystal showing the black/ white contrast at the mosaic block boundaries. The misorientations measured were of the order of 10 minutes of arc.

The resolution is defined by the intrinsic scattering width of the sample since this method relies on the self-selection of the incident divergence. Hence, although this method is termed low-resolution, dislocations can be observed when the scattering widths of the perfect regions are very small, Fewster (1991a).

5.3.2.2. High resolution scanning methods (Lang method):

When the misorientation between the mosaic blocks is very small then the angular sensitivity has to be increased by reducing the angular divergence of the incident beam. To maintain the small divergence yet image large areas it is best to scan the sample back and forth through a restricted divergent beam. This method can be successfully used with a Lang camera in reflection mode using a narrow slit (~200μm) in the scattering plane and the dimensions of the sample in the plane normal to this. A distant X-ray focus is used to achieve a divergence of ~ 0.002^0 (~10"arc), figure 5.3.6a. This method is not without its problems associated with bent samples and long exposure times. The image will in general consist of the medium strength scatter associated with the mosaic blocks and more intense regions associated with the surrounding defects (these extract more intensity out of the divergent beam). However the differing contrast between the mosaic blocks will give an indication of orientations greater than the divergence of the incident beam.

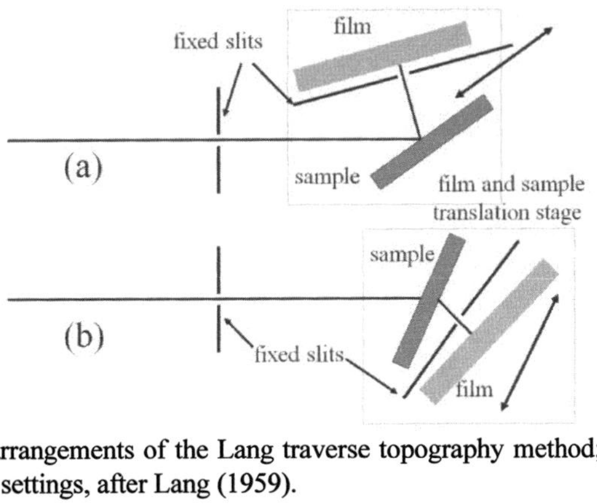

Figure 5.3.6. Two arrangements of the Lang traverse topography method; (a) reflection and (b) transmission settings, after Lang (1959).

Figure 5.3.7. A Lang transmission topograph of a 150 mm diameter Si sample taken with Mo $K\alpha$ radiation. The 700 μm substrate is composed of a heavily boron doped thick wafer with a more lightly doped boron surface layer. The presence of dislocations, (bottom and top right) and swirl (rings) are all evident in this image.

The depth of penetration of the incident beam close to the Bragg condition is severely limited by the scattering process and therefore these reflection techniques will only give information within the top 10μm or so. Transmission topography, figure 5.3.6b, gives a closer representation of a bulk substrate quality however the absorption would seem to be too great to obtain anything worthwhile. For silicon the absorption of Mo $K\alpha$ is not inhibiting and we can obtain reasonably clear images without difficulty. The value of μt for a 700 μm wafer for example is ~1. From section 2.6 we can see that the process described above dominates the contrast in the image, i.e. the defects extract more divergence out of the incident beam and create a more intense image than the perfect regions, figure 5.3.7. However GaAs is highly absorbing for both Cu $K\alpha$ and Mo $K\alpha$ radiations and therefore need to be thinned to ~40 μm to achieve a similar contrast. In section 2.6 we indicated that the absorption for each wave-field and polarisation differs and our notion of a general absorption value is not valid. We can therefore make use of this feature for imaging unthinned wafers where the signal is very weak, but certainly measurable. The sample is set and aligned in the normal way and only the single wave will emerge from the exit surface with a very small angular spread (i.e. the scattering peak is weak and sharp and therefore more difficult to align). The contrast is formed in a very different way since the transmission relies on the perfect regularity of the atoms in the sample. Any deviations from this perfect alignment will almost certainly mean the X-rays will be deviated and absorbed resulting in missing intensity. The contrast is reversed from that of the low μt samples, figure 5.3.8. These methods do reveal useful information but are rather slow and therefore we will consider a somewhat faster method.

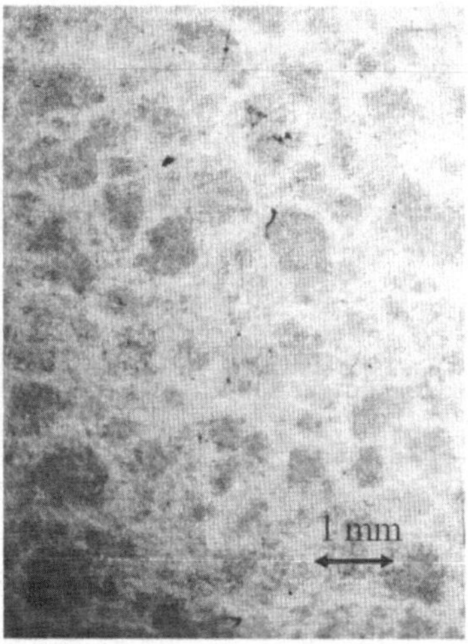

Figure 5.3.8. An anomalous transmission Lang topograph from GaAs with a $\mu t \sim 18$ for Mo $K\alpha$ showing the loss of intensity where there are a high number of defects. The GaAs is semi-insulating and full of mosaic blocks.

5.3.2.3. Multiple Crystal methods for revealing mosaic blocks:

We shall consider a semi-insulating GaAs wafer and examine the mosaic structure using reciprocal space mapping and topography, Fewster (1991b). If we start by undertaking a high resolution rocking curve (the incident beam is well defined, the detector is wide open and the incident angle on the sample varied, ω) then we obtain a profile that is very broad, many times that of the intrinsic calculated profile, figure 5.3.9.

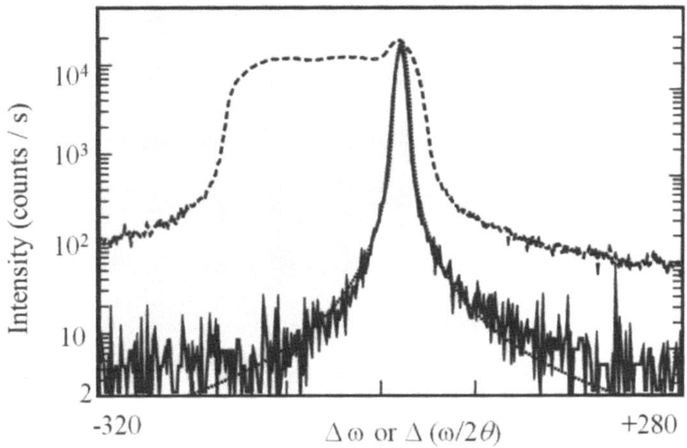

Figure 5.3.9. The scattering from a mosaic GaAs sample; dashed curve the profile with an open detector (ω scan), the continuous curve is that obtained from the multiple-crystal diffractometer (ω / 2θ) compared with that expect theoretically, dotted line.

At this stage we cannot make too many conclusions since we are mixing contributions of bend, mosaic block orientation, strain variations and the intrinsic scattering all into this one profile. The introduction of an analyser crystal, using the geometry of figure 4.3.5 allows us to scan along the plane normal direction that is sensitive to strain, figure 4.3.8b, immediately it becomes clear that this profile matches that expected from the intrinsic profile for a perfect crystal. The major contributing factor to the width of the open detector scan is therefore normal to this direction. If we now centre the probe on the peak of figure 5.3.9, using the configuration of figure 4.3.5, and scan in ω and keep the detector at the scattering angle 2θ (= $2\theta_B$) of the peak, then we can measure the orientation spread over the area of the beam on the sample, figure 5.3.10.

It is important to realise that any measure of angle is a projected angle onto the plane of the diffractometer, therefore we can only use this angle quantitatively if we assume the orientation distribution is isotropic and the sampled region is representative. These latter points are easily confirmed with experiments performed at different ϕ azimuths.

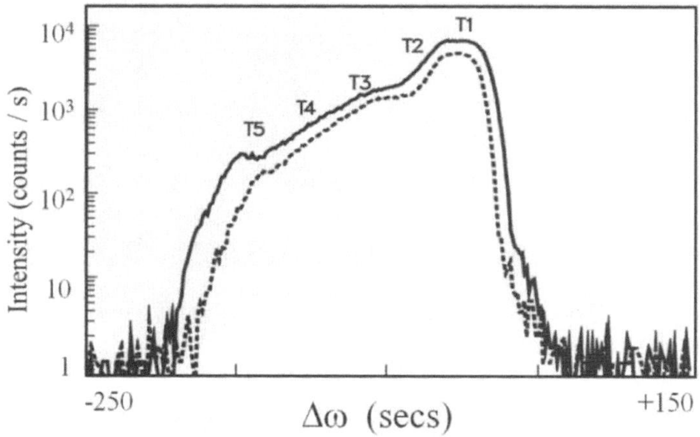

Figure 5.3.10. The rocking scan for a mosaic structure with a multiple-crystal diffractometer. The dashed curve is a repeat scan with a slit reducing the sampled area in the scattering plane, the width does not scale with this reduction and therefore is not due to curvature.

A more complete understanding will be obtained by collecting a high-resolution reciprocal space map with a multiple-crystal diffractometer, figure 4.3.5. The map is obtained by undertaking a series of scans by driving the 2θ and ω axes in a 2:1 ratio and then offsetting ω by a small amount and repeating, figure 5.3.11. The diffraction space map is transformed to a reciprocal space map through equation 4.2.1 and contains all the information of the three scans, figure 5.3.9 and 5.3.10. With this map it is now possible to see whether there are strain variations associated with different orientations. It is again important to recognise that a map is a projection of a small section of the Ewald sphere onto the plane of the diffractometer. Therefore any mosaic blocks with scattering planes inclined to the main axis of the diffractometer will give an error in its ω and 2θ position and therefore there will exist an uncertainty in the value of the derived strain. The extent of the error introduced is covered in the sections 4.2.2 and 5.3.4 on lattice parameter determination.

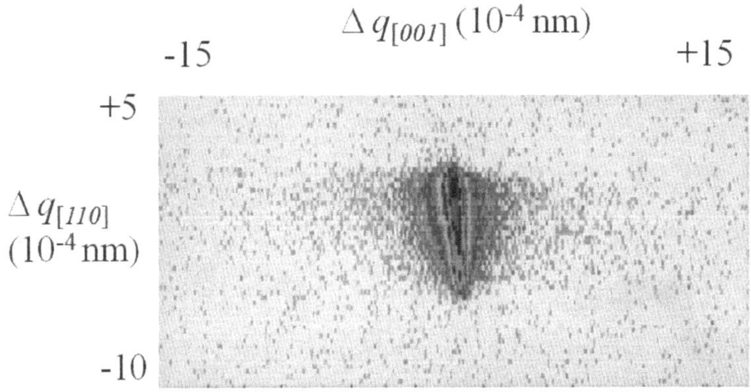

Figure 5.3.11. The reciprocal space map of the mosaic structure described in figures 5.3.9 and 5.3.10.

To overcome all these projection errors Fewster and Andrew (1995a) have introduced the concept of three-dimensional reciprocal space mapping. In this procedure the axial divergence is heavily reduced at the monochromator and analyser, which reduces the intensity, but the diffraction space probe is contained close to the plane of the diffractometer. The projection effects are severely reduced and the data is collected as a series of reciprocal space maps at different tilt values, section 4.2.2. Hence only scattering that is contained in the bounds of the diffraction space probe can be accepted. This is still at very high angular space resolution, where the beam cannot deviate more than ~0.2° out of the plane of the diffractometer. The distribution of intensity will then appear as in figure 5.3.12 for a mosaic GaAs substrate with an AlGaAs layer on top, from which the true strain associated with each mosaic block can be determined. This method has also been used for analysis of polycrystalline (figure 5.7.9) and single crystal enzyme structures, figure 5.3.13, Fewster (1996b).

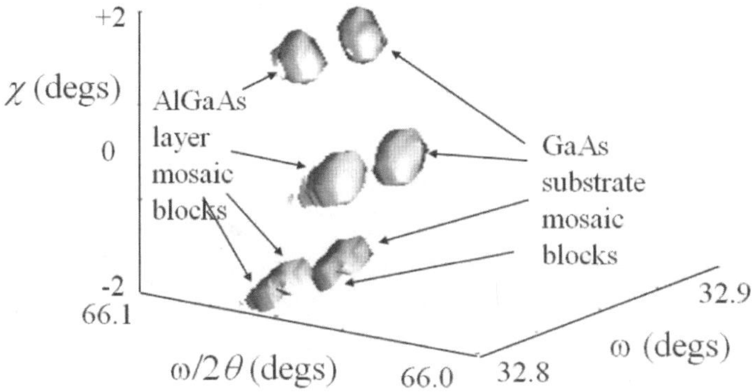

Figure 5.3.12. A three-dimensional reciprocal space map showing the orientation relationship of an AlGaAs layer to that of the GaAs substrate. Note also how the shape is mimicked from substrate to layer.

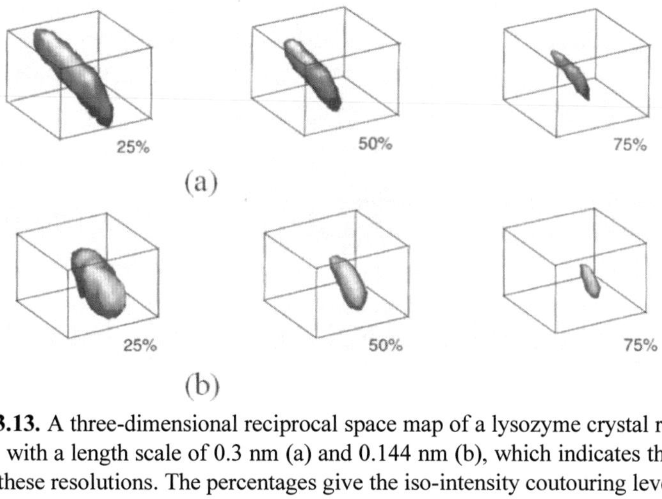

Figure 5.3.13. A three-dimensional reciprocal space map of a lysozyme crystal reflection associated with a length scale of 0.3 nm (a) and 0.144 nm (b), which indicates the crystal quality at these resolutions. The percentages give the iso-intensity coutouring levels.

Of course these measurements give some quantitative values of the degree of strain and the orientation spread, etc., however it does not reveal the dimension of the mosaic blocks. For this particular sample the mosaic blocks are large 0.7 mm in diameter and they can be separately imaged by topography. In general the probing beam although very small in diffraction space is large in real space, approximately 8 mm × 1 mm. Therefore the data collected so far is a contribution from a large region on the sample. However within the probed area we have regions of varying contribution.

A fine-grain photographic emulsion is placed in the scattered beam from the sample, it is possible to observe contrast on a scale comparable to the developed grain size of a few microns. Individual images can be observed if their relative orientation is greater than the divergence of the incident beam. The spatial resolution of this method differs in the axial and scattering planes. The resolution in the scattering plane is defined by the divergence of the incident beam and the wavelength spread, see section 4.2.2, which have a negligible influence on the spatial spreading of the image. The scattering plane resolution is therefore limited by the developed grain size of the emulsion. The axial resolution is influenced by the wavelength spread along this direction and the axial divergence passed by the monochromator as well as the overall geometry. The wavelength spread is small but this effectively allows the beam from the point source to pass and the spatial resolution is dominated by the geometry. The resolution is therefore simply given by

$$R_{axial} = \frac{film\_to\_sample\_length}{source\_to\_sample\_length}(source\_height) \quad 5.3.10$$

In practice it is difficult to place the film closer than about 10 mm from the sample and using a typical source to sample distance of 300 mm and point focus (0.4 mm high), the axial resolution is about 13 μm. This compares with 1 μm in the scattering plane. The shape of the defect images will be smeared to this extend by the instrumental resolution.

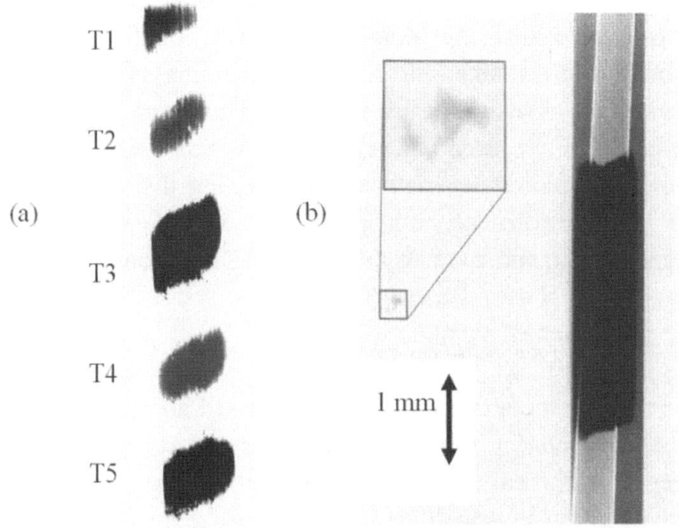

Figure 5.3.14. (a) X-ray reflection topographs at positions denoted by figure 5.3.10 of a GaAs mosaic crystal. Defects at the edge of each block are just visible and a topograph taken whilst rocking the sample creates a continuous image. The topographic emulsion was placed immediately after the sample. (b) a topograph from a lysozyme crystal giving more information on the mosaicity in this 'single' crystal, with the capillary and beam size imaged on the right.

Take figure 5.3.10, where we know that each data point comes from an object that satisfies this orientation and strain value. We have assumed that the strain broadening is small, figure 5.3.9 and therefore placing a photographic emulsion immediately after the sample, T_1 in figure 4.3.5 will create an extended probe (described in figure 4.3.8a as an open detector probe). However because of the nature of the shape of the reciprocal lattice point, figure 5.3.11, the dominant scattering image will relate to the orientation differences. Hence a series of topographic images at various rotations in ω will image the contributing feature. In this case the mosaic blocks are large and do not create diffraction broadening along ω, but each image is of a mosaic block surrounded by defects accommodating the small tilt boundary. In this case the relative tilts are greater than the divergence of the incident beam (~5"arc) and therefore each block can be imaged separately, figure 5.3.14a. Because each image is obtained after sliding the film cassette perpendicular to the ω axis the relative positions of the blocks can be related to the sample position (provided that the axial plane tilt is small). This type of topograph can be obtained at the peak maxima and therefore the exposures can be very short (a few minutes). The image in figure 5.4.14b indicates how this approach can reveal information about the quality of crystals of large biological molecules. Revealing dislocations in these materials is more difficult, but perfectly possible with synchrotron sources, collimation and geometry, Stojanoff and Siddons (1996).

5.3.3. *Characterising the surface quality:*

It is clear that the topography methods discussed above rely on relative contrast of the scattering from perfect regions and defects. A significant advantage of reciprocal space mapping with a very small scattering probe is that we can separate the scattering close to the maximum from a nearly perfect sample into distinct features: Bragg scattering, surface truncation rod or dynamical streak and diffuse, figure 5.3.15. The Bragg scattering comes from the bulk of the sample and is effectively a point in reciprocal space, i.e. the Fourier transform of an infinitely large three-dimensional object. The surface truncation rod is related to the surface, i.e. this is similar to a Fourier transform of an extended two-dimensional plane. The diffuse scattering on the other hand comes from the defects existing in the material, the shape of the scattering could be related to the defect type; however there can be many contributors to this. There will be 'additional' scatter associated with diffraction effects from the new theory, section 2.9 (see Epilogue). From the topography methods above, the images are collected that will include all these contributions. Suppose now we centre our small diffraction space probe on the diffuse scattering and form an image of the scattering over the probed area using a photographic emulsion immediately after the analyser crystal, T_2 in figure 4.3.5. With this arrangement we are obtaining an image of just the defects, with no Bragg scatter but some influence from the additional scatter.

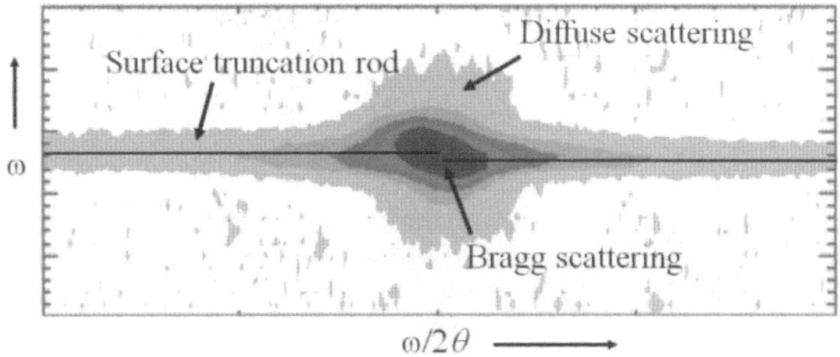

Figure 5.3.15. The main features of the scattering from a nearly perfect bulk crystal (*004* reflection of GaAs).

The defects surrounding mosaic blocks and from surface scratches can be imaged rather easily in this way. Different regions of the diffuse scattering can be associated with different defects, Fewster and Andrew (1993a). By placing the probe very close to the Bragg peak the intensity can be enhanced and most of the defects will contribute to the image, figure 5.3.16. The exposure times can be as short as ~1 h (using a 2 kW Cu $K\alpha$ point focus and L4 nuclear emulsion plates) depending on the number of defects contributing to the image. If few defects are present then the intensity recorded at the detector will be concentrated into small grains and the exposure is shorter, whereas distributed defects will lengthen the exposure time for a given detected count-rate.

Occasionally the surface streak can exhibit some rather strange wiggles. These wiggles can be easily observed with high resolution diffractometry and relate to interplay between the surface and the diffraction planes. This can sometimes be most obvious with a reciprocal space map (small undulations) or a multiple crystal scan normal to the surface plane (the intensity either side of the Bragg peak has a different variation), figure 5.3.15. This variation can be modelled by assuming the surface is undulating in a rather asymmetric form, figure 5.3.17. This example should indicate that the surface can be characterised at all levels and with a fairly complete theoretical model we can reconstruct the surface.

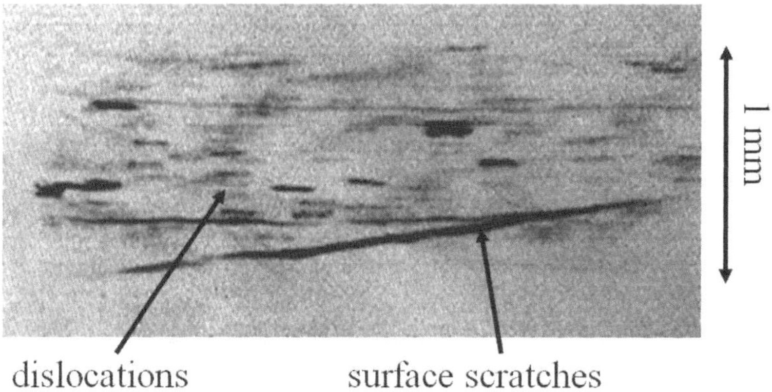

Figure 5.3.16. A diffuse scattering topograph of a bulk sample crystal indicating surface damage and dislocations. The exposure time was 4 hours.

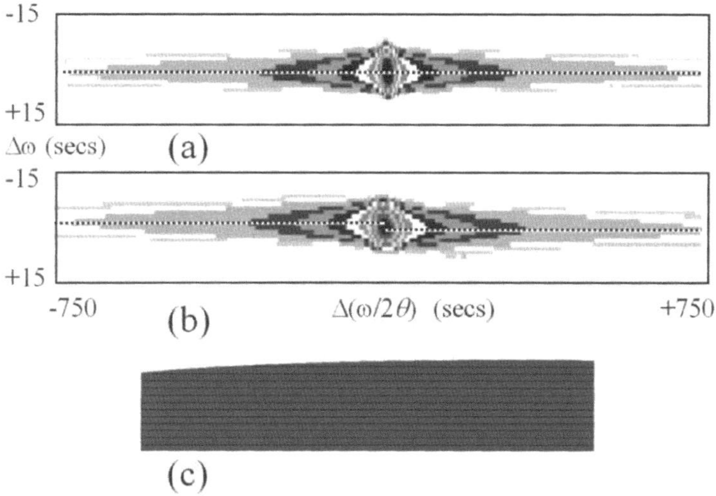

Figure 5.3.17. The simulation of the scattering from a perfectly flat sample (a) and that from a sample with a gradual small undulating surface (b). In this simulation 30% of the surface is perfectly flat with a gradual undulation leading to a maximum of 2.8^0 misorientation over 2% of the surface (c). Experimentally this effect is evident in figure 5.3.15.

5.3.4. *Measuring the absolute interatomic spacing in semiconductor materials:*

So far we have concentrated on the general aspects of quality, but have not discussed aspects of strain within the sample. The strain variation can be determined over the scattering volume with an $\omega/2\theta$ scan using the multiple crystal diffractometer as in the previous example. The $\omega/2\theta$ scan profile conveys information normal to the crystal diffraction planes. The strain variation will then be quantified by comparison with the calculated profile assuming the material to be perfect. However the absolute interplanar spacing in a sample is sometimes very important to know since this can form an internal reference for composition analysis or for determining the true state of layer strain or lattice parameter relaxation.

The assumptions and requirements of any measurement depend on the quality of the sample and the standard to which it is related. We will assume that the wavelength of the peak of the Cu $K\alpha$ spectral line is known, or is at least a transportable standard. Since this depends on a quantised transition in a Cu anode of an X-ray tube this is essentially invariant, although we may argue on the relative value to that of the standard of length defined internationally. Standard materials are very difficult to achieve due to differences in processing, surface preparation, mounting strains, etc. Another requirement is the homogeneity of the sample if the method relies on two measurements as in the method proposed by Bond (1960). A single measurement is therefore ideal. The method described here is that of Fewster and Andrew (1995b).

The principle is simple in that the incident beam and accepted scattered beam directions can be defined very precisely using the multiple crystal diffractometer, figure 4.3.5. The incident beam direction is precisely defined by scanning the analyser / detector assembly through the path of the beam from the monochromator. The width of this profile is very narrow, figure 4.3.6. The centre of this peak defines the condition $2\theta = 0$. Scattering from the sample will now be on an absolute angular scale, provided it is confined to the plane normal to the analyser / detector axis, and that the integrity of the angular movement is precise. The sample does not have to be accurately centred on the diffractometer nor does the incident beam have to pass through the centre of the goniometer axis. Although in general this makes the whole process easier and prevents the beam from being displaced spatially. This reduces the main errors that are very difficult to compensate for or know with certainty, figure 5.3.18.

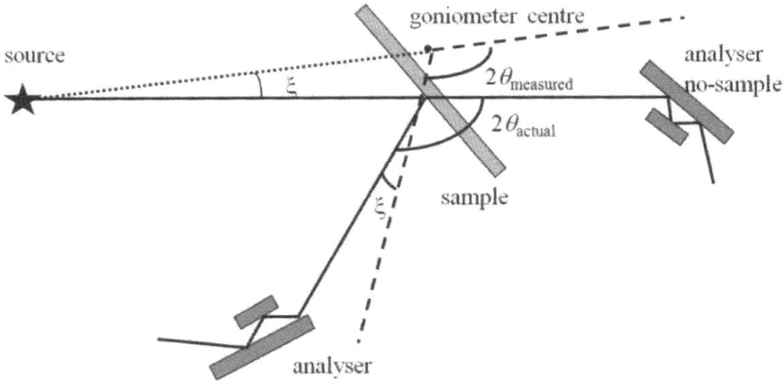

Figure 5.3.18. The principle of the method for determining the interplanar spacing with a single measurement. The zero scattering angle is determined from the direct beam, the X-rays do not necessarily have to pass through the goniometer centre, nor does the sample have to be mounted precisely to achieve 1ppm precision. Dashed line is the assumed beam path.

The only alignment necessary is to ensure that the scattering planes are parallel to the main axes, section 4.3.3.1. The most precise alignment procedure used is to initially rotate the sample in ϕ so that the plane of interest, with an angle of φ to the surface normal is maximised in the plane of the diffractometer. The sample should now be aligned as given in section 5.3.1.2.2 to place ω on an absolute scale. As an approximation the sample can be brought into the beam so that it cuts the incident beam in half. The sample is scanned in ω to ensure the surface is parallel to the beam and the maximum intensity is close to the ω = 0 position. The sample and detector should then be driven to their respective positions to capture the scattering of interest. Then for a given ω value, where there is intensity, the χ axis should be scanned and the aligned position corresponds to the midpoint of the profile. This procedure is done without the analyser crystal and just an open detector, figure 4.3.1b. If the profile is double peaked, this is generally the case unless the alignment is very close, then the intensity can be recovered by driving the ω axis to a lower value from this midpoint, figure 4.3.9b. The alignment is much more reliable when the axial divergence is restricted as much as possible with a slit; this will decrease the spread in acceptable tilt values and sharpen the profile, as well as removing additional contributions from differently tilted regions in imperfect samples. The intensity is then maximized by scanning in ω.

The analyser should now be inserted and apart from driving the instrument axes there should be no touching of this unit, so that the integrity of the measurement is maintained. The 2θ axis is scanned to find the maximum intensity value, which gives the equivalent of a cross hair on the reciprocal space map. To determine the 2θ angle precisely a scan along the scattering vector should be performed or a reciprocal space map collected along $2\theta/\omega$. As before the most precise method of all is to perform a three-dimensional reciprocal space map, when the alignment of the tilt occurs naturally. The corresponding scattering angle 2θ can then be converted to an interplanar spacing, d_{hkl} with equation 2.2.11 after account has been made of the refractive index. The refractive index can be determined by simulation (comparing the Bragg angle and the actual peak position to account for absorption effects) or through this formula:

$$\Delta\{2\theta_B\} = (n-1)\{\cot(\theta_B - \varphi) + \cot(\theta_B + \varphi) + \tan\theta_B\} \quad 5.3.11$$

where n is the refractive index given by equation 2.10.1. θ_B is the Bragg angle and φ is the angle between the scattering plane and the surface plane. Since the angle $2\theta_B$, can be measured to very high accuracy with the multiple crystal diffractometer, the lattice parameter can be determined quite easily to the part per million level for high Bragg angle reflections. This is sufficient to detect deviations from cubic space groups and detect surface damage, Fewster and Andrew (1998).

The advantage of this method is that it can be used to place the reciprocal space maps on an absolute scale and determine the lattice parameters of mosaic blocks.

5.3.5. *Measuring the curvature of crystalline and non-crystalline substrates:*

The curvature of a wafer can be important to establish the state of strain and hence the stress created by the deposition of a thin film. The principle of the method is to establish how the incident angle must be varied to maintain the scattering condition from different regions of the sample. The difference in incident beam angle from two different regions illuminated by X-rays will give the radius of curvature according to

$$R = \frac{x_1 - x_2}{\omega_1 - \omega_2} \qquad 5.3.12$$

The angular separation can be determined to very high precision and if the separation between the two measurement points is reasonable the accuracy of this measurement can be very high. There are a few important points to consider in this measurement. If the sample is translated then the movement has to be sufficiently good to prevent any twisting that will cause a rotation in ω. This can be the major cause of error. The positions on the sample is best defined with a knife-edge, this then selects the position and size of the region depending on how close the knife-edge is to the sample, figure 5.3.19. Since we are only concerned with the change in the incident beam angle the most suitable configuration will use the double-crystal diffractometer to create a parallel incident beam, figure 4.3.1b.

For nearly perfect crystals the angular shift of two peak positions measured at a relatively high scattering angle reflection will yield a reliable result. The sampled region should be as small as possible to reduce peak broadening from the curvature and more precisely define the x position. Precise determinations can be achieved with the arrangement given in figure 5.3.20. The principle of this method is to translate the knife-edge instead of the sample to overcome the uncertainties in the sample translation. This does rely on a large parallel source which can restrict the separation, $x_1 - x_2$. The choice of reflection and the range of $x_1 - x_2$ may limit the accuracy because of the finite beam size. Clearly a lower order reflection will increase $x_1 - x_2$.

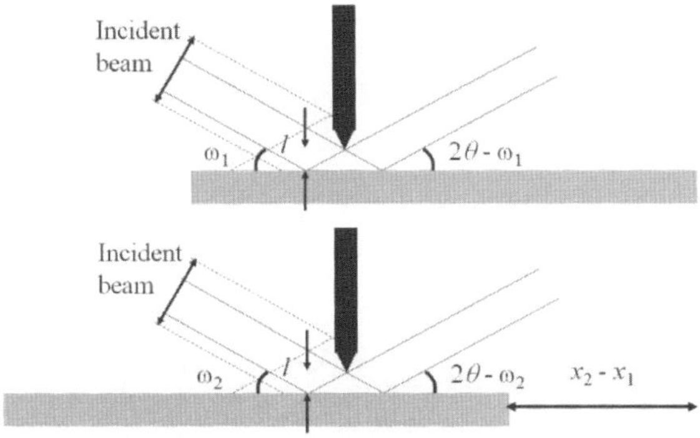

Figure 5.3.19. A method of determining the radius of curvature by defining the sample position with a knife-edge and translating the sample.

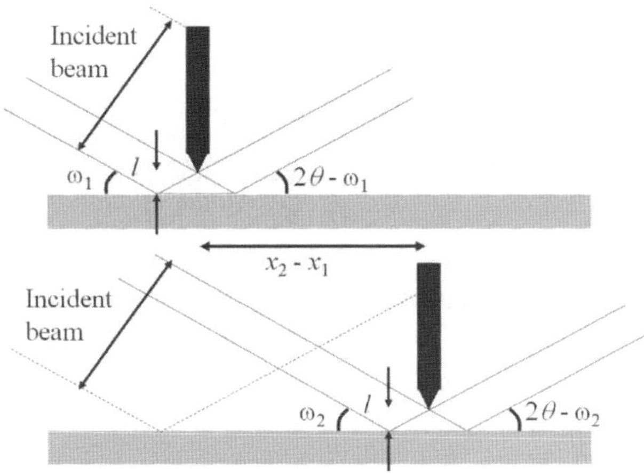

Figure 5.3.20. The method of determining the curvature by translating the knife-edge to overcome any problems an imperfect sample translation stage.

An alternative and very rapid method of determining the curvature is to use the Beam-Selection diffractometer, section 4.3.4. If the double pinhole is used, e.g. 100 μm, then the whole surface can be mapped, figure 5.3.21. The analyser is set at 2θ and the sample is scanned in ω to maximize the intensity. The variation in ω will relate to the radius of curvature through equation 5.3.12.

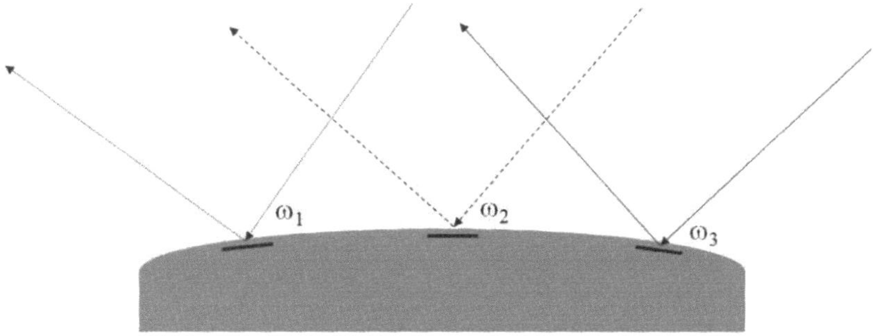

Figure 5.3.21. The method of determining the curvature with the Beam-Selection diffractometer by stepping the sample in *x*.

If the substrate is not crystalline or poorly crystalline then the radius of curvature can be obtained by using the principle described in section 5.3.1.2.2. A relatively large parallel beam from the 2-crystal 4-reflection monochromator incident on the sample is used in conjunction with a knife-edge to define the area on the sample as in figure 5.3.20. A suitable scattering angle, $2\theta = 1.4^0$, is chosen and the incident beam angle ω_1 that gives the maximum intensity is noted. Translating the knife-edge across the sample to the second position and determining ω_2 will then give the radius of curvature through equation 5.3.12. The absolute position of 2θ is maintained constant throughout. Clearly to obtain a reasonably well-defined region on the sample the knife-edge has to be very close to the sample, this will also limit the broadening of the peak from the curvature.

5.4. Analysis of nearly perfect semiconductor multi-layer structures:

The definition of a nearly perfect semiconductor layer was covered in Chapter 1, Table 1.3.1. If the layer is perfect then we would only be concerned with composition as a function of depth and the surface orientation. Orientation has been covered in the last section. For nearly perfect or real samples, we have defects and possibly small tilts and we shall consider these aspects as well in this section. The analysis of periodic multi-layers will be discussed separately since the approach and assumptions are different.

We shall consider firstly the measurement of composition and then thickness building from simple assumptions and how these give rise to errors. The most complete approach is through simulation to extract all this information and this is the desired approach, however some of the initial discussion is very useful for determining approximate working values for modeling the scattering. The most sensitive methods rely on the relationship between strain and composition and this will be the main concentration here, which is scattering from Bragg reflections. Reflectometry as a route to composition measurement will be considered for more imperfect structures and periodic structures.

Sections 5.4.1 to 5.4.4 will consider 'simple structures', which refers to structures with a few layers, although as will become clear this does not suggest the available information is simple. This really depends on how deep the analyst wishes to delve, so these sections should help in making that decision. The following sections 5.4.5 to 5.4.7 consider 'complex structures', which refers to periodic multi-layer analysis through to the analysis of quantum dot structure analysis.

5.4.1. *The first assumption and very approximate method in determining composition (simple structures):*

Any material phase will have certain overall characteristics associated with the periodic repeat unit, i.e. lattice parameters a, b, c, α, β, γ. For example AlAs, GaAs, InP, InAs, GaSb, etc., all have the same arrangement of atoms in their unit cells yet the atomic radii differ and consequently the length a $(= b = c)$ is different for each phase. The angles α, β, γ are all 90^0. These differences in lattice parameter through Bragg's equation give rise to different scattering angles. For nearly perfect layers this difference in lattice parameter is accommodated by elastic distortion, see figure 1.6.2, and more generally in section 1.6.1. The strain in the layer can be expressed as a function of $\Delta\theta$ from the Bragg equation. If the layer is strained to fit to the substrate then the crystal planes originally inclined to the surface plane (by angle φ) will be rotated with respect to the equivalent substrate planes (by angle $\Delta\varphi = (\varphi_L - \varphi_S)$; the subscripts refer to the layer and substrate). Hence

$$\left(\frac{\Delta d}{d}\right) = \frac{\sin(\varphi + \Delta\varphi)}{\sin\varphi} - 1 = -\frac{\sin\theta - \sin(\theta + \Delta\theta)}{\sin(\theta + \Delta\theta)} \qquad 5.4.1$$

d is the inter-planar spacing of the substrate and $\Delta d = (d_L - d_S)$ is the difference in the inter-planar spacing for the reflection measured. The difference in the incident angle from the layer and substrate scattering planes is the sum of these contributions; $\Delta\omega = \Delta\theta + \Delta\varphi$. If we assume that $\Delta\varphi$ and $\Delta\theta$ are small then from equation 5.4.1 $\Delta\varphi \sim \tan\varphi \cot\theta \Delta\theta$ and we can write

$$\left(\frac{\Delta d}{d}\right) = \frac{\Delta\omega}{\tan\theta + \tan\varphi} \qquad 5.4.2$$

Again considering the geometry of the rotation of the crystal planes for a perfectly strained layer on a substrate we write

$$\left(\frac{\Delta d}{d}\right)_{\perp} = \frac{\tan(\varphi + \Delta\varphi)}{\tan\varphi} - 1 \qquad 5.4.3$$

The perpendicular sign refers to the equivalent value normal to the surface. Bringing these terms together we have

$$\left(\frac{\Delta d}{d}\right)_{\perp} = \left(\frac{\Delta d}{d}\right) \frac{\left\{\frac{\tan(\varphi + \Delta\varphi)}{\tan\varphi} - 1\right\}}{\left\{\frac{\sin(\varphi + \Delta\varphi)}{\sin\varphi} - 1\right\}} \sim \left(\frac{\Delta d}{d}\right) \frac{1}{\cos^2\varphi} \qquad 5.4.4$$

An additional influence on the scattering plane rotation can result from a layer grown on a terraced surface, e.g. on a substrate surface that is not parallel to the (*001*) planes. The layer interplanar spacings will be constrained to the step surface (e.g. the (*001*) plane) and also to the step edge such that the expected tetragonal distortion is reached over the expected step length <*L*>. The assumption is that no relaxation has occurred. If the step edge height is *h* then the step length will be <*L*> = $h/\tan\varphi$, where φ is the macroscopic surface to crystal plane angle. The rotation due to this stepped surface is given by

$$\Delta\varphi' = \tan^{-1}\left(\frac{\Delta h}{<L>}\right) = \tan^{-1}\left(\left(\frac{\Delta h}{h}\right)\frac{h}{<L>}\right) = \tan^{-1}\left(\left(\frac{\Delta d}{d}\right)\tan\varphi\right) \quad 5.4.5$$

where $\Delta d/d$ is the strain along the principal crystallographic direction in the layer with respect to the substrate and is equivalent to $\Delta h/h$, Nagai (1974), Auvray et al (1989). To establish an order of magnitude of this effect an AlAs layer on a GaAs (*001*) substrate will give an additional angular separation of 2.4" arc for a 0.25^0 substrate off-cut, 20"arc for 2^0 and 30"arc for 3^0, etc. Clearly as the strain in the layer increases or the off-cut increases the assumptions that no strain relaxation from dislocations occurs becomes invalid and the whole modeling process becomes more complex, see section 5.4.4.

The perpendicular mismatch can be related to the mismatch parallel to the interface, i.e. that of the substrate to the free-standing value for the layer, figure 1.6.2. From equations 1.6.13 and rearranging we can state

$$\varepsilon_{//} = \frac{1-\nu}{1+\nu}(\varepsilon_{//} - \varepsilon_{\perp}) \quad 5.4.6$$

Taking our special case of a cubic (*001*) orientated sample and figure 1.6.2, we can write our strains as

$$\varepsilon_{\perp} = \frac{c_L - a_L}{a_L}$$
$$\varepsilon_{//} = \frac{a_S - a_L}{a_L} \quad 5.4.7$$

Substituting into equation 5.4.6 and multiplying both sides by (a_L / a_S) we have

$$\left(\frac{\Delta d}{d}\right)_{//} = -\frac{1-\nu}{1+\nu}\left(\frac{\Delta d}{d}\right)_{\perp} \qquad 5.4.8$$

where ν is a Poisson's ratio for a particular direction. We could equally well define some distortion coefficient given by $(1-\nu)/(1+\nu)$, Hornstra and Bartels (1978). This is an appropriate expression for isotropic strain in the plane of the interface when we can relate the strain of the layer to that of the substrate.

For cubic space groups the relaxed strain parallel to the interface is equivalent to the strain in the relaxed cubic lattice parameter for the layer, hence

$$\left(\frac{\Delta d}{d}\right)_{//} = -\left(\frac{\Delta d}{d}\right)_{Relaxed} = -\left(\frac{\Delta a}{a}\right)_{Relaxed} \qquad 5.4.9$$

Hence from our X-ray scattering experiment we can determine $\Delta\omega$ and the perpendicular strain using equation 5.4.2 and 5.4.4. The parallel mismatch can be determined from knowledge of the elastic constants. The parallel mismatch relates the atomic layer separation of the free-standing material to the strained value that matches to the substrate.

The X-ray experiment in its is simplest form can be undertaken with a double-crystal diffractometer (figure 4.3.1a). The scattering angle of the collimator crystal should match that of the sample scattering angle to limit the wavelength dispersion. The large detector window has the advantage of collecting the intensity from both layer and substrate when the scattering peaks are well separated (e.g. this becomes more critical when the scattering planes are inclined to the surface plane and small angles of incidence). For scattering planes parallel to the surface plane then a Bragg-Brentano diffractometer can be used, section 4.4.1.3. However the wavelength dispersion can add confusion (effectively figure 3.2.1 is superimposed on each peak). The peak separation increases with scattering angle and therefore the sensitivity is improved at higher scattering angles.

Clearly knowledge of the parallel and perpendicular mismatch and the elastic constants allows us to determine the value of $(\Delta d/d)_{\mathrm{Relax}}$ and this can be compared with the value expected for a given composition.

For most semiconductor systems we can assume that the composition follows Vegard's rule (i.e. the lattice parameter of a mixed phase, which can exist across the full solid solubility range, is given by the sum of the lattice parameters of all the phases present each multiplied by their proportions), Vegard (1921). The elastic parameters should be scaled in a similar manner. Hence for a ternary alloy $A_xB_{1-x}C$ with a known inter-planar spacing d_{ABC} and constituent binary compounds with interplanar spacings d_{AB} and d_{BC} we can write

$$x = \frac{d_{DE}\left[\left(\frac{\Delta d}{d}\right)_{\mathrm{Relax}} + 1\right] - d_{BC}}{d_{AC} - d_{BC}} \qquad 5.4.10$$

where $\Delta d = (d_{ABC} - d_{DE})$, the difference in the inter-planar spacing between the layer and the substrate and $d = d_{DE}$ the inter-planar spacing within the substrate. For a sample that is cubic in its natural free standing state then $a \propto d$ and therefore a can be substituted for d throughout these equations.

The assumptions are that the relative orientation between the substrate and the layer is zero, that the inter-planar spacings and elastic parameters follow Vegard's rule and that the scattering can be simply interpreted in this way, i.e not influenced by dynamical diffraction effects. Also since this is a relative measurement the inter-planar spacing of the substrate (the internal reference) is assumed to be a known value.

In general the deviations of the angular spacing from linearity (this brings together lattice parameters and elastic parameters) is very small and very difficult to measure. Most of the problems are associated with finding a better method of measuring the composition for comparison; some of these methods have been reviewed in Fewster (1993a). The Si – Ge alloy system has been studied very closely by Dismukes et al (1964), who suggest that it can be represented more precisely by a polynomial relationship. There is perhaps a stronger justification for using this relationship since the deviation is nearly 3% in the worst case. The expression is given by

$$a_{\mathrm{Ge}_x\mathrm{Si}_{1-x}} = xa_{\mathrm{Ge}} + (1-x)a_{\mathrm{Si}} + 0.007\{(2x-1)^2 - 1\} \qquad 5.4.11$$

If we assume Vegard's rule then the composition will be underestimated. Vegard's rule is applied to the elastic parameters and is sufficient within the accuracies that these are known.

The lattice parameter assumption concerning the substrate is very important, but as described in section 5.3.4 we can determine this independently and thus remove this uncertainty. Another very important assumption concerns the scattering mechanism itself and this will be discussed in the next section.

The discussion so far has concentrated on the single variable that is an alloy that exists between two binaries. However more complex phase mixtures are of interest and these either require additional information, e.g. photoluminescence determination of the energy gap, or additional X-ray scattering experiments. This whole analysis until now has concentrated on the peak positions with the simple Bragg equation transform from scattering to real space; however the scattering strength is not considered. This deficiency again points us to the power of simulation methods that include all aspects.

5.4.2. *The determination of thickness* (*simple structures*):

Interference fringes observed in the scattering pattern, due to the different optical paths of the X-rays, are related to the thickness of the layers. From section 2.7.2 we derived the Scherrer equation 2.7.10 that relates the width of the diffraction peak to the thickness of the independently scattering region. From section 2.10 we indicated that the reflectometry profile also contains fringing that relates to the thickness of the layers. These appear attractive routes to determining the thickness, however we have to be cautious of these approaches as will be discussed below. Despite this, simulation methods are very reliable.

5.4.2.1. Determining the thickness from the fringes close to main scattering peaks:

From figure 5.4.1a we can see that the profile from a relatively simple structure is basically composed of a substrate peak, layer peak and thickness fringes. We can determine the thickness fringe peak positions $m_1, m_2, m_3, \ldots$ etc., and determine a characteristic length scale, L, from the Bragg equation which gives the simplest transform from scattering to real space. We write this condition for planes parallel to the surface as:

$$\begin{aligned} 2L\sin\omega_1 &= m_1\lambda \\ 2L\sin\omega_2 &= m_2\lambda \end{aligned} \qquad 5.4.12$$

Hence

$$L = \frac{(m_1 - m_2)\lambda}{2(\sin\omega_1 - \sin\omega_2)} \sim \frac{(m_1 - m_2)\lambda}{2\Delta\omega\cos\omega} \qquad 5.4.13$$

Where $\omega_1, \omega_2, \ldots$ correspond to the angular positions of the peaks and $\Delta\omega = \omega_1 - \omega_2$, where ω is the average of the two values. This equation is valid for scattering from planes parallel to the surface.

For the more general case we can derive an expression from the reciprocal lattice construction of figure 4.2.1:

$$\begin{aligned} L &= \frac{(m_1 - m_2)\lambda}{\{\sin\omega_1 - \sin\omega_2 + \sin(2\theta_1 - \omega_1) - \sin(2\theta_2 - \omega_2)\}} \\ &\sim \frac{(m_1 - m_2)\lambda}{\{\cos\omega\Delta\omega + \cos(2\theta - \omega)(\Delta\{2\theta\} - \Delta\omega\ )\}} \end{aligned} \qquad 5.4.14$$

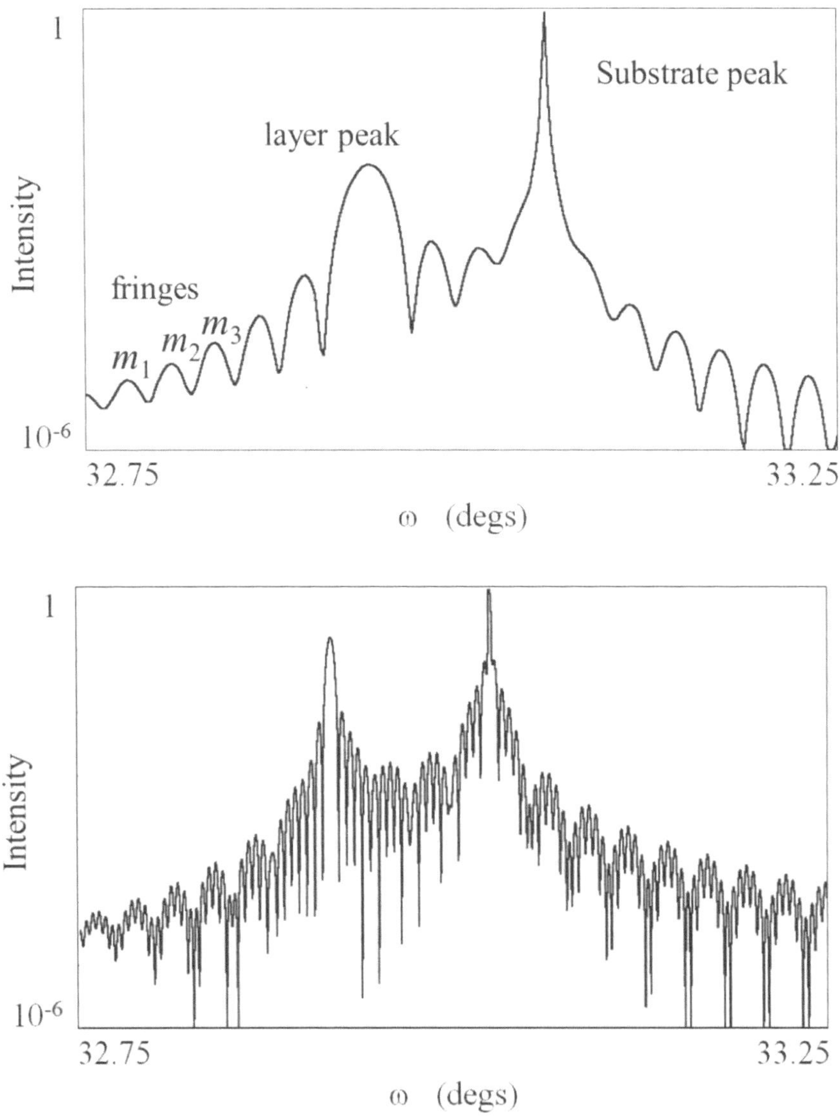

Figure 5.4.1. (a) The scattering from a single layer on a substrate illustrating the peak broadening and fringing from the limited layer thickness. (b) The scattering from a two layer structure indicating the two modulations. Both simulations indicate the degree of complexity in the scattering profile leading to peak-pulling effects and inaccuracies in making direct determination of structural parameters.

Table 5.4.1. Errors in the assumption of equation 5.4.14 due to diffraction effects.

Simulated hickness μm)	Equation 5.4.14 derived thickness (μm)	Error (%)
2.0000	1.9597	-2.02
1.0000	0.9798	-2.02
0.5000	0.4899	-2.02
0.2500	0.2398	-4.08

If we take a series of layers of different thicknesses, simulate the profile and derive the thickness with equation 5.4.13 we can compare the peak separation method directly, Fewster (1993a). The error can be as much as 4%, and varies with thickness, Table 5.4.1. Remember of course that this is the simplest case and therefore making assumptions based on more complicated multi-layers does become problematic and more unreliable.

As with the composition determination from peak separation, section 5.4.3.1, this method will only give an approximate value, a useful guide or an input to a simulation model.

5.4.2.2. Determining the thickness from the fringes in the reflectometry: profile:

Fringes are very evident in a reflectometry profile, figure 2.10.3 and as pointed out by Keissing (1931) these should relate directly to the thickness of the layer causing them. As we can see from Table 5.4.1 the errors due to dynamical scattering effects can be very large and therefore it is not surprising that these errors also appear in the direct determination of reflectometry scans. By way of an example consider a 30 nm layer on a substrate and compare the thickness measured with equation 5.4.14 with that obtained from simulation and see the very large differences, Table 5.4.2.

Table 5.4.2. The errors in the direct interpretation of reflectometry profiles for thickness based on equation 5.4.13 for a sample with a 30 nm layer of GaAs on Si. n is the refractive index.

Mean angle of measurement (degrees)	Thickness determined directly (μm)	Error (%)	Thickness determined with n (μm)	Error (%)
0.4285	0.0442	47.3	0.0291	3.0
0.6626	0.0338	12.7	0.0294	2.0
0.9332	0.0317	5.7	0.0299	0.3
1.2151	0.0309	3.0	0.0302	0.7

The errors decrease rapidly with increasing scattering angle and therefore any measurements based on this direct interpretation should be done at the highest angles possible. The cause of this is mainly related to the refractive index effects that must now be included. If we consider equation 5.4.12 to determine the path length associated with the peak of the m th order fringe, and consider the plane spacing as the thickness of the layer, L, then:

$$2L \sin \omega_{\mathrm{i}} = m_i \frac{\lambda}{n} \qquad 5.4.15$$

ω_{i} and n_{i} are the incident angle for the peak of this fringe and the refractive index inside the layer. The incident angle on the layer surface, ω_0, is related to ω_{i} through Snell's rule, equation 2.10.8 and the refractive index can be related to the critical angle, ω_c, equation 2.10.22

$$n_i = \frac{\cos \omega_0}{\cos \omega_{\mathrm{i}}} = \cos \omega_{\mathrm{c}} \qquad 5.4.16$$

Combining these equations and using the relationship $\sin^2 a = 1 - \cos^2 a$

$$m_i \frac{\lambda}{2L} = n_i \sin \omega_{\mathrm{i}} = \cos \omega_{\mathrm{c}} \left\{ 1 - \frac{\cos^2 \omega_{\mathrm{i}}}{\cos^2 \omega_{\mathrm{c}}} \right\}^{1/2} = \left\{ \cos^2 \omega_{\mathrm{c}} - \cos^2 \omega_{\mathrm{i}} \right\}^{1/2} \approx \left\{ \omega_{\mathrm{i}}^2 - \omega_{\mathrm{c}}^2 \right\}^{1/2}$$

5.4.17

If we now measure the angle at the peak of two fringes that are M orders apart, as in equation 5.4.13, then the layer thickness becomes:

$$t = \frac{M\lambda}{2\left(\left(\cos\omega_c^2 - \cos\omega_2^2\right)^{1/2} - \left(\cos\omega_c^2 - \cos\omega_1^2\right)^{1/2}\right)} \approx \frac{M\lambda}{2\left(\left(\omega_2^2 - \omega_c^2\right)^{1/2} - \left(\omega_1^2 - \omega_c^2\right)^{1/2}\right)}$$

5.4.18

The improvement can be seen when we include this into our estimation of our 30nm layer thickness, table 5.4.2. The effect is very significant for the reflectometry profile. Again simulation is the best method of extracting information from a reflectometry profile.

5.4.3. *The simulation of rocking curves to obtain composition and thickness(simple structures):*

In the previous section we have simply assumed Bragg's equation for each layer and substrate reflection. We could take into account the refractive index but this is a small correction, especially since we are generally comparing the difference in the positions of peaks, hence any error is related to the difference in refractive index correction. However as pointed out in Fewster (1987) and Fewster and Curling (1987), figure 2.3.7, the peak position from single layers less than 0.5 μm or buried layers less than 2 μm do not correspond to the peak position expected from Bragg's equation. These differences can be large, giving errors in the composition up to ~15%. The reason for this is that the build-up of the wave-field in the crystal requires a reasonable sustained periodicity to lock into and for thin layers this has not been established. As with all the arguments expressed so far we cannot isolate aspects in the experimentally determined profile and assign them unambiguously to certain features. All we can state is that some scattering feature is predominately influenced by some layer. Dynamical theory considers the whole process of scattering as wave-fields including all the interactions, and therefore this becomes the most exacting way of describing the scattering. To illustrate the effect, consider figure 5.4.1a and b for single and buried layers.

Both profiles give more peaks than there are layers and result from a mixture of thickness fringes (interference from the lower and upper interfaces of a layer) and interaction between all these contributions. The scattering process is therefore quite complex and the simulation and comparison is the most reliable approach. At this stage we are assuming the material to be nearly perfect. Since we are dividing the structure into layers of constant composition and hence strain, we can simply use the same approach for any strain distribution, for example ion-implanted materials, (Sevidori et al, 1996: Klappe and Fewster, 1993). The method is very general.

To indicate the level of information obtainable from careful and rapid data collection and analysis consider the data from a SiGe heterojunction bipolar transistor, HBT, described below.

5.4.3.1. Example of an analysis of a nearly perfect structure:

We wish to collect and analyse the structural details as quickly and accurately as possible and this example should give an indication of what can be done. This SiGe HBT is a two layer structure (a thin (0.1 μm) $Ge_{0.15}Si_{0.85}$ alloy layer wedged between a Si substrate and a Si cap layer (~0.7 μm). The simplest and easiest experiment is the *004* rocking-curve since the structure is grown on a (*001*) surface. The double-crystal difrractometer, figure 4.3.1b, gave an intensity at the detector at the zero position without the sample of about 4 million counts per second. The setting procedure is as follows:

1. The sample is mounted on the support plate.
2. *An absorbing (nickel, copper) foil is placed between the monochromator and X-ray tube. The detector is at* $2\theta = 0$.
3. *The sample is driven into the X-ray beam to reduce the count rate by half. Since the incident beam may not be passing parallel to the surface, the sample is scanned in* ω *and held at the position of maximum intensity. The sample is driven again to reduce the original intensity by half. The* ω = 0 *is now set and the sample is in the centre of the circles.*
4. *The absorbing foil is removed.*

5. The detector (2θ) and sample (ω) axes are driven to the approximate angles for the Si *004* reflection ($\omega \sim 34.5^0$ and $2\theta \sim 69^0$ for Cu $K\alpha_1$ radiation).
6. The detector is held stationary and the sample scanned very quickly over about 1^0 to find the strong substrate peak reflection. When found the sample axis is fixed at this position.
7. The tilt axis is then scanned over a couple of degrees (χ axis, section 4.3.1.1) and the mid-cord of the resulting profile, figure 4.3.9b, defines the aligned tilt angle. A small drive down in the sample axis will recover the intensity if this profile is double peaked.
8. The GeSi layer will have a larger lattice parameter than the substrate Si and therefore the rocking curve will be centred on the low angle side of the substrate. A 5-minute rocking-curve is then performed and the result is as in figure 5.4.2.

Step 2 to step 4 are not essential but can avoid difficulty in finding the reflection in step 6. The ω angle will have some uncertainty and therefore a larger rapid search scan may be necessary (step 6) and the sample centring will ensure that the scattered beam is closer to the true 2θ angle for the substrate.

We shall now concentrate on two approaches; the very direct method given above and simulation (manual and automatic).

5.4.3.2. Direct analysis from peak separation and fringe separations:

We can assume that for this relatively simple structure that the strongest and second strongest peaks come from the substrate and GeSi layer respectively. The top Si layer should be hidden under the substrate peak if the layers are all strained with respect to the substrate. The peak separation yields a composition of 18.2% Ge in the layer using equations 5.4.2 to 5.4.10. The fringes between the layer and substrate yield a thickness of some layer in the structure of 0.10±0.01 μm using equation 5.4.13. From our knowledge of the structure this will almost certainly be associated with the GeSi layer.

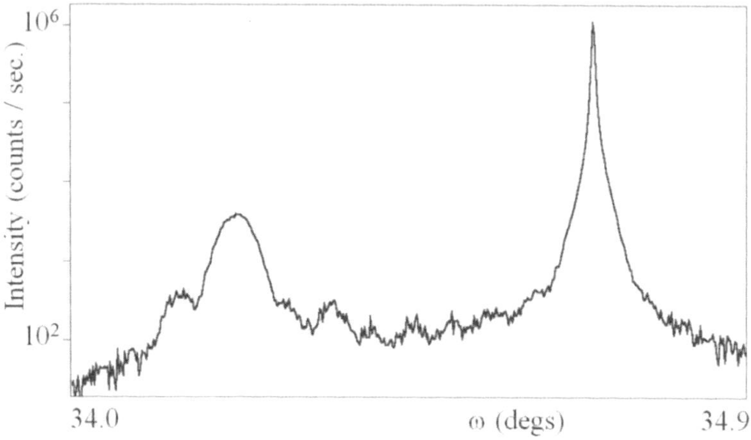

Figure 5.4.2. The scattering from a SiGe / Si structure with a 5 minute scan with the double-crystal diffractometer, figure 4.3.1b, (2-crystal 4-reflection monochromator and open detector).

Now we know that neither of these assumptions is precise, also we do not know the thickness of the cap layer, yet this appears somewhere in the profile.

5.4.3.3. Simulation using an iterative adjustment of the model:

Initially we can either start with the found values from the direct analysis or take the assumed values from that expected from the growth. If we take the latter approach then the peak position of the layer could be a long way from the best-fit value and therefore we should get these roughly overlapping first. The fitting of the thickness will now be more precise since the match is to the broadened shape of the layer and the fringe separation. The Si cap layer will have a much more subtle effect on the profile, since its thickness is ~7× larger than the SiGe layer the associated fringing will be oscillating at ~7× the frequency. The quality of the data may not be too good to fix this parameter too precisely. Clearly if the thickness is considerably smaller than expected, 0.3 or 0.4 μm for example, then the observed fringes from the SiGe will be modified such that their rise and fall can be slightly different giving the appearance of displacing the fringes. However if the cap layer is considerably thicker than expected then the influence on the pattern is insufficient to detect with this data quality for this particular example.

5.4.3.3.1. Linking parameters to cope with complex multi-layer structures:

If there is good control over the growth method then we can simplify the analysis by linking the growth rates of the various phases deposited. This can prove very useful for analysing very complex structures and will also aid automatic approaches to fitting profiles, Fewster (1990). Suppose the growth rate of phase A (e.g. GaAs) and phase B (e.g. AlAs) in layer L are denoted by $R_A(L)$ and $R_B(L)$ and these are consistent throughout growth. Then the timing of the growth for each layer can be directly related to the layer thickness. The "sticking coefficient" of each phase should be known and for many materials this can be considered as unity, should it be less than this but always consistent then this will not influence this approach, just change the effective growth rate. To emulate the growth process precisely we should include transients due to the temperature differences of an open Knudsen cell compared with a closed one in an MBE growth chamber. The high temperature and hence higher growth may occur at the beginning of each layer deposited and this may become significant in the comparison of growth rates for very thin and very thick layers. We therefore have for a mixture of two binary phases

$$t_A = T_L R_A(L) V_1 \qquad \qquad 5.4.19$$
$$t_B = T_L R_B(L) V_2$$

where t_A and t_B are the thickness contribution of binaries A and B. T_L is the time for the layer to be grown and V_1 and V_2 are variables to be refined (in effect the modification to the expected growth rates). Then the total thickness of the layer is given by

$$t_L = t_A + t_B \qquad \qquad 5.4.20$$

and the composition or proportion that is phase B is given by

$$x_L = \frac{t_B}{t_L} \qquad 5.4.21$$

Since this is a volume concentration the assumption here is that the two materials have similar lattice parameters. We have now related the thickness to the time that the shutters are open. This timing is far more precise than knowledge of the various growth rates. To include the transients we can assume an exponential form to reach equilibrium:

$$\begin{aligned} R_A(L) &= R_A(L-1) - [R_A(L-1) - {}_0R_A](1 - e^{-A_0 T_L}) \\ R_B(L) &= R_B(L-1) - [R_B(L-1) - {}_0R_B](1 - e^{-B_0 T_L}) \end{aligned} \qquad 5.4.22$$

${}_0R_A$ and ${}_0R_B$ represent the equilibrium growth rates when the shutter is open and A_0 and B_0 are the characteristic cooling rates of the cells and are further parameters to be refined. Of course since the cells cool from being open they must also warm whilst being closed and therefore the growth rate must also be considered to take into account the equivalent equilibrium growth rates, ${}_CR_A$ and ${}_CR_B$ with the shutters closed:

$$\begin{aligned} R_A(L) &= R_A(L-1) - [R_A(L-1) - {}_CR_A](1 - e^{-A_C T_L}) \\ R_B(L) &= R_B(L-1) - [R_B(L-1) - {}_CR_B](1 - e^{-B_C T_L}) \end{aligned} \qquad 5.4.23$$

where A_C and B_C are the characteristic warming rates. These characteristic rates of cooling and warming should be constant and in a well characterised MBE system known. We therefore have two additional variables to include in any refinement associated with each phase.

For well-controlled growth and large Knudsen cells the effects of transients is less significant. This example does indicate the detail that can be extracted or alternatively an effect that we should be aware of especially in the growth of very short period superlattices. The general idea though of linking growth rates is very useful for rapid evaluation of complex multi-layer systems with mixed phases.

The iterative process can be fast with experience, however this whole process can be greatly aided by automatic fitting of the data.

5.4.3.4. Automatic fitting of the data by simulation:

If we want a fairly "push button" method so we work from the basis that the growth model is the starting value. The fitting of the data is a non-linear problem and full of false minima and so is not a simple procedure. Non-linear least squares procedures have been used successfully when a very good estimate of the model is known already, e.g. when all the direct analysis information is included, Fewster (1990). The basic problem is to somehow find a position close enough to the global minimum (the minimum error possible) or have an algorithm that will allow movement out of local minima (false minimum errors) so that the true best fit to the global minimum is found. The global minimum will contain the parameters associated with the true wanted model of the structure.

There are several approaches to finding the global minimum. Simulated annealing algorithms work on the basis of allowing any solution to move out of local minima and the search will progress by refinement of the model until the global minimum is found (or what it thinks is the global minimum). Genetic algorithms also offer possible routes to the solution of scattering profiles, (Dane, Veldhuis, de Boer, Leenaers and Buydens, 1998). This procedure works on a random process of trial structures that favours the closest fits to the model; the parameters from these fits are then combined randomly to create another set of trial structures. Of course these methods may find the global minimum by chance. If they find local minimum then the problem is overcome by randomly changing some parameters in the hope of moving out of these. The whole fitting process can vary in timing because of the random nature of the algorithm and difficulties in judging the end point, Wormington, Panaccione, Matney and Bowen (1999).

Another approach that we shall discuss here is an algorithm that reduces the presence of local minima and can find the correct solution exceedingly rapidly, Fewster (1993b). This approach proved to be very successful for analysing ion-implanted layers, Klappe and Fewster (1993) and works very well for heterostructures, Tye and Fewster (2000). The underlying assumption is that the amount of detail in the data is initially too great and by fitting the main features an approximate solution close to the global minimum can be found. This method is ideally suited to rocking curve simulation. For reflectometry profiles the occurrence of false minima is again clear and this was the main problem that Dane et al (1998) were trying to solve. However an alternative approach based on a different principle is preferred, Fewster and Tye (2001). This is a segmented fitting routine that assumes that the overall structural form is contained in the first part of the reflectometry curve and so this is fitted first. When this fits well, more of the profile is included bringing in more detail. This approach steers the automatic fitting to the global minimum.

However we are mainly interested in the fitting of rocking curves so we shall concentrate of the method of Fewster (1993b). Consider our profile of figure 5.4.2, the main features are the layer and substrate peak. The algorithm works by simulating the profile with a starting model and then heavily smoothing this calculated and experimental profile for comparison, figure 5.4.3a(i). Using a steepest descent algorithm the agreement between these two profiles is achieved very quickly, figure 5.4.3a(ii). The procedure is repeated with this improved model but with less smoothing. A solution is found in a few seconds on a personal computer, figure 5.4.3b. The agreement is based on a logarithmic comparison of intensities and because of the smoothing the influence of noise, data quality and material quality (peak broadening) have less effect than direct comparisons. The background can be added and fixed or refined.

Obtaining the uncertainties in the parameters from the fitted profiles is complex because little is revealed about the parameter correlations. It is possible to extract these for simple models by looking at pairs of correlations once the fit is obtained, rather similar to finalizing with a least-squares fit and inspecting the correlation matrix. However this assumes that the process to achieve that fit arrives at the right result. A more encompassing stochastic approach that is guided by the prior probability associated with the uncertainty can achieve impressive results. This gives a final uncertainty associated with all parameters and reveals some of the difficulties expressed above, Anderson, Fewster,

Ozsvald and Tye (2010). This analysis method is used in some of the examples described in the section 5.4.5 (complex structures).

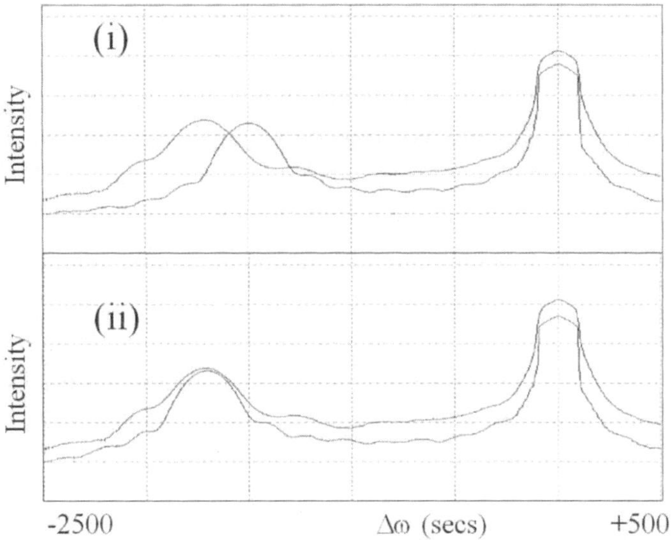

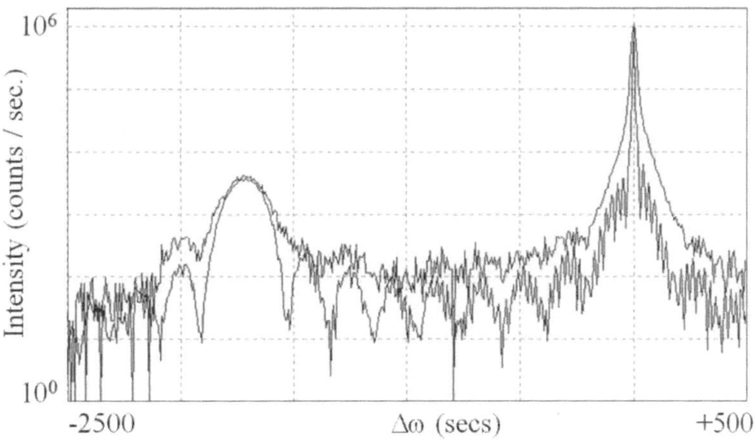

Figure 5.4.3. (a) The principle of the fitting algorithm; (i) the simulation of the expected structure compared with the measured profile after both have been smoothed. (ii) because of the removal of noise and the removal of local minima the first stage of the refinement is very rapid. (b) The final fitted profile after the smoothing has been removed.

The final fitted parameters are 18.51(2)% Ge in the GeSi layer that is 0.101(1) μm thick, capped with a Si layer of 0.70(5) μm. The composition is higher than in the direct analysis and this is the peak pulling effect discussed at the beginning of section 5.4.3. The thickness determined by simulation for this SiGe layer is in close agreement with the direct method. A value is given for the cap thickness, although as discussed it is not well determined in this experiment.

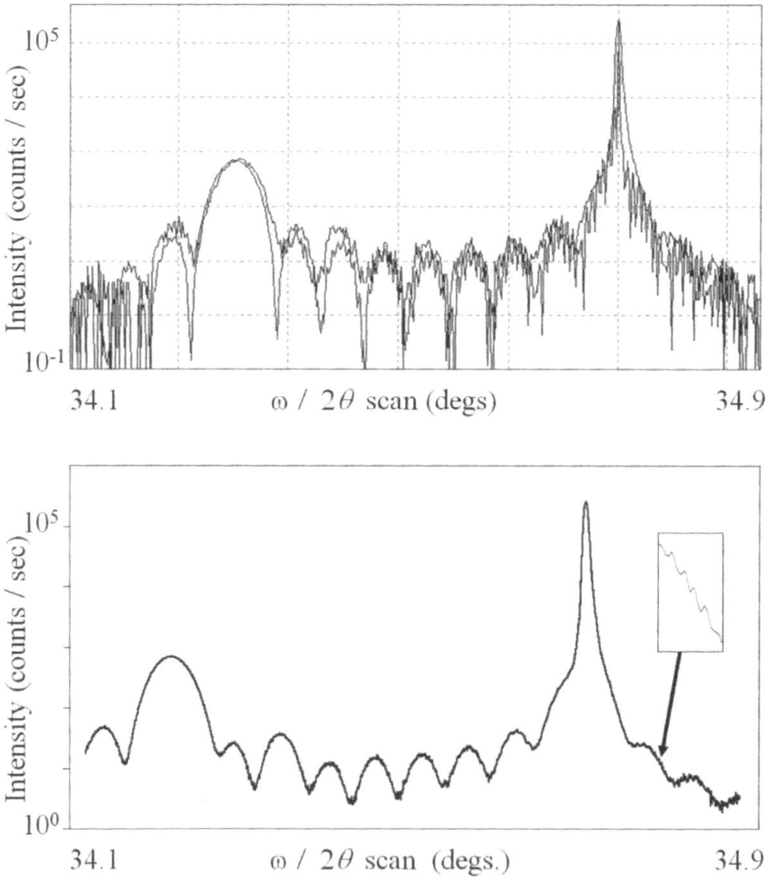

Figure 5.4.4. (a) A repeat of the procedure illustrated in figure 5.4.3b but with a multiple-crystal profile taken in 5 minutes. (b) The same profile but taken over 1.5 h, illustrating the fine fringing that relates to the thick cap layer. The addition of an X-ray mirror would reduce this collection time by approximately a factor of ten.

Let us consider the profile more closely. The layer and fringing appear to fit very well, but close to the substrate peak the fit is poor. We could include some additional or diffuse scattering into the model. This has a $(1/\Delta\omega)^2$ dependence profile with a certain height and width, section 2.8.1; if this is the dominant effect or the diffraction effects illustrated in figure 2.9.7, although in this case this makes little difference. Sometimes this can help to achieve a more rapid and exact fit in some structures although it is not relevant to a greater qualitative physical understanding. We shall now repeat the experiment except with an analyser, taking the same length of time for the measurement, i.e. 5 minutes.

5.4.3.5. Data collection with the multiple-crystal diffractometer:

The experiment set-up is as in figure 4.3.5, initially without the X-ray mirror. The experiment was performed following this procedure:

1. The sample is mounted on the support plate.
2. *An absorbing foil is placed between the monochromator and X-ray tube. The detector is at* $2\theta = 0$.
3. *The sample is driven into the X-ray beam to reduce the count rate by half. Since the incident beam may not be passing parallel to the surface, the sample is scanned in* ω *and held at the position of maximum intensity. The sample is driven again to reduce the original intensity by half. The* $\omega = 0$ *is now set and the sample is in the centre of the circles.*
4. *The absorbing foil is removed.*
5. The detector (2θ) and sample (ω) axes are driven to the approximate angles for the Si *004* reflection ($\omega \sim 34.5^0$ and $2\theta \sim 69^0$ for Cu $K\alpha_1$ radiation). The analyser / detector is substituted with an open detector.
6. The detector is held stationary and the sample scanned very quickly over about 1^0 to find the strong substrate peak reflection. When found the sample axis is fixed at this position.
7. The tilt axis is then scanned over a couple of degrees (χ axis, sections 4.3.1.1) and the mid-cord of the resulting profile, figure

4.3.9b, defines the aligned tilt angle. A small drive down in the sample axis will recover the intensity if this profile is double peaked.

8. An ω scan is performed to precisely locate the peak of the substrate reflection.
9. The analyser / detector assembly is then substituted for the open detector.
10. The 2θ (analyser / detector) axis is then scanned to detect the scattered beam. This will be a very sharp peak and so a small step size is needed for the final location of the peak position.
11. With the two axes set to detect the scattered X-rays they can be driven together. Since we are studying the *004* reflection from a *001* sample the ratio can be maintained at 1 : 2 for ω : 2θ. An ω /2θ scan is carried out over the same range of ω as in the previous experiment and we achieve a perfectly useable profile in 5 minutes. However the fringing is rather weak and is greatly improved by including an X-ray mirror. The data are collected again in 5 minutes and shown in figure 5.4.4a.

The procedural steps 1 to 7 are identical to those needed to collect a rocking curve. Again steps 2 to 4 are not essential. There are alternative alignment methods indicated in figure 4.3.9, which can be used in place of steps 5 to 9, if the swapping between the analyser / detector and open detector is not possible. This is though rather like finding a cross-hair on the peak, which can be iterative or found by reciprocal space mapping, e.g. Δω versus $\Delta\chi$ at fixed 2θ to obtain the mid-chord in $\Delta\chi$.

Clearly the profile is superior and now we can accurately fit the profile close to the substrate. The results become (x = 18.58(18)%, t_{GeSi} = 0.098(1) μm, t_{cap} = 0.717(3) μm). The simulation does not include background or diffuse scattering and assumes that the sample is perfect. The top cap layer, as mentioned previously is rather thick and leads to very high frequency fringing that can be observed with a more careful scan of 1.5 h (small step size and longer counting times) to yield the fringing and thickness, figure 5.4.4b. The advantage of this experimental configuration is now clear. However it may be important to know the reason for the difference in the profiles of figure 5.4.4 and figure 5.4.2.

This all becomes clear when we perform a reciprocal space map, where this sample is discussed further in section 5.4.4.

5.4.3.6. Interface analysis when inter-diffusion exists:

The Silicon-Germanium structure described above has a relatively simple form, although the complexity and detail can be quite extensive. We could also see that the analysis of the main structural features could be obtained rather readily. This next example is of a quantum well laser. The performance was far from expected and this had to be related in some way to the growth of the structure. Initial fast analyses could not reveal the problem and therefore very high quality data were required. For this a limited area reciprocal space map was measured close to the *004* reflection and projected onto the *<001>* direction using the multiple-crystal diffractometer, section 4.3.3. The profile is given in figure 5.4.5. From inspection we can point to the “substrate” peak, the AlGaAs “cladding” peak and the very weak “InGaAs quantum well layer” peak. The assignment of peaks relate to their width and scattering strength (broad and weak means thin layer, strong and narrow mean thick layers) and also their relative positions (larger or smaller lattice parameters normal to the surface plane). The quotes refer to the fact that they represent the dominating scattering associated with these peaks.

Approximate values for composition and thickness can be made from the approaches described previously. However an intermediate broad profile could not be modeled successfully. This is where we now bring in the understanding of growth and recognise that indium can segregate during growth. If this happens the quantum well will be asymmetric, the energy levels will be altered and the emission not what is expected. By introducing a small exponentially falling In concentration towards the surface from the InGaAs quantum well produced an excellent fit to the whole profile. The final model of the structure from the best fit is given in figure 5.4.5. This example gives the degree of sensitivity possible with high quality data and persistence in striving for a perfect fit.

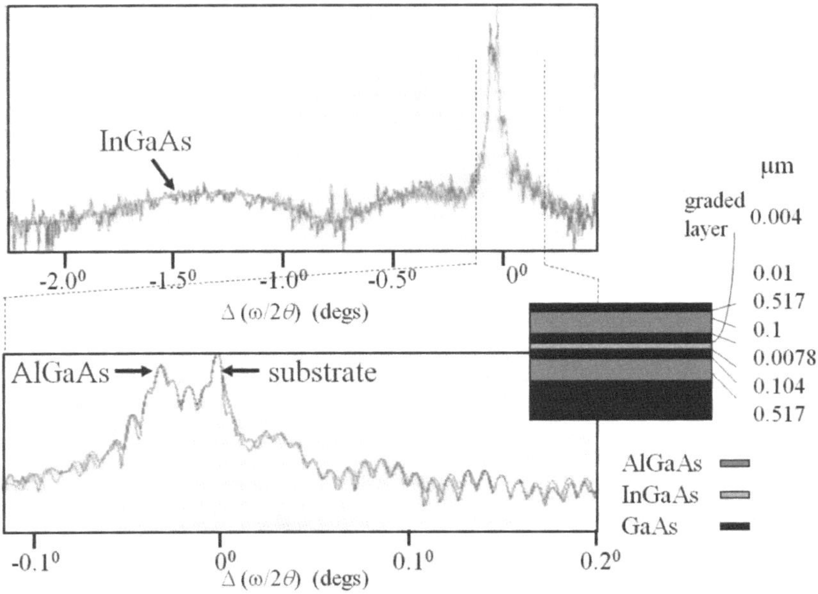

Figure 5.4.5. The detailed comparison of the measured and simulated profiles gives significant structural information including the segregation of In out of the quantum well. The best fit parameters are also given.

The example above indicated a problem with the fitted profile from the start because the peak at ~-0.4^0 that carries considerable information associated with the In segregation. This is a consequence of the interference between the layers. However for simpler structures the effects of inter-diffusion at an interface is less obvious. Each case will reveal these effects in different ways and the sensitivity is best shown by simulation.

In this next example the sample is a 2 layer structure with nominal dimensions; GaAs substrate, 0.015 μm of $In_xGa_{1-x}As$ and $x = 0.15$, and 0.1 μm of $Al_xGa_{1-x}As$ and $x = 0.25$. The diffraction profile can be fitted well with a fairly typical dynamic range of intensity close to 5 orders, on the multiple-crystal diffractometer with an X-ray mirror and projecting a limited area reciprocal space map, figure 5.4.6a(ii). This high-resolution is required to resolve the details of the fringing. The intensity and resolution can be improved with the beam-selection diffractometer, section 4.3.4. The dynamic range is increased to 6.5 orders with a count-rate on the substrate peak of 30 Mcs^{-1}, when an X-ray mirror is used. The data was collected in 7.3 min, figure 5.4.6b. This profile reveals more information, i.e. the peak at ~4500 s and this does not fit the former analysis until some inter-diffusion is included.

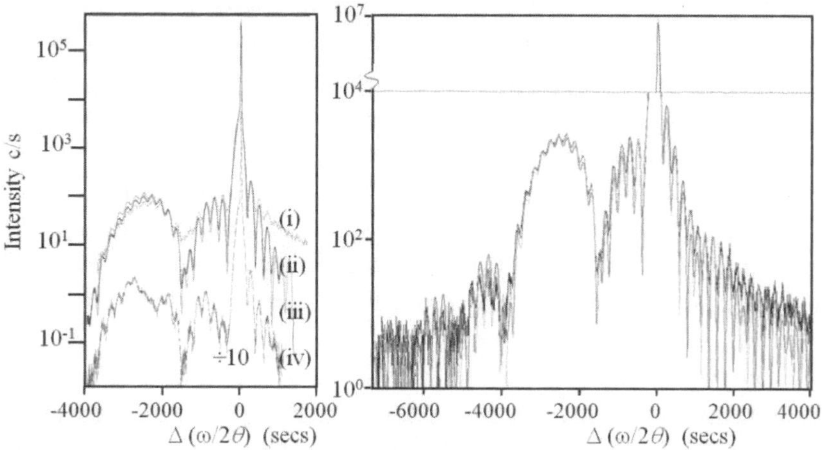

Figure 5.4.6. The experimental and best fit profiles for a structure containing a GaAs substrate and layers of InGaAs and AlGaAs obtained with (a) the double-crystal diffractometer (i), the multiple-crystal diffractometer map projection (ii) and its best fit profile (iii) and a single multiple-crystal diffractometer scan (iv); the large fluctuations in intensity result from the diffraction 'wiggles' (sections 5.3.3, 5.4.6.3). (b) the beam-selection diffractometer with its best fit profile (overlapping) and the simulation assuming an abrupt interface clearly seen as an overestimated intensity at -4500 s.

The best fit to the profile based on data collected over 5 orders of magnitude, given in figure 5.4.6a, gave a structure corresponding to; GaAs substrate, 0.0157 μm of $In_xGa_{1-x}As$ and $x = 0.145$, and 0.0808 μm of $Al_xGa_{1-x}As$ and $x = 0.258$. With the increased dynamic range and more resolvable features, figure 5.4.6b, the best fit structure becomes; GaAs substrate, 0.0130 μm of $In_xGa_{1-x}As$ and $x = 0.145$, and 0.0773 μm of $Al_xGa_{1-x}As$ and $x = 0.258$, with InAlGaAs graded approximately linearly between the two layers over 0.0059 μm. The overall thickness remains the same with the two fitted profiles, however the natural segregation of In towards the growing surface clearly exists in this example.

5.4.3.7. The analysis of quaternaries; a mixture of three phases:

The best sensitivity to composition is based on the state of strain induced through the effect of differing covalent radii of the constituents. Up until know we have concentrated on a two-phase mixture, i.e. InGaAs being a mixture of InAs and GaAs, however a three-phase mixture that is sometimes termed a quaternary like InAlGaAs or InGaAsP, then the combined strain does not unravel the composition of the three phases without additional information. The additional information could come from photoluminescence for example, or if the scattering strength of the different phases is sufficient the intensity fit to the profile combined with the peak positions can yield the information quite easily. It can be useful to select the reflection that best reveals this information. Good choices are often from reflections whose intensity is based on the difference of scattering strengths (the weaker reflections); the *002* and *006* reflections can be useful for this analysis rather than the *004* that is popularly used. Or a combination of these reflections can be used.

An interesting example is SiGeC, which is a three-phase mixture where the carbon content is kept to below 1% and is used to compensate the strain associated with high Ge concentrations. The covalent radius of C is very small and because of the low concentration used, the profile appears to be dominated by Ge. The Ge peak will appear more intense than expected for its apparent composition, if C was not present. In this example, Zhang *et al.* (2005), combined reflectometry, which is insensitive to strain but sensitive to the Ge scattering, and the *004* reflection that is sensitive to strain. This breaks the correlation and gives a reasonable precision for the Ge and C content. The Ge concentration can be determined with reflectometry, sections 2.10 and 5.4.5.5.2, and in this case could be measured to ~0.1 % in x for $Ge_xC_{1-y}Si_{1-x-y}$,provided the C content is known to within 1 % in y. This C concentration can be obtained from fitting the *004* profile, which also assumes that the covalent radius of C is known, figure 5.4.7. Zhang et al used the covalent radius of the cubic form of SiC, which was consistent with the results from composition measured with secondary ion mass spectrometry (SIMS).

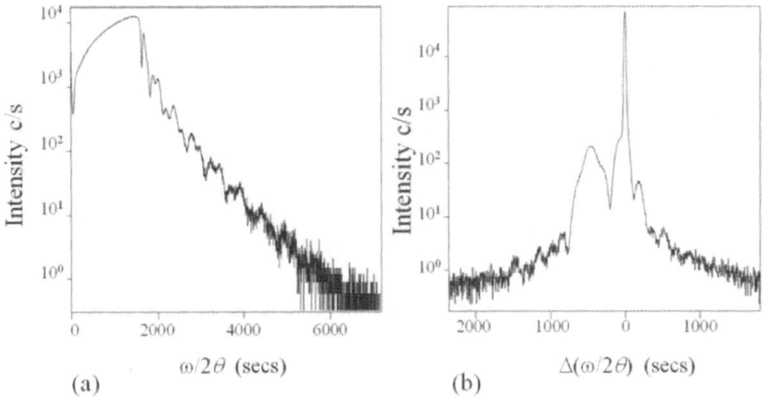

Figure 5.4.7. (a) The reflectometry scan from $Ge_xC_{1-y}Si_{1-x-y}$, which is predominantly sensitive to the Ge concentration provided the C concentration is known to within 1 % (typically these structures have C < 1 %). (b) the scattering profile close to the *004* reflection taken on the multiple-crystal diffractometer. This combination made it possible to extract the C and Ge concentration in this structure through fitting both profiles independently.

5.4.3.8. Rapid analysis of composition and thickness values:

All the methods described so far require scanning that limits the speed of data collection. The ideal situation would be to have minimal sample alignment when mounted on the diffractometer and fast data collection time, i.e. the whole process should be short to give rapid feedback. This is possible, but there are compromises in that they become more sample specific, which is often not a problem as a feedback for tuning the composition of a growth system. From the last section the beam-selection diffactometer can be considered a rapid method because of the very high intensities and alignment of the tilt is usually not required, figure 4.3.13.

If we compromise on the detail then the Static diffractometer described in section 4.3.6 will give the greatest speed of data collection.

Again the alignment is minimal, because the sample if set in approximately the correct position it will select the divergent beams from the source that satisfy the scattering conditions, Fewster (2005). The reflection chosen for analysis requires 2θ - ω to be very small and positive. With reference to figure 4.3.18 (the reciprocal space map for the *113* reflection of a SiGe structure with the static diffractometer), we can see that the detector captures most of the information along 2θ and therefore an ω-scan will reveal the Cu $K\alpha$ doublet for the substrate (assuming this is the dominant contributor to the profile). This pin-points the ω value for the Cu $K\alpha_1$ peak, so when set, the detector will collect the whole 2θ profile instantaneously, figure 5.4.8.

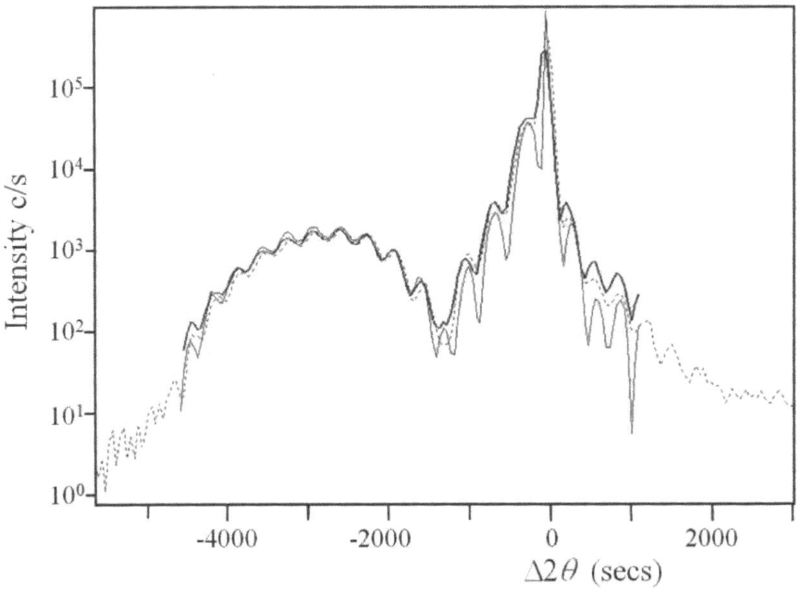

Figure 5.4.8. The scattering profile from the same structure described in figure 5.4.6, section 5.4.3.6 obtained with the Static diffractometer in 1 s (dark line). After the sample was mounted on the stage the automated alignment procedure took ~ 10 s, i.e. the whole process took approximately 15 s. The best fit profile is also included (grey line), which matches precisely that obtained with the multiple-crystal diffractometer. The dash line is the profile from extending the 2θ range by scanning.

The simulation is slightly more involved, since the wavelength distribution should be included for the best precision, however this is not a deterrent to its use because the results obtained with the Multiple-Crystal diffractometer, section 5.4.3.6 and figure 5.4.6, are used for the composition and thickness parameters in the simulation given in figure 5.4.8.

5.4.3.9. Mapping the composition and thickness over a wafer:

The advantage of rapid analysis methods as discussed in the previous section 5.4.3.8, is that it then becomes possible to spatially map inhomogeneities over a wafer. The 1 s data collection time illustrated above represents the average over a dimension of ~ 40 μm × 14 mm, so if a ~ 100 μm × 100 μm double slit is used to reduce the spatially sampled region to small values, then the intensity drops by a factor of 50 to achieve comparable data quality. The data quality can be compromised considerably yet still yield very useful information. An example of the usefulness of this mapping is given for a semi-processed SiGe based wafer; the information required was to have information regarding the Ge concentration and level of strain relaxation, if any, in certain regions. The interpretation of the strain relaxation will be covered later in section 5.5.3. The point of concern referred to an 80 μm pad that should be representative of the composition elsewhere, but may be strain relaxed.

The nature of the data collection is such that the actual relationship between the illuminated area and where the information emanates is more dependent on the focus size; this can be illustrated by modeling the whole configuration. As can be shown in figure 4.3.16 each part of the scattered intensity (i.e. along 2θ) comes from a different position on the sample as the sample extracts the associated incident beam trajectory. Therefore the probed region associated with each 2θ angle corresponds to the focus size, which is ~40 μm for a long fine focus X-ray source in the scattering plane. A similar argument can be given for the scattering out of the scattering plane (axial direction), and in this case the pinholes do result in some averaging. A simulation of the position on the sample from where the dominant scattering comes from for an InGaAs sample that created a 'static' profile, figure 5.4.9a, is given in figure 5.4.9c for an illuminated area given in figure 5.4.9b.

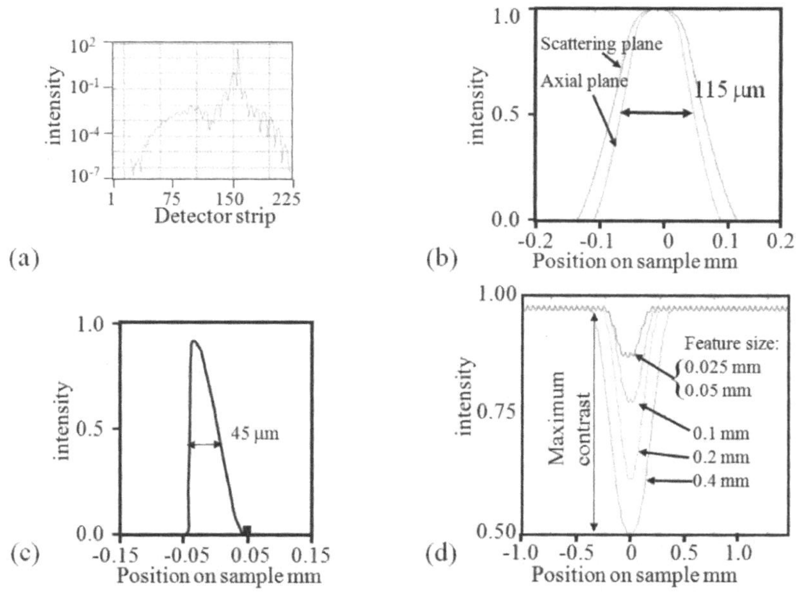

Figure 5.4.9. (a) gives the *113* reflection profile from an InGaAs/GaAs HEMT structure using the 'static' diffractometer. (b) is the distribution of intensity across the sample for a 100 μm double-pinhole and (c) gives the scattered intensity distributed across the sample, which predominately comes from the substrate. (d) illustrates the change in contrast for various narrow feature sizes and the consequent contrast using a 100 μm double pinhole and 50 μm steps, when maximum scattering contrast is 0.5.

Another important aspect is what can be observed in terms of feature size, i.e. how small a region can be detected. Various examples of the effect on contrast is given in figure 5.4.9d. Making use of these possibilities we have mapped a large Si wafer to isolate a small 80 μm square pad and determine the state of strain. The idea behind this approach is to set the detector to capture the whole GeSi/Si profile, figure 5.4.10a and the peak intensity used to map the wafer, so that when it is moved laterally in x and y the regions that have a Ge layer will absorb the intensity and the Si peak intensity is reduced, figure 5.4.10b. This was compared to an optical image of the wafer and with this the local region associated with the pad could be isolated. The contrast close to the pad was performed by monitoring the Ge peak, figure 5.4.10c. Once the pad was located we knew that the probe was accurately positioned and a reciprocal space map could be obtained, figure 5.4.10d (left hand image). The reciprocal space map took ~30 min and the SiGe layer was found to be fully strained to the Si substrate. The composition and thickness of the layer were determined through simulation to be 22.0(0.2) % Ge and 0.045(0.003) μm thick. In another region on the sample there was strain relaxation, figure 5.4.10d (right hand image).

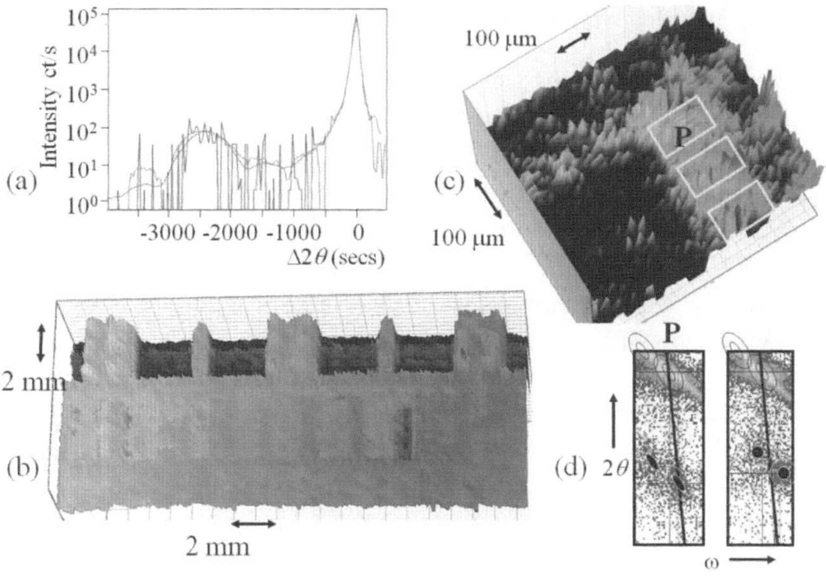

Figure 5.4.10. (a) gives the *113* reflection profile from a SiGe (darker spiky line) from the 80 μm square pad (40 s capture time), along with the smoothed profile and the simulated best fit. (b) the mapping of the Si peak intensity reveals processed features. (c) the mapping of the Ge layer peak that reveals the pad region. (d) left-hand reciprocal space map reveals that the pad is fully strained, whereas the right-hand map from another area reveals that the layer is partially strain relaxed.

With more powerful sources such as synchrotrons, wafer mapping can become quite rapid. For example Zeimer et al (2007) have studied the distortions in GaAs/InP/AlAs alloy structures with bonding and heat-sinking included. This makes it possible to examine the curvature and strains under heating, cooling and optical pumping the laser structures locally; thus giving a detailed interpretation of the changes under operational conditions. Much of this work was achieved at 100 × 100 μm^2 dimensions, however this can be reduced further. The parallel 'white-beam' impinges on the crystal orientated to the *004* reflection at ~8 keV (i.e. close to the laboratory Cu $K\alpha$ energy) and the CCD detector at 1 m away captures the Bragg scattering. As the sample is scanned in x and y, the sample can be probed for curvature and strains and indicate the influence associated with device heating.

5.4.3.10. Imaging with Tomography:

Another wafer mapping approach is based on X-ray tomography originally proposed by Ziedes des Plantes (1932), and now has developed into computerised tomography. This method reveals a more direct image of the device information and the basic principle is quite straightforward. A laboratory based system is often composed of a small point source, the sample to be imaged, and a 2D detector, whereas a synchrotron configuration relies on a large parallel source, which removes the distortion and makes the interpretation simpler. An exposure is taken at each rotation step, from which the image in 3D can be reconstructed (the algorithms are similar to those used in medical imaging CT systems). These methods usually rely on absorption contrast, i.e. the imaginary part of the refractive index given in equation 2.10.4. For soft tissues this contrast is weak, but a beam close to a boundary between regions can have differing path lengths, which will result in a phase difference and therefore contrast; this is associated with the real part of the refractive index, which compares with the dominant contrast in diffraction.

The contrast in tomography is largely related to the differences in electron density, and therefore loosely with composition for similar path lengths. The spatial resolution is strongly related to the source size, or parallelism of the incident beam, and the detector pixel dimension. The contrast will be a mix of the X-ray wavelengths used, the composition of the sample and the quality of the detector. A good photon counting detector will limit the noise levels giving a very large dynamic range. The pixel dimension is very relevant to parallel incident beams, but for point sources the detector can simply be moved further from the sample to achieve a natural magnification; therefore the pixel dimension is less relevant, but the source is important.

The work of Wilkins et al (1996) used phase contrast tomography with a micro-source and a large distant detector to achieve stationary impressive stationary results on biological specimens. An example of the synchrotron parallel beam is given by Heflen et al (2007), enabling detailed investigation of bonding contacts in integrated circuits with a detector pixel size of 1.6 μm.

5.4.3.11. General comments on imaging:

There are clearly many options in imaging, from the direct absorption tomography, where the contrast relies on electron density differences, to phase-contrast tomography to reveal edges of regions in weakly absorbing materials, section 5.4.3.10, through to mapping curvature and inferred strain at high spatial resolution, sections 5.3.5 and 5.4.3.9, and composition and thickness. These methods can be very useful in revealing inhomogeneities across wafers. The detail can of course be almost too much to usefully absorb, but even a limited wafer map, e.g. 9 points, can give some reassurance that the reported results are representative of the whole.

5.4.4. *Taking account of sample imperfections (simple structures):*

If the SiGe HBT structure, discussed earlier in sections 5.4.3.1 to 5.4.3.4, was perfect then the profile obtained with the double-crystal diffractometer (figure 4.3.1b) should be almost identical to that from the multiple-crystal diffractometer (figure 4.3.5). However these two profiles are not the same (figure 5.4.2 compared with figure 5.4.4), especially close to the layer and substrate peaks: there is significant broadening at their bases. As was illustrated in figure 4.3.8, the double-crystal diffractometer integrates intensity across the detector aperture, i.e. a large angle in 2θ, whereas the multiple-crystal diffractometer captures a very small region of diffraction space. The diffraction space map, or reciprocal space map (this differentiation is given in section 2.5), will reveal these differences.

5.4.4.1. Obtaining a reciprocal space map:

The description here will be for obtaining reciprocal space maps with the multiple-crystal diffractometer, where the experiment is that as described in section 5.4.3.5, with the monochromator and analyser (i.e. procedure 1 to 10). Rather than conducting a single $\omega/2\theta$ scan we should perform several of these with slightly different settings. Continuing on from the action number 10 above, and since we know that the intensity close to the substrate is broader than it should be a suitable next step is:

1. Scan in ω over a reasonable range over the substrate peak to see how far the intensity spreads.
2. Use this angular range as a guide to the spread in the width (ω) of the scattering to be captured by reciprocal space mapping.
3. The reciprocal space map is then collected over the angular range given by the $\omega / 2\theta$ scan of the previous experiment and the ω spread of step 1.
4. The resulting reciprocal space map can be collected either at a similar or shorter count-time to the previous experiment depending on the detail of the analysis. For this purpose it was collected with 1 s per point. There were 1000 steps in $\omega / 2\theta$ and after each scan the ω angle was offset by 0.0015^0. This was repeated 60 times to give a respectable reciprocal space map in 17h, figure 5.4.11.

The reciprocal space map shows significant diffuse intensity around the substrate peak and the layer peak. We can take a guess at the cause of this additional scattering or we can build on our picture by placing the instrument probe (i.e. setting the ω and 2θ angles) to be on these "wings."

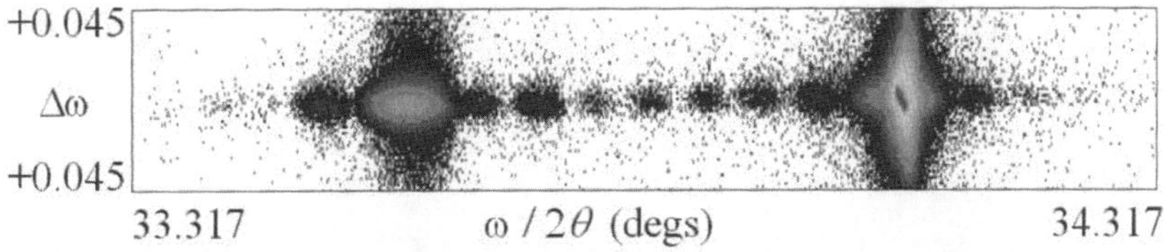

Figure 5.4.11. The reciprocal space map of the SiGe / Si sample, indicating the significant diffuse scattering close to the substrate and layer reflections.

5.4.4.2. Multiple-Crystal topography:

If we now place a topographic emulsion after the analyser we will see which parts of the sample contribute to this scattering, figure 5.4.12. What we observe is a crosshatch of lines that lie along the *<110>* type directions. We can assume from this that defects are forming at the GeSi / Si substrate interface and these are 60^0 dislocation that are partially relieving the internal stresses in the sample. The image width of these lines in the scattering plane is about 7 μm and in the axial direction 100 μm. These lines are predominately the strain-fields from single dislocations. We can understand the whole scattering shape to arise from the local stretching of the Si substrate lattice parallel to the interface plane; this reduces the lattice parameter normal to the interface through the Poisson effect, figure 2.8.3. Consequently the scattering angle is increased for this region and the tips of these "wings" are at higher angles than the main peak.

The spread in the ω direction arises from the curvature of the scattering planes close to the dislocation. The very high strain sensitivity of this experimental configuration means that the distortion is detected a long way from the dislocation core (i.e. many microns). This scattering is only that which comes from between the perfect region and the outer limits of the dislocation strain-field and explains the very small angular spread of the scattering. The analysis close to the layer peak can be carried out in a similar way.

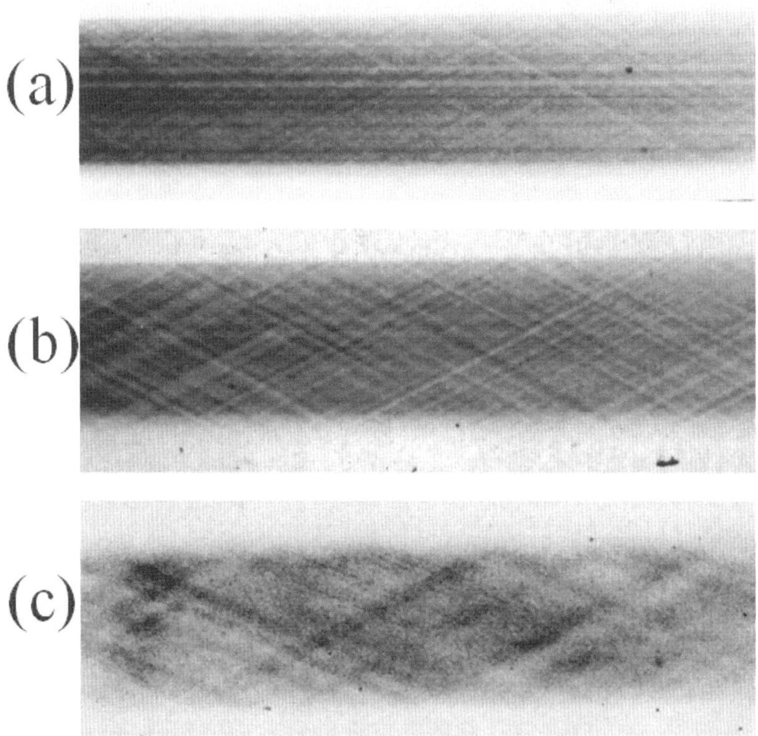

Figure 5.4.12. A series of topographs (a) and (b) with the emulsion placed immediately after the sample at the layer and substrate peak respectively (the later is a 5 minute exposure L4 emulsion). The topograph (c) is obtained on the diffuse scattering region close to the substrate with the emulsion placed after the analyser (2 h exposure). Note the cross-hatch of lines parallel to the <*110*> directions typical of relaxing layers.

Since the individual dislocations are clearly observed we can count the number per unit length and determine the degree of strain relaxation. We have to make an assumption on the type of dislocations that exists in the sample and in Si these are almost entirely 60^0 dislocations with a Burgers vector, b, 45^0 to the interface plane. Each dislocation will therefore take out a line of atoms equivalent to $b_{//} = d_{<110>}/2 = a/\{2\sqrt{2}\}$. In this case we can measure m (~12) dislocations in a length M (~230 μm) in two orthogonal directions, giving a dislocation density of

$\rho_{disl} \equiv \dfrac{2\mathrm{m}}{\mathrm{M}^2} = 4.54\times10^4$ cm^{-2}. The mismatch and dislocation density in the plane of the interface is given by

$$\left(\frac{\Delta d}{d}\right)_{//} = \frac{\mathrm{m}b_{//}}{\mathrm{M}} = \frac{\mathrm{m}\dfrac{a}{2\sqrt{2}}}{\mathrm{M}} \qquad 5.4.24$$

$$\rho_{\mathrm{disl}} = \frac{2\left(\dfrac{\Delta d}{d}\right)_x\left(\dfrac{\Delta d}{d}\right)_y}{\left|b_{//}\right|^2} \qquad 5.4.25$$

In this case the mismatch is ~2×10^{-5}. The actual mismatch of free-standing $Ge_{0.19}Si_{0.81}$ and Si is ~ 7.93×10^{-3}, giving the relaxation % as

$$R = \frac{2\times10^{-5}}{7.93\times10^{-3}} \sim 0.26\% \qquad 5.4.26$$

This is the only convenient way to measure very small relaxations, since the influence of the scattering pattern as we will see later is inadequate to change the peak separations. For higher relaxation a peak shift is detectable and the method of counting dislocation lines fails because the individual lines overlap and cannot be resolved.

Another feature of the reciprocal space map is that the layer and substrate peaks do not lie on the same $\omega/2\theta$ scan. This is because the layer is tilted with respect to the substrate. The effect is very small, only 7.2 s difference in this case. However we ought to be aware of the difference since this can impact our analysis. The tilt angle adds or subtracts directly from our peak separation in the rocking curve with no analyser. In our previous rocking curve analysis we had not taken this into account and this we shall do now.

5.4.4.3. Taking account of tilts in rocking curve analyses:

Returning to our GeSi sample we can determine the influence of tilt on our composition determination. In this particular example the tilt is very small; from the reciprocal space map we can see that it amounts to 7.2 s. From figure 5.4.2 we can see that a layer tilted with respect to the substrate by $\Delta\varphi$ will decrease or increase the peak separation $\Delta\omega$ depending whether the tilt is towards or away from the incident beam direction. Therefore the average angular separation from two rocking curves of the same set of scattering planes with opposite beam paths (a rotation of π in ϕ) will have the correct value based on a simple analysis, section 5.4.1. Similarly the difference in angular separation $\Delta\omega$ for the two rocking-curves will correspond to $2\Delta\varphi$. We therefore have two choices: include $\Delta\varphi$ into the model of the structure or model both rocking curves and take some average. Clearly the better way would be to include the tilt into the model of the structure since all the parameters will be influenced (composition and thickness).

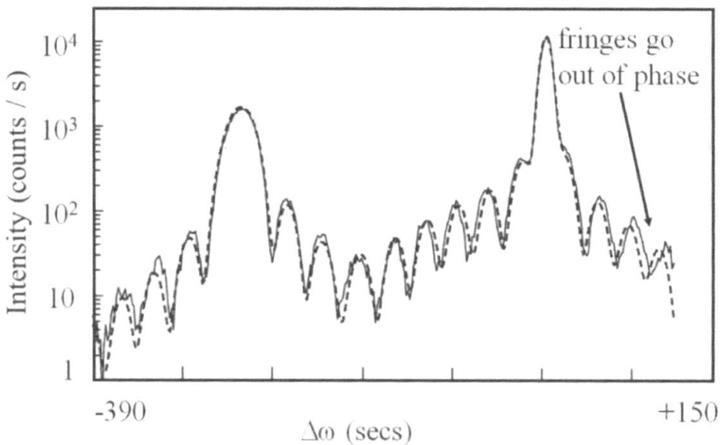

Figure 5.4.13. The measured and simulated profile indicating some of the problems in achieving an exact fit.

In the GeSi example considered above the difference in the two rocking-curves is barely discernible, because we are observing a 7 s tilt in a separation of 1850"arc between the layer and substrate. However when the tilt is a significant proportion of the difference, we have to take care in the interpretation. The most exacting measurement is achieved with the multi-crystal diffractometer, figure 5.4.4 unless the tilt is too large and one simple $\omega/2\theta$ scan will not capture both the layer and substrate peaks. This will be covered in section 5.4.4.4.

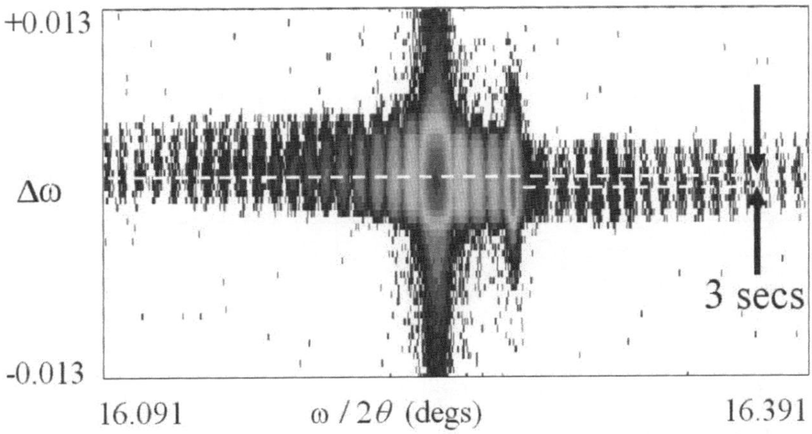

Figure 5.4.14. A reciprocal space map indicating the subtle kinks in the dynamical streak that leads to the small displacements in the fringing evident in figure 5.4.13.

Before we leave the problem of analysing tilts in nearly perfect structures we should consider a fairly common problem where a perfect fit can be difficult to achieve. Consider the rocking curve of an AlAs / GaAs superlattice on a GaAs substrate, figure 5.4.13. The simulated scattering pattern looks almost a perfect fit, however the fringes appear to go out of phase outside the region between the substrate and layer peaks. We also find that the problem exists when we use multi-crystal optics. The reciprocal space map, figure 5.4.14 reveals a rather strange phenomenon, the fringes appear to have their maxima aligned along the surface normal direction outside the region of the substrate and layer peaks, whereas they lie along a line between the two peaks. The layer and substrate have a small relative tilt. This phenomenon can be simulated when we use a very exact model taking into account the full instrument function response and coherence of all the scattering contributions in the reciprocal space map, section 2.8.2. The simulated reciprocal space map illustrating this effect is given in figure 5.4.15. Except in the most precise analyses we can ignore these effects and take some reassurance that these shifts will reflect a small tilt. The shift of these tilts can be determined geometrically from knowledge of the tilt angle.

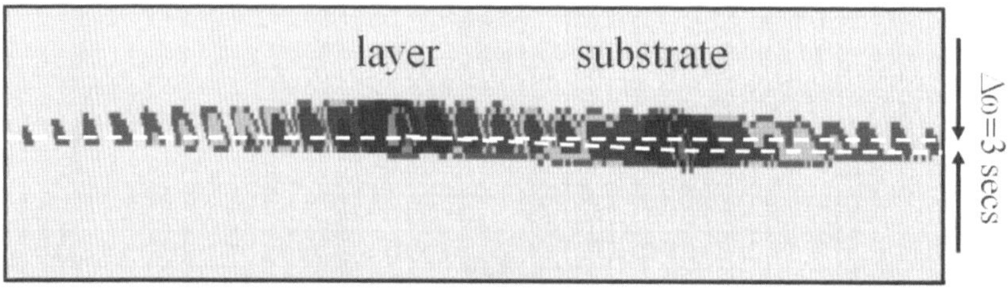

Figure 5.4.15. The simulation of a reciprocal space map of a structure containing a very small layer misorientation with parallel surface and interfaces.

5.4.4.4. Modeling the extent of the interface disruption in relaxed structures:

Using the above analysis we can obtain a good fit to the multi-crystal diffraction profile whereas that from the double-crystal diffractometer shows discrepancies. We have already indicated the source of this effect from topography, i.e. dislocation strain-fields close to the interface of the substrate and GeSi layer. From the strain model and scattering theory described in section 2.8.2 we can build a better picture of our sample by estimating the extent of the distortions around this interface.

In this particular example we will project all the intensity of the reciprocal space map onto the *<001>* direction, this then takes out the lattice rotation effects and tilts but includes all the associated diffuse scattering. This will not be identical to the rocking-curve, since the rocking-curve effectively projects the intensity that exists along a line inclined to the direction normal to *<001>*, see figure 4.3.3. However we should be able to account for all the intensity in this profile in terms of thickness, compositions and interfacial strains. A full simulation of the reciprocal space map will give the plane rotations and tilts, etc.

The projected profile is given in figure 5.4.16a. The initial composition and thickness values are taken from the above analysis. Clearly if we are extending the interfacial disruption then this will influence the apparent thicknesses. The best-fit profile given in figure 5.4.16a was obtained with the depth of interfacial disruption into the layer and substrate as 4.4 nm and 77 nm. The strain has been assumed to

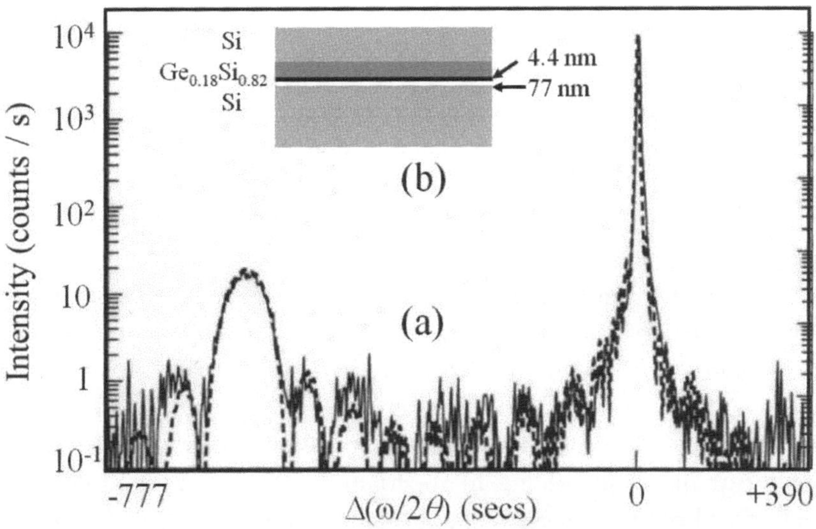

Figure 5.4.16. The profile obtained by projecting the reciprocal space map to include all the diffuse scattering. The fit to the full profile (dash line), including the diffuse scattering gives the full structure including the distortion at the interfaces.

take on an exponential form and this disruption depth represents the full extent of this gradation in strain. The thickness and compositions were modified slightly to give a picture of the sample given in figure 5.4.16b. The shape of the profile close to the interface can therefore be understood in terms of the distortion associated with dislocations close to the interface.

This above analysis just reveals some average of the defect disruption at the interface, and information about the dislocation density can be revealed if it is assumed that they are of a particular type and are randomly distributed. For a more detailed analysis bringing together correlated positions and mixtures of dislocations, the reader is referred to section 2.8.2 and the cited references.

5.4.5. *Analysis of periodic multi-layer structures (complex structures):*

A periodic multi-layer structure is composed of a sequence of layers that are repeated one or more times to create different physical properties from the individual layers. If the properties of interest see this repeat sequence behaving as a whole then this is a superlattice. Periodic multi-layers create very special properties either as Bragg reflecting stacks for opto-electronic devices, or as superlattices that have strong wave function coupling to produce band-folding and new possibilities in quantum effect devices. The periodicity also has a large influence on the X-ray scattering and presents opportunities for extracting information quite directly and for studying interfaces in greater detail.

The structural information required from periodic multi-layers is very similar to that discussed above except that additionally we are interested in variations in periodicity or departures in periodicity and the abruptness and roughness of interfaces. The variation in periodicity and interface quality would influence the reflecting power in a Bragg stack or the quality of the band folding in superlattices. The concentration in this section will centre on these additional parameters.

5.4.5.1. The analysis using direct interpretation of the scattering pattern:

If we simulate the scattering from a periodic structure with dynamical theory we will see, depending on the thicknesses involved, a series of satellite peaks associated with the average composition. We can consider this as a modulation of the composition and the strain about some average value.

From equation 5.4.13 we see that the thickness can be obtained from the separation of the fringes in the scattering profile. Therefore there should be fringes associated both with multi-layer periodicity and with the overall thickness of the periodic structure. We can determine the period by combining various combinations of satellite positions

$$\Lambda = \frac{(M_i - M_j)\lambda}{2(\sin\omega_i - \sin\omega_j)} \qquad 5.4.27$$

If the structure is composed of a binary (e.g. GaAs) and a ternary (e.g. $Al_xGa_{1-x}As$) and the thicknesses of the contributing layers are t_{BC} and t_{ABC} then the average composition within the period Λ is given by

$$\bar{x} = x \frac{\frac{\bar{a}}{a_{ABC}} t_{ABC}}{\Lambda} \sim \frac{x t_{ABC}}{\Lambda} \qquad 5.4.28$$

where

$$\Lambda = t_{BC} + t_{ABC} \qquad 5.4.29$$

The approximation sign in equation 5.4.28 arises from the fact that the composition ratio x is related to the unit cell and not to a common unit of length throughout the structure. We have therefore had to relate the number *"A"* atoms per unit length and translate this to the number of equivalent *"A"* within the period with average lattice parameter $\bar{a}$. However in the initial analysis we could assume that the unit cell repeat parallel to the growth direction for the two layers are the same. If the periodic structure is composed of two materials of known composition (e.g. two binaries, AlAs and GaAs) then we can consider the "average" structure to be composed of an alloy of the two materials (e.g. $Al_xGa_{1-x}As$). The average composition is hence given by

$$\bar{x} \sim \frac{t_{AC}}{\Lambda} \qquad 5.4.30$$

where t_{AC} is the layer thickness of the phase AlAs for example.

We can simply determine a value for the period, equation 5.4.27 and the thickness of the individual layers, equations 5.4.29 and 5.4.30 for a periodic structure of known compositions in the individual layers if the average composition is determined. The average alloy can be determined approximately from the angular separation of the "average" layer peak and that of the substrate in the rocking-curve, section 5.4.1.

From these equations we can derive some basic information directly from the scattering profile.

5.4.5.2. The analysis using kinematical theory:

From the above analysis we can obtain some information, although it does not yield the composition in complex (e.g. three layer repeat) superlattices or interfacial quality. To obtain this information we need to model the intensities. Kinematical theory can prove to be very convenient and fast in extracting this information, Fewster (1986, 1988, and 1998). The basis for this approach is to consider the repeat as a unit cell and determine the associated intensity using the kinematical theory approximation.

Firstly we determine the structure factor for the period. The size of the unit cell repeats parallel to the interface can be assumed or measured, but equated to the normal unit cell repeat. However the unit cell perpendicular to the surface is now the repeat period of the superlattice. We can determine this period directly, but of course it may not be equivalent to an integer number of unit cells of a_{BC} and a_{ABC}. In reality it will be composed of several combinations depending on the quality of the superlattice and therefore the period is given by

$$\Lambda = (na_{BC} + ma_{ABC})K_1 + ([n+1]a_{BC} + [m-1]a_{ABC})K_2 + ([n-1]a_{BC} + [m+1]a_{ABC})K_3 + ([n+1]a_{BC} + ma_{ABC})K_4 + (na_{BC} + [m+1]a_{ABC})K_5 + ([n-1]a_{BC} + ma_{ABC})K_6 + (a_{BC} + [m-1]a_{ABC})K_7 + \ldots\ldots\ldots\ldots \qquad 5.4.31$$

where

$$K_1 + K_2 + K_3 + K_4 + K_5 + K_6 + K_7 + \ldots\ldots = 1 \qquad 5.4.32$$

and a_{ABC} and a_{BC} represent the lattice parameters along this direction and m and n are integers. The integers m and n are values that give the closest fit to the period.

$$\Lambda \sim na_{BC} + ma_{ABC} \qquad 5.4.33$$

We therefore have to calculate the structure factor several times to account for the incommensurability in the structure, depending on the superlattice quality and precision we wish to achieve. If the integers are large then the importance of including a high number of terms is less important because the ratio of the contributing layers are not too seriously affected. The index of reflections associated with each satellite become rather large ~ $(m+n)$ and because the period will almost certainly be incommensurate with the lattice repeat periodicity this index is not exact. It therefore becomes easier to index the satellites as *–2, -1, +1,*

+2, etc. with respect to the "average" or "zeroth order satellite" peak for the superlattice.

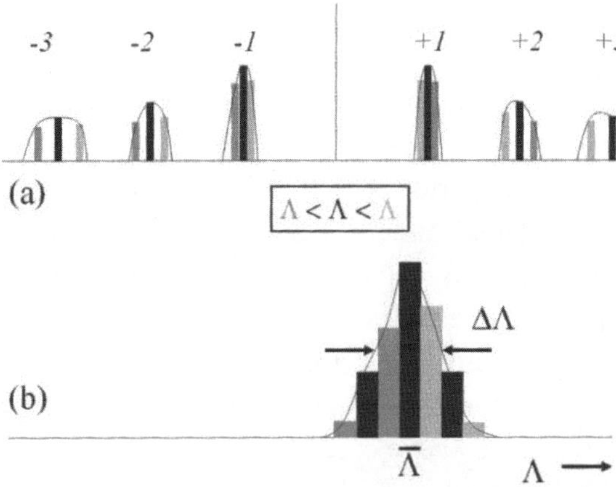

Figure 5.4.17. (a) A simple explanation of how the distribution of periods in a structure will give rise to the broadening of satellite reflections. The distribution of periods can be approximated to a Gaussian form (b).

We now have to establish the magnitude of the K_j values and this will depend on the variation in the period, which is also related to the interfacial roughening, Fullerton, Schuller, Vanderstaeten and Bruynseraede (1992). This can be determined directly, Fewster (1988). From a rather simplistic viewpoint we can imagine a structure composed of several different periods that overall still has some average structure and an "average" scattering peak. Those parts of the structure that have a longer period will have satellites at smaller angles than the average position and those parts that have a shorter period than the average will scatter at larger angles than the average, figure 5.4.17a. For satellites further away from the "average" peak the angular difference to the "average" satellite position increases and the satellites broaden. If we now differentiate equation 5.4.27 above for the period to obtain the variation in the period we obtain

$$\Delta\Lambda = \frac{(M_i - M_j)\lambda\Delta\omega_M}{\cos\omega(\Delta\omega)^2} \qquad 5.4.34$$

where $\Delta\omega$ is the distance between two satellites i and j, and $\Delta\omega_M$ is the angle within one satellite reflection for a difference in period of $\Delta\Lambda$.

The satellite order is M (in this notation $M = 0$ is the "average" peak order). This can therefore be measured directly by plotting the satellite order against the full-width-at-half-maximum intensity and deconvolving the zeroth order peak width. The zeroth order peak width should be unaffected by period variations and should represent the broadening effect of the data collection method and sample broadening effects. This method assumes that the period variation is random and not systematic. Since this can be measured directly from the width of the satellites this information can be included in the modelling of the satellite intensities. The period variation is assumed to be Gaussian, and the number of unit cells to create a repeat unit is distributed about the mean value, figure 5.4.17b. The deconvolution of two Gaussian profiles is also rather trivial and straightforward, in the above case we can write

$$\Delta\omega_M = \left\{\Delta_m\omega_M^2 - \Delta_m\omega_0^2\right\}^{\frac{1}{2}} \qquad 5.4.35$$

where $\Delta_m\omega_M$ is the measured peak width and M the satellite order.

The structure factor is calculated using equation 2.3.9 for all the possible repeat units that straddle the measured period and these are added coherently (maintaining the phase relationships). The intensity is obtained from equation 2.7.14 taking into account the absorption of the X-rays above the periodic structure. The angular position for the satellite is given by

$$\theta_j = \sin^{-1}\left(\frac{M'\lambda}{2\bar{a}} + \frac{M\lambda}{2\Lambda}\right) \qquad 5.4.36$$

where M is the satellite order with respect to the "zeroth order satellite" peak and M' is the order of the reflection for the average lattice parameter in the period, $\bar{a}$. Now of course the average lattice parameter $\bar{a}$ is given by

$$\bar{a} =$$

$$\frac{(na_{BC} + ma_{ABC})}{(n+m)}K_1 + \frac{([n+1]a_{BC} + [m-1]a_{ABC})}{(n+m)}K_2 + \frac{([n-1]a_{BC} + [m+1]a_{ABC})}{(n+m)}K_3 \qquad 5.4.37$$

$$+\frac{([n+1]a_{BC} + ma_{ABC})}{(n+m+1)}K_4 + \frac{(na_{BC} + [m+1]a_{ABC})}{(n+m+1)}K_5 + \frac{([n-1]a_{BC} + ma_{ABC})}{(n+m-1)}K_6$$

$$+\frac{(a_{BC} + [m-1]a_{ABC})}{(n+m-1)}K_7 + \ldots\ldots\ldots$$

We can therefore very easily determine the intensity associated with each satellite and compare these with the measured intensities. As discussed in section 2.7.2 on the assumptions of the kinematical theory the shape is determined by the thickness and this will smear the intensity that we have just evaluated. The simplest approach is to determine the integrated intensity associated with each satellite and compare this with various model structures. The advantage of this approach is that we can ignore the shape of the peaks and this gives us the possibility to collect the data using many different methods.

Since this approach is not restricted by the contents of the periodic repeat unit we can very easily include variations in composition and several layers in each period. The variation in composition could be interfacial spreading for example and this can take on any shape (Gaussian, linear, exponential or erf) appropriate to how the grading is formed. Also the advantage in just concentrating on the integrated intensities of the satellites is that we can use high intensity low-resolution diffractometers as described in section 4.4.1.

The structure factors of all these various contributions are added as before and then multiplied by its complex conjugate to give the intensity after including the absorption and instrumental aberrations, equation 2.7.14.

This kinematical theory model of the scattering makes several assumptions. The major assumption is that the structure behaves as some "average" layer with a modulated perturbation of the strain (lattice parameter variation) and composition. What we show in the next section is that these assumptions are only valid over a certain range of thicknesses.

5.4.5.3. Considerations regarding the analysis of periodic multi-layers with dynamical, kinematical and optical theories:

Modeling with dynamical theory is the most exact way of simulating the scattering from periodic structures. Fewster (1993a, 1998) has considered when the kinematical theory breaks down and this will be briefly summarised here and the effects that occur.

We have already established that the peak position of the layer, in this case the "average" layer, cannot be a reliable measure of the composition or strain value because of the dynamical pulling effect, section 5.4.3. So clearly we have to simulate the profile to obtain a reliable value. Also the whole structure should be simulated and not the average layer composition for the most precise work since the equivalent "average" layer peak does not occur at the same position as the periodic structure "average" layer peak. This error is of the order of 4% for a 50×[AlAs (5 nm) + GaAs (5 nm)] superlattice on GaAs for example. Let us assume now that we obtain some "average" composition then as stated above we have to take account of the difference in lattice parameters and elastic parameters of the individual components in the superlattice. The composition determined is that for the average lattice parameter and will therefore vary with the ratio of the layer thicknesses. Consider the ternary / binary system then the composition we need to include in the simulation is

$$x = \frac{\bar{x}\Lambda a_{ABC}}{t_{ABC}\bar{a}} \sim \frac{\bar{x}\Lambda}{t_{ABC}} \qquad 5.4.38$$

Since in the dynamical theory we are concerned with inputting the composition and layer thickness we can derive the average lattice parameter

$$\bar{a} = \frac{\Lambda}{n'+m'} = \frac{\Lambda}{\left\{\frac{t_{BC}}{a_{BC}} + \frac{t_{ABC}}{a_{ABC}}\right\}} \qquad 5.4.39$$

where n' and m' are non-integer and the a parameters refer to the direction normal to the layers. The simulation of the rocking-curve on the basis of this "average" layer compared with the simulation of the full superlattice structure is given in figure 5.4.18. Clearly for the most precise work we have to be aware of the errors introduced in working from some average composition.

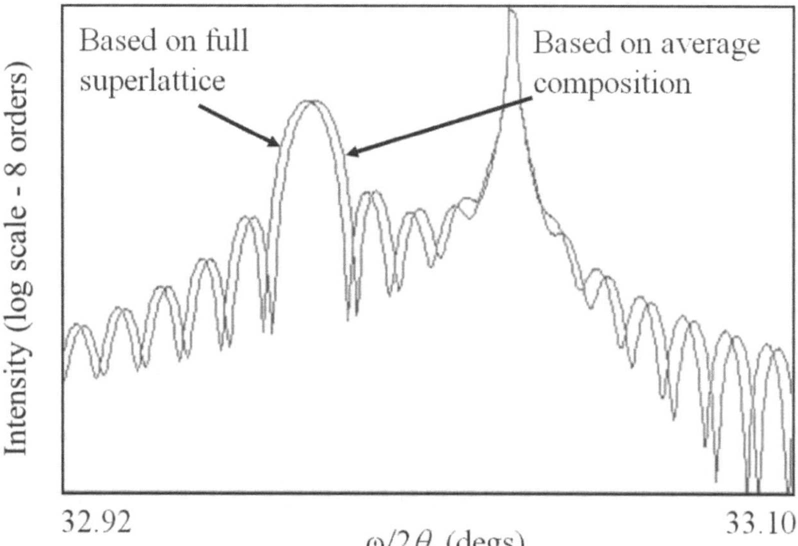

Figure 5.4.18. The difference in simulated profiles from the average of the superlattice structure and the actual superlattice structure.

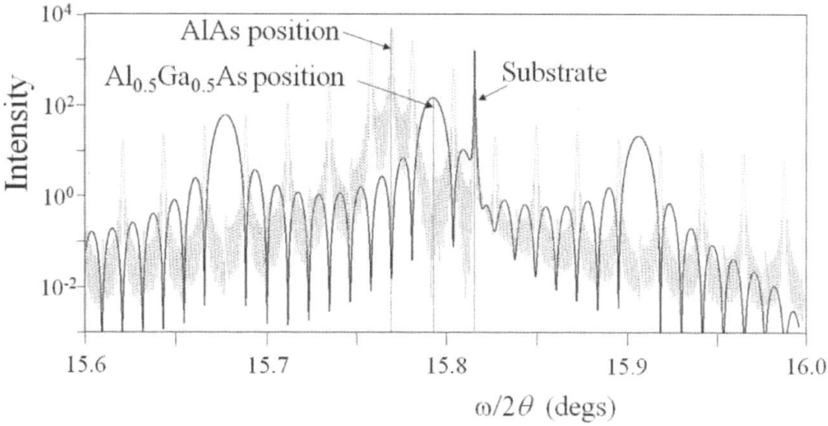

Figure 5.4.19. As the period increases the assigning of the "average" superlattice peak becomes difficult. Both these simulations have the same average composition, equal thicknesses of AlAs and GaAs, yet the longer period structure (0.4 μm grey dots) gives the most obvious position corresponding to the position of AlAs instead of the $Al_{0.5}Ga_{0.5}As$ position as in the 0.04 μm period structure (black line).

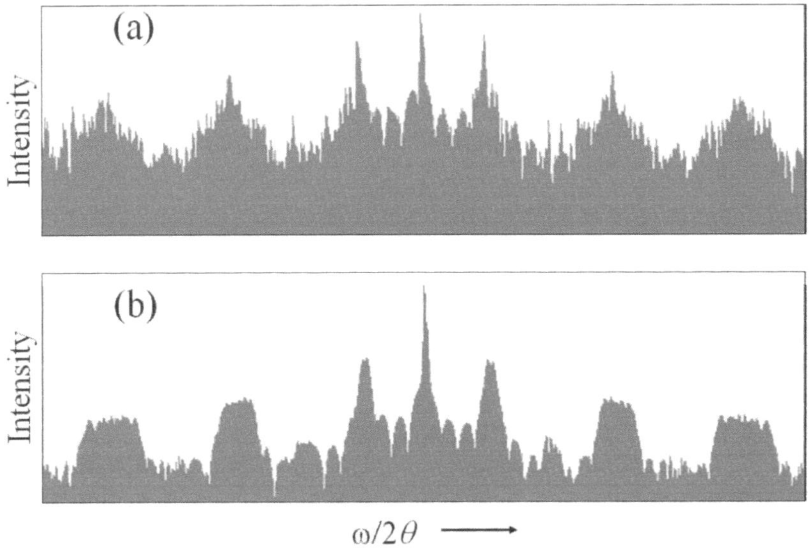

Figure 5.4.20. The influence of random period variations (a) compared with decreasing periods as a function of depth (b), for the same average period throughout the structures.

Perhaps a more dramatic effect occurs when the period is gradually increased. The concept of an average peak disappears, figure 5.4.19. Also for very few periods the satellites become broadened and the characteristic length should be determined from the minima and not the maxima since this becomes equivalent to a simple interference problem, section 2.7.2. Other features to note are the complex interference phenomenon in some of the profiles; these can shift the satellite peaks positions at the 1 % level. Again this may not be significant for the precision of the analysis required but it is important to be aware of the limit of using a more direct analysis. Clearly modelling the scattering with dynamical theory is the most exact method, however the amount of information required to obtain an exact fit can be more readily found by a feeling of the sensitive features that are more easily observed from a kinematical theory viewpoint.

The variation in the period that was discussed in section 5.4.5.2 can also be confirmed using the dynamical theory simulation, figure 5.4.20. The case for large variations has been given for a random variation and a systematic variation. The satellites do increase in width with satellite order for both types however the shapes of the satellites are very different, the random variation are more Gaussian, whereas the systematic variation is much squarer in shape. These calculations do not include randomness laterally, which would smear the high frequency fringes.

We can now give some basic guidelines for the applicability of the kinematical and dynamical theories.

- For >3 periods of 4 nm $< \Lambda <$ 40 nm the kinematical theory (using the integrated intensities of the satellites) gives reliable results.
- For < 3 periods or $\Lambda <$ 4 nm and $\Lambda >$ 400 nm the dynamical theory should be used
- The direct interpretation is a good guide to subsequent simulation or for deriving approximate working values.

These are very approximate guides and in general should be simulated to ensure that these regions are valid for the information required and material being studied.

5.4.5.4. Analysis of periodic structures with reflectometry:

Reflectometry is sensitive to the composition through variations in the structure factor F_{000} as a function of depth below the surface, section 2.10. In this section we are assuming, as in the whole of section 5.4, that the samples are nearly perfect and this influences our measurement and interpretation. The sensitivity to composition is poorer than the experiments close to Bragg peaks, becuase reflectometry is much more dependent on the magnitude of the intensity rather than the angular displacement (strain effect). However reflectometry can prove very useful for the determination of the period in superlattice structures and interface quality providing care is taken in the measurements.

In the previous measurements the diffuse scattering due to imperfections is generally small, but close to the *000* origin of reciprocal space (the reflectometry region) these effects can be large and can influence the interpretation. However the multiple-crystal diffractometer can separate the diffuse from the main signal. A bent sample can cause problems with this measurement because different regions will scatter at different incident angles. The intensity can be recovered by a small reciprocal space map and projecting along ω; if there is significant diffuse scatter this will be included and could change the intensity fall-off and change the estimated roughness, see figure 2.10.3.

The setting up of the experiment for obtaining a reflectometry profile follows a slightly different approach to that in section 5.4.3.5. Although that procedure is sufficient for that analysis, the following is more precise and preferred for a more careful experiment. This procedure puts ω and 2θ onto an absolute scale.

1. *The analyser / detector assembly is scanned through the direct beam to establish* $2\theta = 0$ *from the peak position.*
2. *The sample is placed on the stage and translated until the beam is approximately cut in half, with* $\omega \sim 0$.
3. *The analyser/detector assembly should then be offset to* $2\theta \sim 0.3^0$, *i.e. just above the critical angle.*
4. *The* ω *angle should then be set to* $\sim 0.15^0$ *and then scanned over* $\pm 0.15^0$, *and set to the peak. This peak position will be exactly* 0.15^0, *and from now on the* ω *and* 2θ *angles should be driven together in the ratio of* 1:2.
5. *A scan with these two angles coupled in this way will reveal a reflectometry scan on an absolute scale.*

From equation 5.3.11, we could see that the influence of the refractive index is larger for smaller scattering angles and this will influence the satellite positions as well as the fringe spacing effects, section 5.4.2 and table 5.4.2. The direct interpretation clearly suggests that simulation of the profile is the only reliable method to interpret the profile.

5.4.5.5. Example analyses of nearly perfect periodic multi-layers:

In many ways the analysis of a nearly perfect periodic multi-layer does not differ from the multi-layer structures described in section 5.4.3 and the whole analysis can be approached with dynamical theory. For ease of analysis we can assume that each repeat unit is identical and hence the layer thickness can be linked, unless we include some systematic or random variation in the period. We will consider three analyses using the kinematical, optical and dynamical theories, and the complications of highly asymmetric periods, very complex structures and interface considerations.

5.4.5.5.1. Analysis based on the kinematical approach:

The positions of the satellites are defined by equation 5.4.36 above and the intensity is given by the kinematical equation 2.7.4 with a correction for the absorption in 2.7.14. The constant of proportionality in equation 2.7.4 is effectively the overall scale-factor and this has to be determined. The structure factor can therefore be directly related to the measured intensity. Now the structure factor in this case is as defined in equation 2.3.9 and the unit cell corresponds to the repeat unit. We can therefore add in aspects of interfacial grading by changing the proportion of partial occupancy, whereas the incommensurability is included by coherently adding several structure factors corresponding to different unit cell (period) combinations, as expressed in equation 5.4.31. The measured intensity should correspond to the integrated intensity after background removal. The process of fitting the experimental intensity to the calculated intensity now becomes an iterative process, but unlike the dynamical model we are unconcerned with the profile shape and therefore the number of data points to be matched is related directly to the number of satellites measured.

To simplify the process the number of parameters should be reduced or at least estimated for rapid convergence. The average period is known and should represent the centre of some (e.g. Gaussian) distribution of periods, figure 5.4.17b. The width of this distribution can be determined from the period variation, equation 5.4.34. If the lattice parameter of the whole structure is constant parallel to the interfaces throughout its depth then we have just thickness and composition to include throughout one period. If strain relaxation exists then the modeling requires a combination of reflectometry and this analysis to achieve a complete picture, Birch, Sundgren and Fewster (1993). In this example we will assume the layers are not relaxed. We should consider the analysis of the interfaces through an understanding of the growing surface, figure 5.4.21. During deposition it is highly likely that the composition is switched before an exact number of unit cells or atomic layers are completed, also the low surface diffusion will limit the ability for all dips and hillocks to interact. Therefore we must expect some roughness, which when averaged laterally appears as a spreading in the interface. The interface will also in general not correspond to an integer number of building blocks (unit cells or atomic layers) therefore we have an incommensurate structure and this should be accounted for by including several combinations (with a minimum of two for each interface).

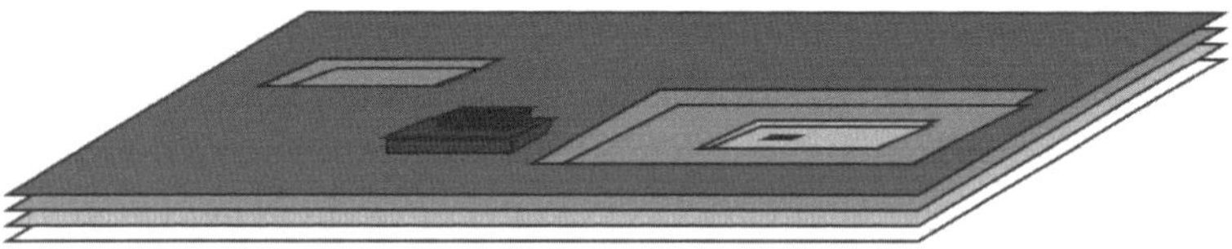

Figure 5.4.21. The variation in atomic layer completion from limited surface diffusion leading to surface roughening and incommensurate layer structures.

We have now narrowed the problem to a description of the composition (that relates to the strain, through the lattice parameter and Poisson ratio, equation 5.4.4 to 5.4.10) within the period. These parameters can now be varied in discrete steps, since there are discrete numbers of atomic layers until the best agreement between the experimental and calculated intensity is achieved.

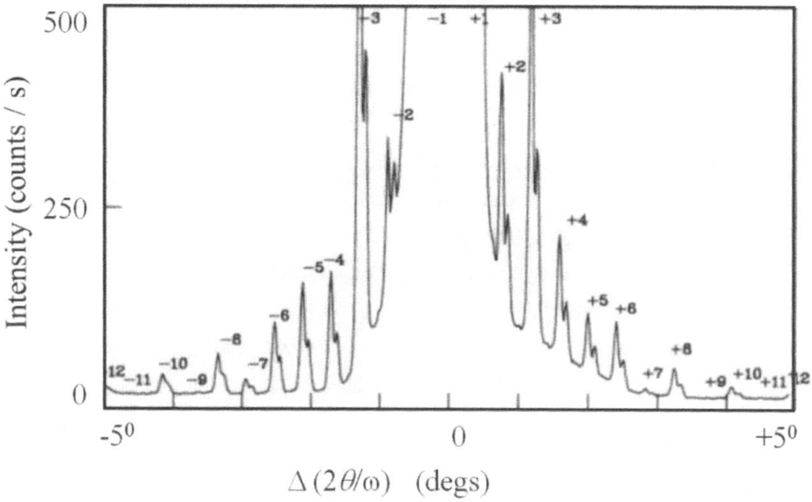

Figure 5.4.22. A 30 minute data collection scan of a periodic structure using a slit based diffractometer close to the *002* reflection from GaAs. The structure is a GaAs / $Al_xGa_{1-x}As$ superlattice on GaAs.

In this example we shall just consider an AlGaAs / GaAs superlattice and derive the composition and thicknesses of the individual layers and determine the interface extent. Since we will initially be concentrating on the kinematical theory for rapid analysis this method is restricted to the bounds defined in section 5.4.5.3. This example will not use high-resolution instrumentation at all (although there are advantages in doing so) to indicate what can be achieved with what is conventionally called a "powder" diffractometer, section 4.4.1.

One significant advantage of this approach is that we can use low-resolution diffractometry with all the advantages of high intensity. Also since this is only applicable to periods less than about 40 nm the problems of peak overlap are a less significant problem. We can consider the satellites as the coefficients of a Fourier expansion and therefore to obtain good detail about the intermixing at the interfaces we should measure to as higher order as possible. We also should choose a suitable reflection that enhances the satellite intensities and this would depend on whether the contrast between the different layers is dominated by differences in scattering strength or strain. For AlGaAs / GaAs the strain effects are not too strong, however the scattering contrast between them is significant. From equation 5.4.2 we see that the strain effects become more prominent for higher 2θ values, whereas the scattering contrast will depend on the relative phases of the scattering components in the structure factor, equation 2.3.9. For GaAs and AlAs the *002* reflection is dominated by the difference in scattering strength of the Ga and As and the Al and As scattering factors respectively. For GaAs this effect is small, whereas for AlAs this effect is large. This reflection has good contrast that is very convenient for (*00l*) surface planes, where a simple scan along ω / 2θ (scanning both axes) in the ratio of 1:2 will capture the full diffraction profile.

The data illustrated in figure 5.4.22 was collected in about 30 min, and after taking account of the background the integrated intensity for each satellite was determined. The full width at half maximum was also measured for each reflection and plotted as a function of satellite order, figure 5.4.23. The angular positions of all the satellites were also determined. The period was determined by combining all the satellite positions and orders, equation 5.4.27. This also indicated the possible uncertainty in this value, equation 5.4.34. Data was collected close to the *006* reflection to obtain the strain difference between the "average" superlattice peak and the substrate peak. The high strain sensitivity of this relatively high angle reflection is perfectly adequate for measuring the composition based on the peak separation assumption, section 5.4.1. The lattice parameter difference between GaAs and $Al_xGa_{1-x}As$ for $0 < x < 1$ is rather small and therefore we can take the simple form of equation 5.4.28, hence we have two unknowns related by a linear equation (the period and average composition have already been determined above). This is where we now include the intensities of the satellites, which are sensitive to the strength and form of this modulation. In this example a very small interfacial grade was included to determine the individual thicknesses, followed by a refinement of the interfacial grade through iteration. The results are given in Table 5.4.3 and the intensity comparison is given in Table 5.4.4.

Table 5.4.3. The parameters determined from two quick scans for a simple superlattice structure using the method of direct interpretation and modelling the satellites in the kinematical approximation.

Property	**Value**
Period	25.61(0.1) nm
Period variation	0.53(0.06) nm
GaAs thickness	11.55 nm
$Al_xGa_{1-x}As$ thickness	13.46 nm
Composition x	0.352
Grade GaAs to $Al_xGa_{1-x}As$	0.54 nm
Grade $Al_xGa_{1-x}As$ to GaAs	0.43 nm

We could include more complex combinations, three layer repeats and quaternaries, etc., by just cycling over all possible arrangements. In the case of the quaternaries a combination of two or more groups of satellites may be appropriate to find the unique solution.

Table 5.4.4. The square root of the measured intensities ($F_0 = \sqrt{(I)}$) compared with the calculated structure factor (F_c) from the superlattice, whose parameters are given in Table 5.4.3. The R-factor for the fit with no grade and with grade are 18.7 % and 6.6 % respectively, where

$$R = \frac{\sum_N |F_0 - F_c|^2}{\sum_N |F_0|^2}.$$

Satellite order	F_0	F_c no grade	F_c grade
-12	116	304	103
-11	23	7	34
-10	213	361	191
-9	23	121	83
-8	290	410	294
-7	163	293	238
-6	392	443	394
-5	475	566	544
-4	484	455	465
-3	1168	1127	1209
-2	522	415	460
-1	4507	3773	4269

0	Not measured	9779	11140
1	3659	3193	3614
2	571	605	671
3	1016	989	1061
4	567	527	539
5	395	471	452
6	405	468	415
7	147	217	177
8	302	406	292
9	40	62	50
10	193	340	180
11	65	41	37
12	128	272	92

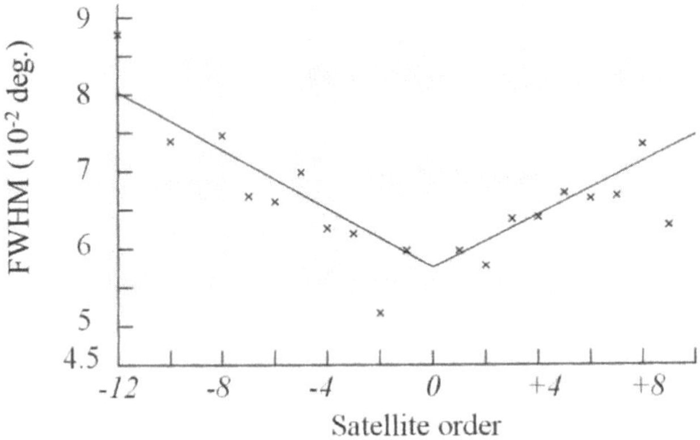

Figure 5.4.23. The variation of the width of the satellite profiles from figure 5.4.22 as a function of satellite order. The full-width-at-half-maximum intensity is obtained after stripping the Cu $K\alpha_2$ component and fitting to a Gaussian function.

5.4.5.5.2. Analysis based on the optical theory with reflectometry:

The kinematical approach presented above could equally well be used close to the *000* reflection, i.e. from a reflectometry scan. However in this section we will use the optical theory, section 2.10, to model a high quality superlattice with data collected using a multiple crystal diffractometer and an X-ray mirror, section 4.3.3. This is an iterative procedure and with the help of direct interpretation or using automatic fitting techniques the analysis can be fairly rapid.

The example in this case is an AlAs / GaAs superlattice that illustrates the sensitivity to the variation in the GaAs to AlAs thickness ratio. The period is very closely determined to the sub 0.1 nm level with this many satellites. The measured profile and the best-fit profile are given in figure 5.4.24, along with the calculated curves for the cases where the GaAs / AlAs thickness ratio has been changed, with thickness changes of 0.05 nm. The differences are quite significant giving a good indication of the sensitivity for this type of superlattice.

The composition sensitivity however is not so good as in the case of scattering close to higher angle reflections. For example changing the composition from AlAs to $Al_{0.9}Ga_{0.1}As$ has a barely perceptible change in the reflectometry profile, whereas close to the *004* reflection the change is very obvious and dramatic: the whole superlattice pattern is shifted with respect to the substrate peak.

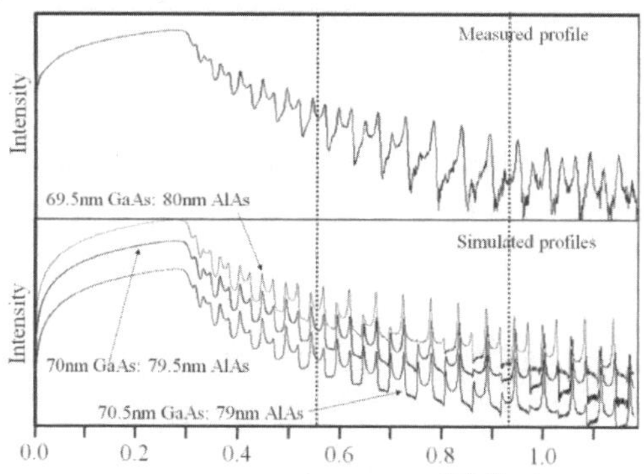

Figure 5.4.24. The sensitivity of the reflectivity profile to the thicknesses in a periodic multi-layer structures. The best-fit model corresponds to the centre profile of the simulated scans. The data was collected with the multiple-crystal diffractometer.

From figure 2.10.3 we can see that any roughening at the interfaces will change the rate of fall in the specular intensity and therefore gives a fairly rapid way of evaluating the interfacial spreading normal to the interfaces. The interfacial smearing normal to the interfaces added in these calculations is 1.27 nm (equivalent to 4.5 monolayers) and is consistent with typical values obtained from a large range of AlAs / GaAs samples, Fewster, Andrew and Curling (1991), and the example in reasonable agreement with table 5.4.3.

5.4.5.5.3. Analysis based on the dynamical theory simulation:

This approach is very well suited to analysing superlattices with long periods for the reasons discussed in section 5.4.5.3. The procedures are very similar to those given in section 5.4.3.4 for any structure. We shall just illustrate the fit to the structure discussed in the previous section with a single profile and to a reciprocal space map to illustrate the contrasting information. The data were collected with a multiple crystal diffractometer with an X-ray mirror, section 4.3.3. The central profile was fitted first using an iterative process and the full instrumental contribution: the measured and calculated reciprocal space maps can be seen in figure 5.4.25. The agreement is very close with the obvious but very weak additional scattering not included. The important aspect to note here, is that even for highly perfect material, where an excellent fit from a single profile will show perfect agreement, the inspection of the reciprocal space maps illustrates differences. Also it should be noted that the "average" peak for the superlattice does not exist. The most dominant peak in the profile, apart from the substrate peak corresponds to a position for the pure AlAs binary strained to the substrate, as in figure 5.4.19b. This is an example where dynamical theory is essential to interpret the data correctly.

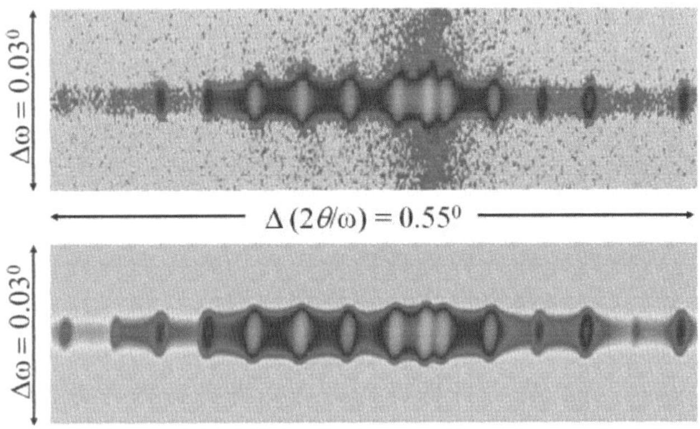

Figure 5.4.25. The experimental and simulated reciprocal space maps of a near perfect AlAs / GaAs periodic multi-layer structure. The simulation of the diffuse scattering is not included in this model. The scale is on a log(log(intensity)) scale to accommodate the full dynamic range for display purposes.

We now have a slight conflict: if the structure requires dynamical theory for its interpretation, which is based on fitting profiles, then high resolution instrumentation is required. As discussed above the move towards high-resolution will result in a reduction in intensity and the detailed analysis can become problematic. One such problem is the analysis of very asymmetric period structures, i.e. those composed of a thick and thin layer repeat. An example is given in section 5.4.5.5.4. Another interesting problem is the analysis of periodic structures with a very large numbers of layers, and this will be considered with an example in section 5.4.5.5.5.

5.4.5.5.4. Analysis of an asymmetric repeat unit (thick and thin layers):

Consider a structure composed of relatively thick layers periodically interspersed with atomic layers, e.g. delta doping, then the diffraction pattern will have clear satellites, e.g. Hart et al (1993, 1995) see figure 5.4.29. Obtaining the composition and thickness of this ~atomic layer is not simple. This type of structure can be considered as composed of three frequencies; the overall period, the thicker layer and the thin layer; it is the interaction between these that give rise to the satellite spacing (the overall period) and the intensity modulation of the satellites. The intensity modulation due to the thick layer is virtually coincident with the periodicity, i.e. the satellites would be of similar magnitude from this alone. The very thin layer will have a variation or envelope function characteristic of its thickness that will be very broad. The centre of this broad envelope will correspond to the 'average' composition within the period. This leads to the conclusion that the composition and thickness of the thin layer are highly correlated, and it is breaking this correlation that is of particular interest here.

To break this correlation of thickness and composition an intense high-resolution diffractometer makes this easier. The example here is a structure composed of {2 nm $In_{0.15}Ga_{0.85}N$ + 19 nm GaN} × 10 on a 3 μm GaN layer on a sapphire (*000l*) substrate, and on top there is a cap of 15 nm of GaN. This is not as extreme as the delta-doped layers, but is illustrative of the problem. The high-resolution instrument with the most intense beam discussed in chapter 4, is the beam-selection diffractometer (section 4.3.4) and is used here. The *220* three bounce analyser crystal was used with an attenuator to remove the Cu $K\alpha_2$ component, figures 4.3.11 and 4.3.12. The incident beam was from a line focus and an X-mirror. The *0002* peak intensity from the GaN in this structure was 7 Mct/s and gave the profile given in figure 5.4.26.

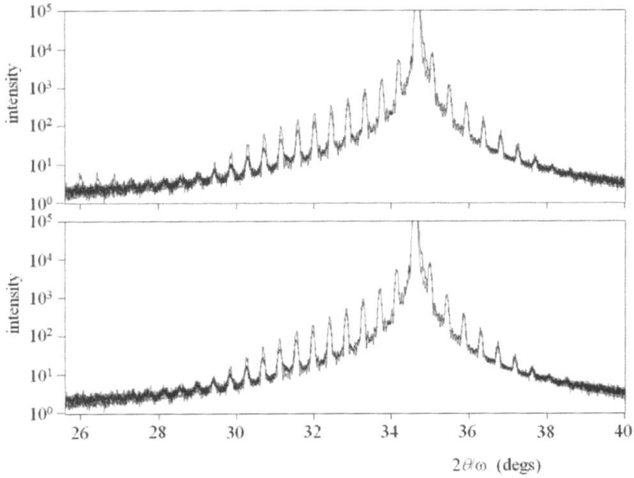

Figure 5.4.26. The profile close to the *0002* reflection of GaN from a structure nominally composed of Al_2O_3 substrate + 3 μm GaN + {2 nm $In_{0.15}Ga_{0.85}N$ + 19 nm GaN}×10 + 15 nm GaN (E Thrush, Aixtron / Univ. Cambridge, UK). (a) is the profile including the best fit assuming no grading in the structure and (b) with grading. See text for the interpretation.

The dynamic range of the beam-selection diffractometer makes it possible to capture useful information from 25^0 2θ to 40^0 2θ. The intensity of the satellites have nodes at 28^0 2θ and 39^0 2θ, which represents the period of the envelope function associated with the InGaN layer. In figure 5.4.26a, the best fit to the profile shows that the intensity below 28^0 2θ is over-estimated and results from the assumption that the interfaces are perfect and abrupt. The measured satellite peaks broaden as function of order and this is indicative of fluctuations in the periodicity, which clearly relates to the quality of the interfaces averaged vertically and laterally, figure 5.4.17. The period variation was estimated to be 0.40±0.07 nm from the measured profile through equation 5.4.34 as in figure 5.4.23, and then included in the simulation of the profile. The resulting fit is given in figure 5.4.26b, where it can be seen that the satellites beyond the nodes are suppressed. The final parameters were {1.7 nm $In_{0.157\pm0.002}Ga_{0.843\pm0.002}N$ + 19.5±0.1 nm GaN} period and a 165.2±0.1 nm GaN capping layer.

These uncertainties are derived based on a stochastic refinement procedure taking into a count all the correlations between parameters, section 5.4.3.4, and assume that the elastic parameters and relationship between composition and strain are reliable. For GaN based compounds, Moram and Vickers (2009) have compiled the spread in lattice parameters, elastic parameters and relationship with Vegard's rule (Vegard 1921): if these are considered to define the limiting bounds (i.e. the 'Poisson ratio' within 10 %, lattice parameters to within 0.004 %), then the confidence level for the In composition is ±0.03. This is the absolute worse case and reflects the range of estimates in the elastic parameters, some of which give confidence bounds well below 10 %. This just suggests that we should be aware when choosing a reliable reference and the rigour in which it was established.

5.4.5.5.5. Analysis of structures with large numbers of layers:

Some structures are very complex, e.g. vertical cavity surface emitting lasers (VCSEL), but can still be analysed if done carefully. These structures are composed of groups of periodic stacks that reflect the light generated from a few quantum wells, and if the periodicity matches the condition for maximum reflectivity of the light generated then considerable enhancement of the emission is possible. The structures can have many hundreds of layers and the consequent diffraction pattern can be very complex, figure 5.4.27, and the modeling will require considerable expertise and persistence. The number of parameters in this example is large; there are 450 layers, and it is necessary to bring together thoughts on growth rates, section 5.4.3.3.1, some direct analysis, section 5.4.5, which is combined with automatic fitting. Automatic fitting with oversight can give an indication of the improved confidence levels as the refinement progresses; and is the best approach at present. The agreement of the fitted profile to that measured is given in figure 5.4.27, and the detailed solution is given in Kidd (2003). The experimental approach is also important in this study, because the growth is on a plane inclined to the [*00l*] direction by 10^0, so a mixture of the *117* reflection and *006* reflections were used. The latter was set so that the [*00l*] direction was in the axial plane, thus making use of the axial divergence to capture much of the scattering along the surface normal. This also results in a projection effect.

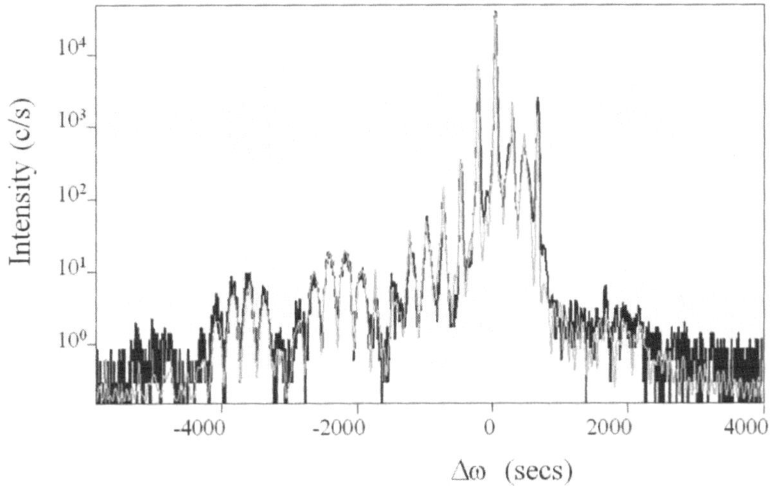

Figure 5.4.27. The *006* profile from a 450 layer VCSEL structure and the final best fit profile, courtesy of P Kidd.

5.4.5.5.6. Lateral information from interfaces:

There are two possible reasons for the interface smearing in s_z in nearly perfect structures, and these are inter-diffusion and roughness. The former will show no intensity in s_x, whereas the latter will show intensity that has characteristic forms given in figure 2.11.7, depending whether the roughness is correlated from interface to interface. As an indication interfaces in the example structure given in figure 5.4.24, has strong lateral streaks in s_x indicating roughness and correlated interfaces. Extracting a lateral length scale requires a model of the interface, and in this case it is reasonable to assume a castellated form (figure 2.11.4c), based on the reasoning given in figure 5.4.21. The importance of deciding on the most appropriate scattering model is discussed in section 5.7.6.1.

5.4.6. *Extracting even more information from periodic structures* (complex structures)

This section gives an indication of what can be extracted by taking the analysis further than discussed so far. Some of the discussion will be left incomplete and qualitative, but it will give an indication of what is possible to achieve with laboratory X-rays sources.

5.4.6.1. Structure with faults, a missing layer:

This was an unusual case in that a GaAs / AlGaAs periodic structure was emitting light under bias at a wavelength that was significantly different from that expected for the thickness of the GaAs quantum wells. Photoluminescence also indicated that the wavelength corresponded to a larger well width that could have arisen if the Al Knudsen cell shutter of the MBE system failed to open at some stage. The fringing close to a satellite has a spacing that is related to the overall thickness of the periodic structure, and their intensities are approximately symmetrical. The diffraction profile of the *-1* satellite did indicate that the structure was not as intended. Substituting a well layer with the thickness of a well and barrier layer, influenced the fringes quite dramatically and the fault could be pinpointed to a specific position in the structure by moving it around until the fringes fitted, Fewster (1993a). The fit was performed with dynamical theory and the fault was situated after 14 periods, followed by 44 periods, which was subsequently confirmed by cross-section transmission electron microscopy.

5.4.6.2. The nature of grading or roughening at the interfaces:

From the overview of the interfacial roughening, section 2.11.1, and the deeper probing of the perfection of periodic structures by studying the high order satellites, section 5.4.5.5.1, we can see that considerable information is available. Since the higher order satellites are revealing greater details on the interface shape, analogous to the coefficients of the Fourier transform, then any spreading of the satellites along s_x will relate to the length scales at that information point in the interface. It should therefore be possible to model the whole interface roughness, in depth and laterally without having to define a model. That was the intention of the work illustrated in figure 5.4.28. The interpretation is only qualitative but clearly shows the increased broadening of the satellites in s_x with order. The high order satellites define the higher frequencies in the modulation, i.e. the abruptness of the interface, and a shortening of the lateral roughness dimension at the beginnings of the interface transition. For lower order satellites this dimension is larger, where the two components and hence their coverage have similar proportions. This interpretation would be consistent with the interface model given in figure 5.4.21.

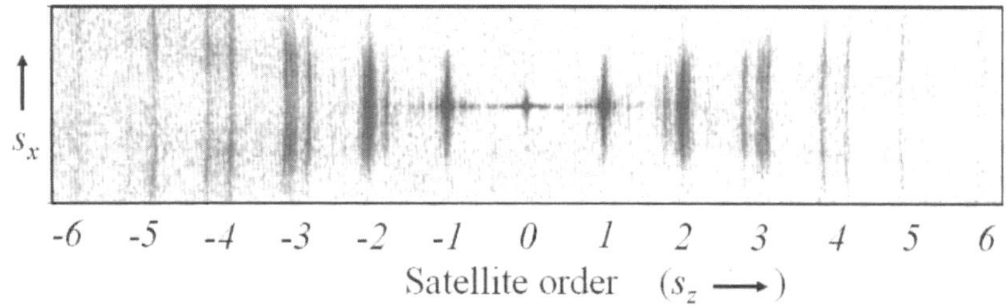

Figure 5.4.28. The data from a periodic structure that shows satellite intensity spreading along s_x that is indicative of the interface shape, which corresponds to the form given in figure 5.4.21. The data is represented on a log(log) scale with a full angular range along s_z of ~4^0 and along s_x of 0.4^0, both on the ω scale. This is the *002* reflection using Cu $K\alpha_1$ for a {AlAs (1.84 nm) + GaAs (8.76 nm)} × 20 periodic structure on a GaAs substrate and with a 10.2 nm cap of GaAs (sample courtesy of Tom Foxon).

5.4.6.3. Diffraction tail 'wiggles':

Experimentally we could see from figures 5.3.15 and 5.4.14 that the diffraction streak or fringing do not appear on a simple line normal to the surface (sometimes termed the crystal truncation rod). The current proposed explanation is that the relative orientation of the surface plane and the crystallographic planes varies, but this is not definitive. These effects can be much more dramatic in periodic structures, e.g. figure 5.4.29.

At this stage there is limited information that can be extracted from interpreting this data, but it is very important to be aware of these effects. For example if a small diffraction space probe is used as that of the multiple-crystal diffractometer a limited area reciprocal space map

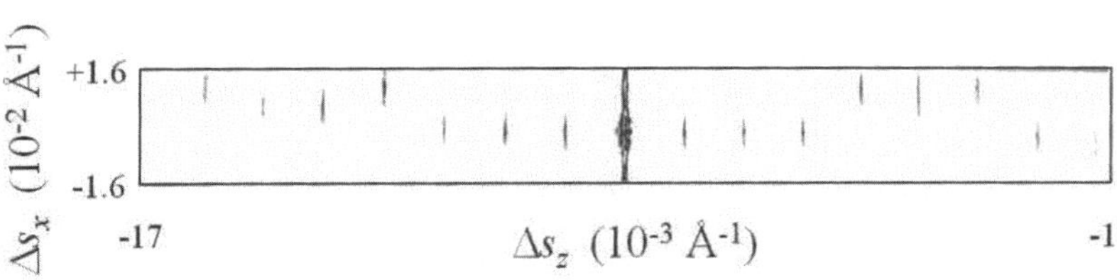

Figure 5.4.29. The 'wiggles' present in a 100 period Al delta-doped GaAs structure. The reciprocal space map is expanded considerably in the s_x direction to show the effect. From Hart et al (1995).

should be captured and projected onto the s_x direction. It is assumed that in the axial direction the 'wiggles' are captured due to the axial divergence: with 3-dimensional reciprocal space mapping, section 4.2.2, these diffraction features can be seen to deviate in the $s_x s_y$ plane. Another aspect to consider is that if a lower resolution probe is used so that no projection is necessary, then the satellite positions will vary.

5.4.7. *Analysis of quantum dot structures* (complex structures)

The analysis of self-assembled quantum dots is considered in this section. The dots are assumed to be distributed randomly and their strain-fields do not overlap laterally. The strain-field of the dots normal to the interfaces can lead to vertical correlation of the dot positions, and the example given here is for that case. In some systems, notably PbSe dots formed on $Pb_{1-x}Eu_xTe$ ($0.05<x<0.1$) layers, there is very interesting ordering; not only in the interface plane but also vertically, Springholz, Holý, Pinczolits and Bauer (1998). The resulting dot arrangement compares to a FCC lattice with the 3-fold axis parallel to <*111*> surface normal direction. This information is revealed in reciprocal space maps of the satellites around the *222* reflection.

5.4.7.1. Analysis of buried dots:

We discussed some of the basic theoretical concepts of in-plane scattering, section 2.11.2 with a specific section 2.11.3 on in-plane scattering from buried quantum dots. In-plane experiments with the multiple-crystal diffractometer, section 4.3.3, will allow the measurement of the precision lattice parameter of planes perpendicular to the surface plane and the shape of the scattering close to Bragg peaks.

A point focus source was used along with a narrow cross-slit after the monochromator to restrict the axial dimension and limit the axial divergence for a good reflectometry profile. A narrow slit is placed in front of the detector that is just wide enough to capture the incident beam, this will define the $2\theta = 0$ position in the plane normal to the sample surface, since the sample is mounted at $\chi = -\pi/2$ such the surface is the scattering plane, figure 4.3.2. The sample was then set to cut the beam in two, by translation and rocking α_i to ensure the surface was parallel to the incident beam as defined in figure 2.11.8. When the slit was removed from in front of the detector, the direct beam was blocked. The sample was then scanned in α_i to obtain a low resolution reflectometry profile and α_i was set to 0.3^0 close to the bottom of the observable reflectometry profile. At this position the beam is entering and can scatter from within the sample. To find the scattering from a plane normal to the surface an open window detector is ideal before invoking the analyser crystal. The analyser is then substituted for an open detector and the $2\theta = 0$ position is set according to section 5.4.5.4, point 1.

This particular example yielded the in-plane lattice parameters along four different directions and indicated that the whole structure had an average in-plane lattice parameter of that expected from the (*001*) GaAs substrate. Therefore no lattice relaxation was detected, however significant diffuse scattering can be seen and interpreted to give the asymmetry of InGaAs quantum dots grown on GaAs, Fewster (1999). By calculating the strains associated with randomly distributed dots in the plane but vertically correlated, the diffuse intensity can be calculated through equation 2.11.21 to yield the shape and composition Fewster, Holý and Andrew (2001). Clearly the dynamic range is sufficient to observe clear shapes and features within the diffuse scattering. The matching of the simulated reciprocal space maps to the experimental results can be achieved by eye, figure 5.4.30. The sensitivity to

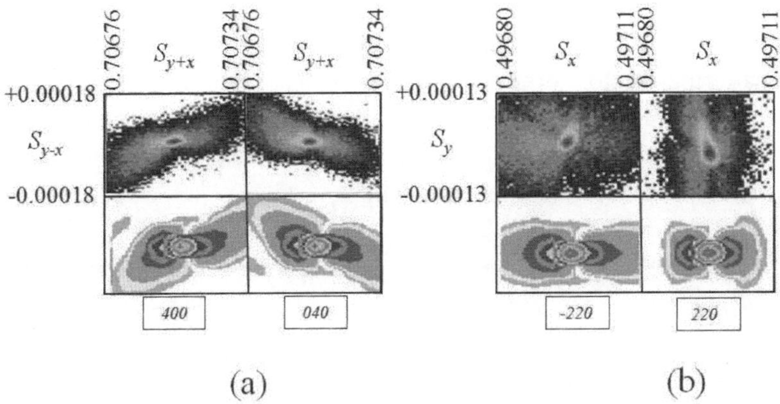

Figure 5.4.30. The diffuse scattering measured from a stack of $In_xGa_{1-x}As$ quantum dot layers imbedded in a GaAs matrix, with the in-plane scattering geometry and the multiple-crystal diffractometer with the best fit simulated reciprocal space maps. (a) the *004 and 040* reflections and (b) the *220* and *-220* reflections. The best fit was obtained using a half-ellipsoid of $In_{12\pm3}Ga_{88\pm3}As$ of height 3±1 and dimension 30±10 and 10±5 in the *220* and *-220* directions respectively.

parameter changes is perhaps the best way to build confidence in the results. Some immediate inferences can be made to guide the fitting, if all the quantum dots are assumed to be the same size and half ellipsoid in shape with the same composition throughout. The layer in-plane axis will be along *220*. Knowing the depth of the dots is an important input. An estimation of the reliability is given in the figure caption as an indication of the usefulness of this method. It has been extended to a single layer of dots from which a fit could be made but the reliability of the information is not so good, Fewster, Holý and Zhi (2003).

5.5. Analysis of mosaic structures (textured epitaxy):

The analysis of mosaic structures is very similar to that discussed for mosaic bulk samples. Generally the mosaicity is associated with relaxed layers with imperfect interfaces, but this is not always the case. Consider the reciprocal space map of an unrelaxed mosaic layer on a mosaic substrate, figure 5.5.1. There are three distinct peaks that we can associate with the substrate reflection and similarly for the layer. The important aspect to remember is that the alignment is based on bringing the scattering vector from one of the blocks into the plane of the diffractometer. If each block is tilted with respect to the aligned block the others cannot all satisfy this requirement. This leads to the projection errors in extracting reliable information.

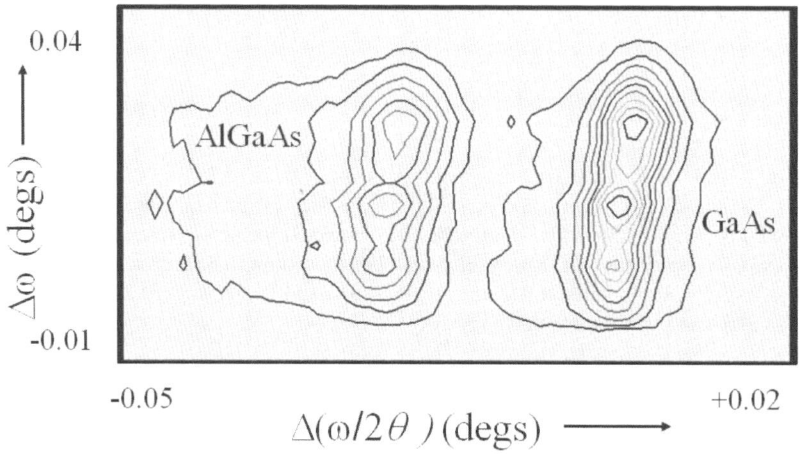

Figure 5.5.1. A reciprocal space map of an AlGaAs layer grown on a mosaic GaAs substrate. The large axial divergence projects all the contributions onto the same plane. A three-dimensional reciprocal space map of the same sample is given in figure 5.3.12.

If now we undertake a series of maps at different tilt values we can observe this pattern change quite significantly as different blocks move into and out of the optimum scattering condition. If we restrict the axial divergence further we can isolate the scattering from each mosaic block build up a three-dimensional reciprocal space map of the scattering, figure 5.3.12. Details of collecting this data are given in section 5.3.2.3. From our reciprocal space map we can see that a simple multiple crystal scan along the surface normal (along $2\theta/\omega$ in this case since this is the *004* reflection from a (*001*) orientated sample) will result in a profile identical to that from a perfect non-mosaic sample. We can therefore apply the simple techniques of interpretation of composition and thickness, etc., given above.

Let us consider some of the advantages of the multiple-crystal concept. Clearly we can analyse any material over very large regions of reciprocal space with very high resolution; i.e. the instrumental artefacts do not significantly smear the data. The probe can be moved throughout accessible regions of reciprocal space and the intensity can be mapped and from Chapter 2, the scattering process can be simulated, and include all the instrumental artifacts. We can also extract information directly from the reciprocal space maps. There are some additional benefits with this geometry, the angle 2θ can be placed on an absolute scale and hence the lattice parameter can be measured absolutely to within a part per million. The low divergence of the spatially large (millimetres in dimension) incident beam makes it suitable for topography. A photographic emulsion, or small pixel area detector, can be placed directly after the sample or after the analyser crystal. In the latter position specific areas in reciprocal space can be imaged (for example diffuse scattering) or more specifically with the former mosaic blocks as in figure 5.3.14.

5.5.1. *Analysis of partially relaxed multi-layer structures (textured epitaxy):*

As discussed in section 1.6 the growth of thin high quality epitaxial layers is limited by the internal stresses that eventually lead to plastic deformation if they cannot be contained by elastic distortion. From section 2.8.2, the theoretical description of the scattering clearly also becomes rather complex. However we can obtain considerable insight by assuming the scattering behaves in a simple way (i.e. Bragg's equation can be applied directly to any diffraction peak and the correlation lengths can be determined from the inverse of the reciprocal space feature length).

If we follow the process during the onset of relaxation then initially the dislocations are isolated with their own strain-field that extends to some limit depending on the stiffness of the material or region that can relieve this strain, for example the surface. Following the discussion of Kidd and Fewster (1994) we can assume that the dislocation strain-fields form isolated regions that will have a lateral spread related to the thickness of the layer (the top surface of the layer is the strain-relieving boundary). Therefore each dislocation will have nearly identical strain-fields in lateral extent with perfect regions of material between them. We therefore have a distribution of correlation lengths associated the shape of the dislocation strain-field and the perfect crystal. For the very early stages of dislocation formation the former is small (comparable to the layer thickness) and the latter is large. We can therefore assume that the scattering is independent so that we can simply add two shape functions with different length scales.

The result is shown experimentally with the simulated reciprocal space map in figure 5.5.2. As the thickness of the layer increases the thickness of the defect region enlarges and narrows the diffraction broadening, however the region between defects decreases so that the width of the layer peak (the longer length scale) will increase. Eventually the defect density increases to such an extent that the strain-fields overlap and the extent of the strain-field normal to the interface is restricted and therefore contained close to this interface. The distortion will lead to defects breaking up the structure locally and will lead to tilts and twists. The consequence of this is that the layer peak can exhibit broadening from the limited length scale and the different shaped regions having different orientations and lattice plane distortions. For a more rigorous analysis the scattering should be based on modeling the whole strained region with the distorted wave Born approximation, discussed in section 2.11. In the limit the defects will completely relieve the internal stresses in the layer and the layer will become totally relaxed, i.e. the layer lattice has the dimensions of that for bulk material of the same phase composition. In reality this is rarely achieved since the driving force is very small as the stress is relieved. Of course we should not consider that the distortion is entirely within the layer and assume the substrate is completely unaffected. In the following determination of the relaxation and composition we will take the substrate distortion into account and then consider how this information can be obtained with different assumptions. We shall then consider how we can begin to extract information on the microstructure.

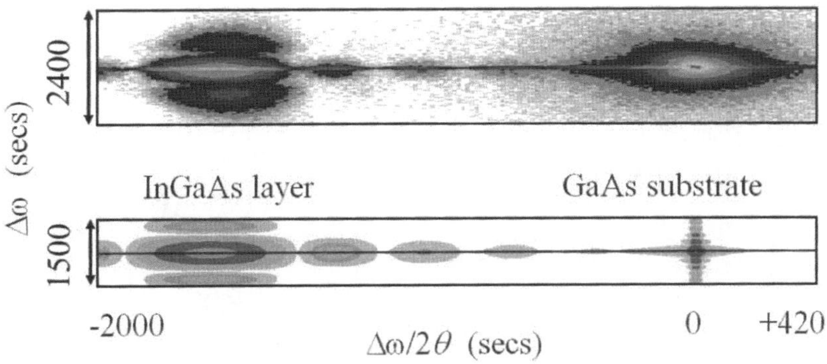

Figure 5.5.2. The experimental and calculated reciprocal space maps of an $In_{0.1}Ga_{0.9}As$ layer on a GaAs substrate. The layer was just beginning to relax and the diffuse scattering, associated with isolated dislocations close to the layer peak can just be observed. The simulation includes several columns of 70±10 nm wide and 70 nm thick with a distribution of distorted regions at the interface ranging from 5 to 20 nm. The total distorted area was ~10% of the total area and the perfect intervening regions were ~700 nm wide. The strain relaxation in the dislocated regions is evident in the experimental map but is not included in this calculation; this gives the fringe movements along $2\theta/\omega$.

5.5.2. *Measuring the state of strain in partially relaxed thin layers:*

The state of strain is fundamental to deriving the composition in thin layers. In partially relaxed layers the state of strain cannot be determined from the equations relevant to those for nearly perfect epitaxy, section 5.4.1. The interatomic spacings above and below a partially relaxed interface are not equivalent and therefore have to be determined independently. We can define the interatomic spacings in the layer and substrate as an orthogonal set d_x and d_y in the plane of the interface and d_z normal to the interface plane. The determination of these spacings will be described with reference to figure 5.5.3, which is a subset of figure 4.2.1 when we consider regions close to two layer reflections and two substrate reflections.

The observation of any reflection can be characterised by an incident and scattering angle, i.e. ω and 2θ respectively, after the scattering plane normal has been brought parallel to the diffractometer plane with ϕ and χ. The interatomic spacing normal to the sample surface is given by the

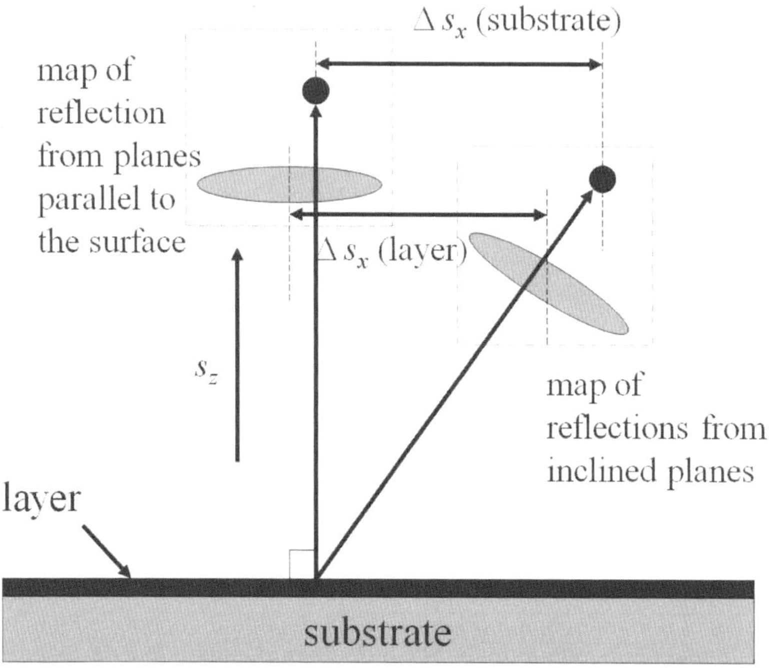

Figure 5.5.3. A schematic of the measurements to be extracted from two reciprocal space maps to determine the relaxation of a layer on a substrate.

reciprocal of the differences of two measured positions $\Delta s_z = s_{z1} - s_{z2}$ in equation 4.2.1

$$d_z = \frac{1}{\Delta s_z} = \frac{\lambda}{[\sin\omega_1 + \sin(2\theta_1 - \omega_1)] - [\sin\omega_2 + \sin(2\theta_2 - \omega_2)]} \qquad 5.5.1$$

For the interatomic spacings in the plane of the interface we can write

$$d_x = \frac{1}{\Delta s_x} = \frac{\lambda}{[\cos\omega_1 - \cos(2\theta_1 - \omega_1)] - [\cos\omega_2 - \cos(2\theta_2 - \omega_2)]} \qquad 5.5.2$$

and similarly for d_y.

Hence if we measure the angular positions of two reflections for the layer and similarly for the under-layer, or substrate, then we can determine the difference in the interplanar spacing parallel and perpendicular to the interface.

For two reciprocal lattice points separated by Δs_z It is clear that if we know the angles ω and 2θ on an absolute scale then we are defining the location of the origin of reciprocal space with Miller indices *000* and we should set $\omega_2 = 0$ and $2\theta_2 = 0$. We can therefore determine the interplanar spacing parallel and perpendicular to the surface using one reflection for each azimuth. The determination of the 2θ angle on an absolute scale is relatively easy, section 5.3.4. The ω absolute angle is less precise, in general, however since this is a relative angle with respect to the surface it will bring in cos(ω) errors to the measured interplanar lengths and for small errors this will not be significant. This angle basically just defines what direction is perpendicular to the surface and only becomes significant when this is related to the assumed orientation associated with the phase composition. We can simply overcome this by defining the perpendicular and parallel interplanar spacings along an appropriate direction, e.g. relate it to the [*001*] direction for a nominal [*001*] orientated sample.

5.5.3. *Obtaining the composition in partially relaxed thin layers:*

This section considers the determination of the interplanar spacings parallel and perpendicular to the interface of interest; the general principles are given above and the various practical methods are given in sections below. However now we should refer to equation 5.4.10, but modify it slightly so that the interplanar spacings are with respect to the layer in the fully relaxed state, rather than being related to the substrate

$$x = \frac{\{d_M\}_{Relax} - d_B}{d_A - d_B} \qquad 5.5.3$$

where *d* are suitably chosen parameters characteristic of the composition, i.e. the unstrained (fully relaxed) interplanar spacings for the alloys forming the layer. In the cubic case, as discussed previously, this could be the unit cell dimension, *a*, whereas in the hexagonal case this could be *c*, and similarly for any other symmetry. The subscripts *A* and B represent the phase limits of the alloy A_xB_{1-x} and $\{d_M\}_{Relax}$ is the determined interplanar spacing along the same direction when all the strain has been relieved, i.e. the equivalent bulk value. Equation 5.5.3 assumes Vegard's rule; however whatever the relationship between composition and interplanar spacing the same principle applies. For example equation 5.4.11 accounts for a small modification to this rule, yet there is a simple relationship between composition x and a lattice parameter for the $Si_{1-x}Ge_x$ alloy system.

Our concentration now is to determine $\{d_M\}_{Relax}$. From the previous section 5.5.2 we will assume that we have determined the interplanar spacing in three orthogonal directions (normal and in the plane of the interface). Suppose now that the interplanar spacings in these three directions for the two limits of the alloy extent are related by certain constants, such that

$$\begin{aligned} {}_1d_x : {}_1d_y : {}_1d_z &= {}_1K_x : {}_1K_y : {}_1K_z \\ {}_2d_x : {}_2d_y : {}_2d_z &= {}_2K_x : {}_2K_y : {}_2K_z \end{aligned} \qquad 5.5.4$$

These K parameters relate to the choice of the reflection in the measurement, e.g. a choice of reflections from a cubic [*001*] orientated layer may be *004* (along z), *444* (along x) and *–335* (along y): this will give $K_z = 4$, $K_x = \sqrt{(4^2+4^2)}$ and $K_y = \sqrt{(3^2+3^2)}$. We will have to assume that these K parameters vary for the alloy in a well-defined manner and it is not unreasonable to assume that they follow the same relationship with composition as the interplanar spacings (i.e. Vegard's rule in general). For the cubic case all these K values are unchanged and for hexagonal structures these will follow the c/a ratio that can vary between phases.

The d values in equation 5.5.4 are all the fully relaxed values and therefore we must derive these from the strained values and the appropriate elastic parameters. We shall concentrate on the interplanar spacing normal to the interface and derive the $\{d_z\}_{Relax}$, now:

$$\varepsilon_{xx} = \frac{d_x - \{d_x\}_{Relax}}{\{d_x\}_{Relax}} \qquad 5.5.5$$

and similarly for ε_{yy} and ε_{zz}. Combining these relations, equation 5.5.4 and equation 1.6.5, we obtain after some manipulation

$$\{d_z\}_{Relax} = \frac{\left\{d_z - \dfrac{F_1 K_z}{F_3 K_x} d_x - \dfrac{F_2 K_z}{F_3 K_y} d_y\right\}}{\left\{1 - \dfrac{F_1}{F_3} - \dfrac{F_2}{F_3}\right\}} \qquad 5.5.6$$

Or if the 'engineering' approximation of the Poisson ratio can be used then this becomes

$$\{d_z\}_{Relax} = K_z \frac{1-\nu}{1+\nu}\left\{\frac{d_z}{K_z} + \frac{\nu}{1-\nu}\left(\frac{d_x}{K_x} + \frac{d_y}{K_y}\right)\right\} \qquad 5.5.7$$

ν is the Poisson ratio for the phase along the direction normal to the surface. The parameters F_1, F_2 and F_3 (or ν) are unknown (since the alloy composition is unknown at this stage) although we know the values at the limits of the two phases. So we relate these parameters F_i (or ν) to the composition and iterate, i.e.

$$F_{A_x B_{1-x}} = xF_A + (1-x)F_B \qquad 5.5.8$$

If we initially assume the parameters as the mean value, i.e. for $x = 0.5$, then a first estimate of $\{d_z\}_{Relax}$ can be used to determine a composition with equation 5.5.3. We can then refine the parameters that now are all associated with the composition iteratively by substituting the new value of x into equation 5.5.6 or 5.5.7 until the parameter or Poisson ratio change is insignificant. Similarly the K values will follow the same relationship and should be included in the iterative process.

5.5.4. *The measurement of the degree of relaxation and mismatch in thin layers:*

These parameters can be useful for estimating the number of stress relieving defects. Until now we have treated the substrate and layer independently because the state of strain gives the composition, etc., whereas the mismatch and degree of relaxation just relate the strain in the layer to that of the substrate or layer below. The layer strain along x in the plane of the interface is

$$Strain = \varepsilon_{xx} = \left\{ \frac{d_x - \{d_x\}_{Relax}}{\{d_x\}_{Relax}} \right\} \qquad 5.5.9$$

and similarly for ε_{yy}. This strain should be considered in two orthogonal directions to fully characterise the misfit at the interface. The strain normal to the interface is given by

$$Strain = \varepsilon_{zz} = \left\{ \frac{d_z - \{d_z\}_{Relax}}{\{d_z\}_{Relax}} \right\} \qquad 5.5.10$$

The degree of relaxation is simply the ratio of the difference in the actual interplanar spacing parallel to the interface to that for the layer in the completely relaxed state.

$$R_x = \frac{d_x - {}_S d_x}{\{d_x\}_{Relax} - {}_S d_x} \qquad 5.5.11$$

${}_S d_x$ is the interplanar spacing along the direction x (parallel to the interface) for the layer below that of interest or the substrate. We can similarly write an expression for R_y. The value of the relaxation in general will vary from 0 (no relaxation: the layer is perfectly strained to fit the under-layer) to 1 (fully relaxed: the strain has been completely relieved). This relaxation value is a convenient simple input for modelling the strain to be included in the simulation of the scattering profile. The strain is determined from the relaxation parameter through equations 1.6.17 and 1.6.18.

The defect density can be derived from the difference in the interplanar spacing parallel to the interface. If all the defects have the same contribution to the strain relief and the component of their Burgers vector parallel to the interface is known then from equation 5.4.25 we can determine their density. Of course we must be aware that the defects can pile up especially at high levels of relaxation, or there could be a mixture of dislocations types that may exist bringing doubt into the estimation.

We shall cover various approaches to the determination of composition and relaxation including the use of reciprocal space maps, rocking curves, etc., and compare their precision with examples.

5.5.5. *The determination of relaxation and composition with various methods:*

In this section we shall give examples of structures that are partially relaxed that also include some tilting and analyse them using various options. The most reliable method will be to interpret the data using reciprocal space maps and calculating the parameters from the absolute angles determined from the peak positions. We can then make some assumptions and gradually work our way through the procedures to single peak determination of composition of relaxed structures. Collecting reciprocal space maps should not necessarily be considered a slow process and results obtained from data collection in a few minutes are perfectly useable. Rocking-curve measurements, assuming that the substrate lattice parameter is known can be used as an internal standard and will also be considered.

5.5.5.1. Determination by reciprocal space maps on an absolute scale:

This method is suitable for a multiple-crystal diffractometer with a monochromator and an analyser. Suppose we wish to obtain the most precise value for the parallel and perpendicular interplanar spacings then we should measure combinations of reflections that have the largest separations in s_x, s_y and s_z. This is purely a geometrical consideration. However we must include some indication of the volume of the sample that is being assigned to these determined parameters. The X-rays penetrate the sample and are attenuated depending on the general photoelectric absorption and extinction effects; these topics are discussed in detail in Chapter 2. Generally if we are determining the parameters in a thin layer, then we sample the whole layer and obtain a reliable average. The substrate will almost certainly be distorted close to the interface due to the strain-field of the dislocation, yet deep below the interface the parameters will be representative of the bulk parameters. To obtain the interplanar spacings of the substrate, in order to extract the degree of relaxation, we should compare results from reflections with a similar probing depth, which should ideally be optimised to be close to the interface. For example for a [*001*] orientated GaAs wafer nearly all the intensity is scattered in the top 5 μm for the *004* reflection and similarly for the *444* reflection, however for the *224* reflection this depth is 2.7 μm, Fewster and Andrew (1998). These depths are determined by simulation and noting the rate of change in the integrated intensity as a function of thickness (effectively regions of the substrate below depths that have little effect on the integrated intensity are not contributing significantly to the scattering).

Let us now consider the accuracy of determining the interplanar spacing for different reflections. As discussed in section 5.5.2, the 2θ angle can be determined with high precision (i.e. the zero can be set precisely) whereas the ω angle is at best an order of magnitude less precise since this relates to the sample surface. Therefore if we wish to determine the strain perpendicular to the interface the rotation in ω should be normal to this direction for the reflection we use; hence scattering from planes parallel to the surface satisfy this condition, therefore the *002*, *004*, *006*, etc., for a [*001*] orientated surface are best. We should now apply our criteria of largest separation in s_z, since we are on an absolute scale we use the chosen reflection in combination with the *000* "reflection." Remember of course we can arbitrarily define the direction of s_z along the main plane normal of the substrate, without introducing too much error and this can define the ω angle zero. We therefore can write equation 5.5.1 on an absolute scale as

$$d_z = \frac{\lambda}{[\sin\omega + \sin(2\theta - \omega)]} \qquad 5.5.12$$

To determine the interplanar spacing parallel to the interface the errors introduced by the uncertainty in ω are minimised for reflections from planes normal to the surface. We have described how these in-plane measurements are conducted; section 5.4.7.1 and these are generally unsuitable for any rapid method. Complications do arise also because this will give a projected angle. However since we are concerned about the uncertainty in ω, from our measurement of d_z we have now defined ω and with a precision goniometer this is now as precise as the 2θ determination, although the absolute ω = 0 position is uncertain. Therefore a combination of any two reflections with differences in s_x will give the interplanar spacing parallel to the interface (or more precisely along a direction normal to s_z). If we measure the *444* reflection with respect to the *004* reflection for a [*001*] GaAs sample then to maintain 1ppm accuracy in the interplanar spacing parallel to the interface (assuming a perfect measurement) ω should be accurate to within 0.1^0 of the absolute value. We therefore obtain d_x from equation 5.5.2, where the subscript '1' refers to the *444* reflection and '2' to the *004* reflection for example. To account for any anisotropy the measurement should be performed along both the *x* and *y* directions. The *d* values are then substituted into equation 5.5.6 and with equation 5.5.3 the composition is obtained.

The relaxation along the *x* and *y* directions are determined with equation 5.5.11, since all the components are known. If we can make assumptions concerning the relaxation per defect then the interfacial defect density can be derived from equation 5.4.25 with the appropriate Burgers vector contribution.

5.5.5.2. Determination by using a series of rocking curves and analyser scans:

This method is suitable for a multiple-crystal diffractometer. It is clear from the above analysis that we only need the ω and the 2θ values for two reflections and the clearest method is to undertake a reciprocal space map including the layer and the substrate, or some other layer reference reflections. Now of course we can obtain the ω value directly from the rocking curve, provided that the layer peak is well defined and the ω reference is defined as described in section 5.5.5.1. If we drive the ω rotation to the layer peak position then with this open detector the instrument acceptance is given as in figure 4.3.8a. If we invoke the analyser, which is already defined on an absolute scale, then a simple 2θ scan will give the second angle we need. The procedure then simply follows that given in the previous section.

This method is less accurate than conducting reciprocal space maps because an inaccurate location of the ω peak position, which can be very broad, will introduce additional errors in the 2θ value. Consider the example of a heavily relaxed GeSi layer on Si, where the rocking curve gives a very poorly defined layer peak, figure 5.5.4a. In contrast the reciprocal space map gives a clear peak for a good determination of the ω angle, figure 5.5.4b. A perfectly measurable reciprocal space map can be obtained in 5 minutes with the multiple crystal diffractometer, *220* symmetric reflections throughout and line focus Cu X-ray source with an X-ray mirror. Clearly the reciprocal space map is the most reliable approach when the layer peak is broad and flat. However for reasonably well defined layer peaks this can be a very quick and useful approach.

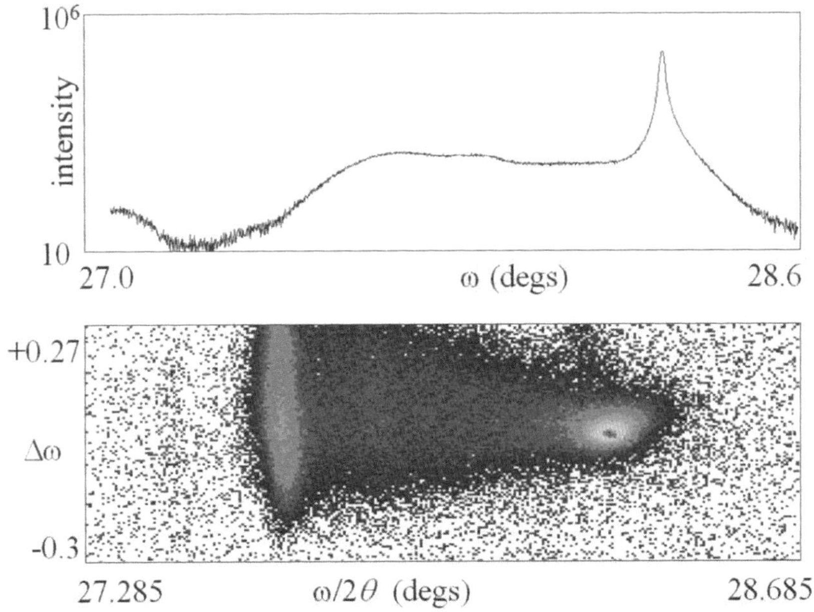

Figure 5.5.4. The difficulties associated with determining the ω angle from a rocking curve (a) compared with a reciprocal space map (b) for a heavily relaxed SiGe layer on a Si substrate.

5.5.5.3. Determination by reciprocal space maps on a relative scale:

This method relies on the substrate, or some other internal reference, to have known ω and 2θ angles. The layer reflection of interest is then collected on the same map and the ω and 2θ angles are determined very simply from

$$\omega = \omega_S + \Delta\omega$$
$$2\theta = 2\theta_S + \Delta 2\theta \qquad 5.5.13$$

The procedure then follows that given above, as though this information was determined on an absolute scale. It must of course be remembered that the reference reflection may not be representative of the substrate material deep below the interface. The substrate or reference under-layer may be strained due to the interface distortion and consequently the resulting composition and degree of relaxation may well be in error. This is best shown with an example, Fewster and Andrew (1998).

Table 5.5.1. Composition and relaxations in a thick InGaAs layer on GaAs; determined assuming the lattice parameter of the substrate is a reliable internal standard and then determining the same values on an absolute scale.

Method	a_z	R_x (%)	a_x	R_y (%)	a_y	$a_{unstrained}$	X (%)
Relative							
Substrate	5.65368		5.65368		5.65368		
Layer	5.67546	108	5.67420	63	5.66550	5.6728	4.75
Absolute							
Substrate	5.65367		5.65386		5.65397	5.65386	
Layer	5.67573	90.3	5.66978	90.7	5.66996	5.67148	4.39

A series of reciprocal space maps from a thick buffer layer of $In_{0.05}Ga_{0.95}As$ grown on GaAs has been analysed assuming the substrate reflection is a true representation of the bulk GaAs lattice parameters. The data was extracted from two *004* reflections for the [*110*] and [*1-10*] azimuths and the *444* and *4-44* reflections. The consequent analysis gives a value of composition and relaxation given in Table 5.5.1. The same procedure gives quite a different result if we make no assumptions about the substrate and determine the lattice parameters on an absolute scale. Clearly the degree of relaxation changes from a value that is above 100% in one direction to something more realistic as we go from the relative method to the absolute method. The composition is also different when determined by the two methods.

5.5.5.3.1. Using rapid mapping methods for multi-layers:

This section considers the compromise between obtaining information on the layer strain and speed of data collection. We covered the 'static' diffractometer in section 4.3.6, that presents an opportunity to collect a reciprocal space map very rapidly as discussed in section 4.3.7. The 'static' diffractometer will include wavelength dispersion (figure 4.3.18 where the Cu $K\alpha$ doublet is resolved) although this doublet can be eliminated with the primary optic of a channel-cut, section 4.3.5, which can add clarity to the reciprocal space map. To optimize the data collection time and resolution it is best to use a reflection close to grazing exit, although we can compromise with a convenient reflection: for GaN based compounds the *21-34* reflection from a (*0001*) surface orientation is the best option if a primary channel-cut optic is used. Scattering from planes significantly inclined to the surface planes will also improve the sensitivity to differences in the in-plane strain because of the larger s_x values for comparison, whereas scattering planes closer to the surface plane are better for obtaining layer/substrate tilting, because the rotation is more along ω. The *10-15* reflection that is commonly used for studying GaN materials, will require a large map to accommodate all the s_z information because of the effect illustrated in figure 4.3.20 and the necessity for a small step-size in ω to ensure the smearing of the 2θ capture does not bring uncertainty to the peak location.

We will assume that the substrate, or pseudo-substrate, of GaN has known lattice parameters. Also we can assume that the row of satellites associated with the periodic structure (PS) is normal to the surface plane and this defines the s_x reference line; so with the accepted s_x(GaN), s_z(GaN) for the GaN, we can determine the lateral strain of the periodic structure to that of GaN, i.e.:

$$\left(\frac{\Delta d}{d}\right)_x (PS) = \frac{-\Delta s_x}{s_x(\text{GaN})} \qquad 5.5.14$$

Δs_x is derived from the peak separations in the map, using equations 4.2.1. Now of course we only know these approximately, because ω is ill-defined, however since the line of satellites must all have the same s_x value we can correct for this uncertainty by calculating the miss-setting in ω, given by α:

$$\alpha = \tan^{-1}\left(\frac{s_{x1} - s_{x2}}{s_{z1} - s_{z2}}\right) \qquad 5.5.15$$

Where s_{xi} is the measured value for a satellite: the greater the separation of the satellites the more precisely this angular offset can be determined. This gives the following:

$$\Delta s_x = \{(s_x(\text{GaN}) - s_{x1}) - (s_z(\text{GaN}) - s_{z1})\tan\alpha\}\cos\alpha \qquad 5.5.16$$

And:

$$\Delta s_z = \{(s_x(\text{GaN}) - s_{x1}) - (s_z(\text{GaN}) - s_{z1})\tan\alpha\}\sin\alpha + \frac{s_z(\text{GaN}) - s_{z1}}{\cos\alpha} \qquad 5.5.17$$

Therefore if Δs_x is negative the lattice plane spacing parallel to the interface of the periodic structure is smaller than the GaN pseudo-substrate, then the first layer or two of the periodic structure has strain relaxed and their associated lattice plane spacing normal to the interface has increased, i.e. Δs_z is positive assuming typical elastic parameters.

One single map will not isolate tilt between layers and this can bring errors to the evaluation of strain. A second map 180^0 in ϕ from the first will isolate the strain and layer tilt components. It really depends on how much information is wanted, since a further two maps at the 90^0 and 270^0 azimuths in ϕ will reveal the full relaxation, if it is anisotropic, as well as the full tilt. It should also be remembered these rapid methods rely on the substrate being a good reference for any absolute lattice plane spacings. For obtaining an indication of the defects associated with the plane spacings these rapid methods can be very useful. The data collection time with the channel-cut + mirror primary optic with the *21-34* reflection is ~16 min which is identical to the 'static' diffractometer using the *21-33* reflection that relies on a slit: both give strain reproducibility < 0.001. The same analysis with the *10-15* reflection with channel-cut + mirror, for the same reproducibility takes ~ 25 min to collect the data.

5.5.5.3.2. Wafer mapping to extract relaxation and composition:

If the sample is not laterally homogeneous then the region of analysis becomes important and this can be achieved with spatial mapping with a small probe volume. This is possible with the ‘static’ diffractometer, section 4.3.6, and the details and an example are given in section 5.4.3.9.

As discussed in the last section, separating strain and tilt from a single map is not possible without assumptions, although the tilt is usually the result of significant relaxation, but combining maps is impractical. An alternative approach is to use a reflection with planes approximately parallel to the surface and use the ‘beam-selection’ diffractometer, section 4.3.4 with a double-pinhole colimator. The analyser will define the 2θ angle, and so any rotation of the planes will be indicated by a peak shift with an ω scan; this scan can be obtained in ~30 s with a range of 0.3^0, which is sufficient to capture typical tilt angles. Spatially mapping over the wafer will yield the tilts, distortions and bends very rapidly, as previously discussed in section 5.3.5.

5.5.5.4. Determination by rocking-curves alone:

This method is suitable for instruments with a monochromator and open detector, for example a double-crystal diffractometer, section 4.3.1. A rocking curve just gives the difference angle $\Delta\omega$ and therefore we have to find some way of obtaining the 2θ angle for each reflection. This method relies on the substrate reflections having known ω and 2θ angles.

Suppose initially we consider the layer and substrate to be perfectly aligned with each other, i.e. there is no tilt, then

$$\left\{\frac{\Delta s_z}{\{s_z\}_S}\right\}_H = \left\{\frac{\Delta s_z}{\{s_z\}_S}\right\}_0 = const \qquad 5.5.18$$

where the subscript H refers to an arbitrary reflection and 0 to a reflection from planes parallel to the surface. $\Delta s_z = s_z - \{s_z\}_S$ where s_z refers to the layer and $\{s_z\}_S$ refers to the substrate. This assumption clearly relates to the fact that $\Delta d/d$ is constant along a given direction. From equation 5.5.1 we can state that

$$\{\Delta s_z\}_H = \left\{\frac{\Delta s_z}{\{s_z\}_S}\right\}_0 \{s_z\}_H = \sin\omega_L + \sin(2\theta_L - \omega_L) - \sin\omega_S - \sin(2\theta_S - \omega_S)$$

5.5.19

Now since we are dealing purely with relative positions there is doubt about the surface normal direction and this has to be defined or chosen to be parallel to a convenient substrate scattering plane normal. We can determine the tilt of the layer with respect to the substrate by collecting two rocking curves at azimuths 180^0 apart in ϕ.

$$Tilt = \varphi = \frac{1}{2}\{\Delta\omega_0 - \Delta\omega_{180}\} \qquad 5.5.20$$

Equation 5.5.19 then should be

$$\{\Delta s_z\}_H = \left\{\frac{\Delta s_z}{\{s_z\}_S}\right\}_0 \{s_z\}_H = \sin\{\omega_L + \varphi\} + \sin(2\theta_L - \{\omega_L + \varphi\}) - \sin\omega_S - \sin(2\theta_S - \omega_S)$$

5.5.21

Rearranging this equation we obtain the 2θ angle for the reflection of interest; the subscript, L, to this layer reflection has been omitted.

$$2\theta = \sin^{-1}\left\{\left(\frac{\Delta s_z}{\{s_z\}_S}\right)_0 s_z - \sin\{\omega + \varphi\} + \sin\omega_S + \sin(2\theta_S - \omega_S)\right\} + \omega + \varphi$$

5.5.22

Therefore we have all the parameters necessary to determine the scattering angle 2θ and ω is found through equation 5.5.13 and because we are defining the surface normal perpendicular to a set of substrate scattering planes, i.e. $2\theta_S = 2\omega_S$ from equation 5.5.21:

$$\left(\frac{\Delta s_z}{\{s_z\}_S}\right)_0 s_z = \sin\{\omega + \varphi\} + \sin\{\omega - \varphi\} - 2\sin\omega_S \qquad 5.5.23$$

A combination of a rocking curve from a set of planes inclined to the surface plane for $\varphi \neq 0^0$ and two rocking curves for $\phi = 0^0$ and 180^0 from planes parallel to the surface will then give us all the information to obtain the composition and relaxation, etc., along the x azimuth. To account for any anisotropy we should include a similar set along y. Additional precision can be achieved by including widely spaced reflections from inclined planes as well as from planes parallel to the surface. All the discussion concerning problems with relative measurements must be born in mind. It should also be clear that this method cannot achieve the accuracy of the reciprocal space map method, as demonstrated in Table 5.5.1.

5.5.6. *Studying dislocations and defects:*

The above methods of determining the relaxation in thin layers are only suitable for significant strain relaxation. When the density of dislocations is small, as in the sample described in section 5.4.4.2, then with topography we can count the number and obtain the dislocation density. However it is sometimes useful to take topographs of samples when the defect density is clearly well above the level to observe single dislocations.

5.5.6.1. Analysis of relaxation with topography:

The microstructure of materials at various stages of relaxation can reveal the nature of the process. We shall consider a series of $In_xGa_{1-x}As$ layers, where $x \approx 5$ %. As discussed previously the present level of resolution achievable with topographs is related to the developed emulsion size (approximately 1 μm, although the diffractometer resolution may increase this value). In figure 5.5.5a there are four topographs taken from four different thickness of layers with the same intended 'In' composition. It is clear that the contrast increases with layer thickness. This can be interpreted as the amalgamation of very small regions with large tilts to create larger regions with similar tilts, suggesting a mosaic grain growth similar to work hardening, Fewster and Andrew (1993c). An estimate of the characteristic length scales and the measured tilts and their spread, from a combination of topography and reciprocal maps, are given in figure 5.5.5b. This example combines the analysis of reciprocal space maps with topography and indicates the tools available for this sort of analysis in understanding the evolution of the microstructure.

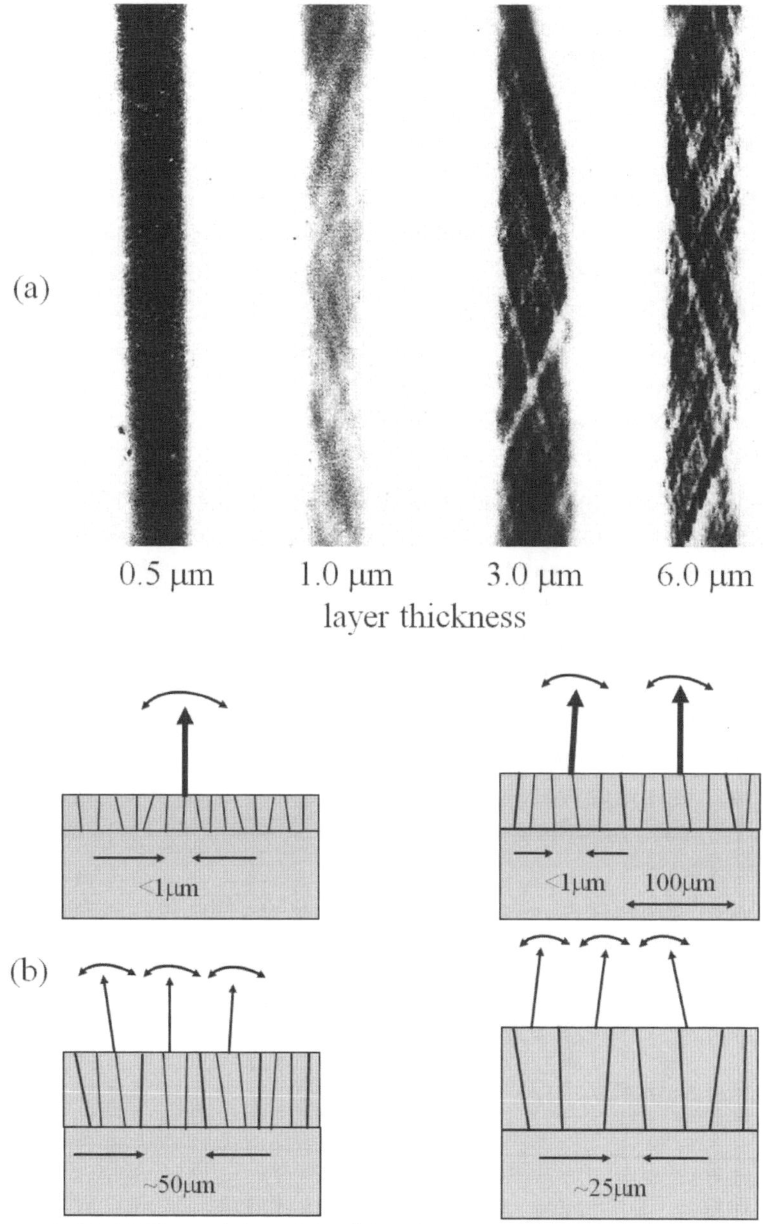

Figure 5.5.5. (a) The evolution of the dislocation network in $In_{0.05}Ga_{0.95}As$ on GaAs as a function of layer thickness and (b) the proposed model of the microstructure based on these topographs and the relaxation measurements.

5.5.6.2. Simulating structures with defects:

In section 2.8.2 we described an extension to the dynamical model to take into account the influence of defects at interfaces. The strain associated with the misfit dislocations and associated defects will extend into the layer above and below. At low levels this can be characterised by the shape of the diffuse scattering close to the substrate and layer peaks. At high levels this can be very pronounced and can lead to significant diffuse scattering streaks parallel to the interface plane. This example indicates the process of extracting information by modelling the full profile.

The sample was a 3 μm $In_{0.05}Ga_{0.95}As$ layer on a GaAs substrate. On top of this layer there was a superlattice of alternating $In_{0.10}Ga_{0.90}As$ and GaAs layers, whose average lattice parameter should be matched to the layer below. The materials problem was to ascertain the strain in the superlattice, was it isotropic, etc. This required a full analysis to ensure that the relaxation in the 3 μm layer was complete and the material quality was reasonable. The method of determining the relaxation was carried out as described above, section 5.5.5.2 and this was included in the simulation. The data were collected with a limited-area reciprocal space map close to the *004* reflection and projected onto the [*00l*] direction. This recovered the full intensity associated with tilting, the limited mosaic size effects and also the diffuse scattering associated with the interfaces.

Since the degree of relaxation was well above the few percent level the X-ray wave-fields are uncoupled at the interface and this was shown to immediately influence the relative heights of the layer and substrate peaks. Matching peaks alone assuming the material is perfect gave a thickness value of 2 μm, however when the wave-fields are uncoupled in the simulation the value was close to 3 μm. The diffuse scattering contribution alters the shape of the peaks at the base of the substrate and layer peak, and by including the strain distribution this can be modeled. This region is divided into regions of constant strain, which are considered to scatter independently (the exponential variation in the strain is split into 10 layers above and below the interface). The shape is

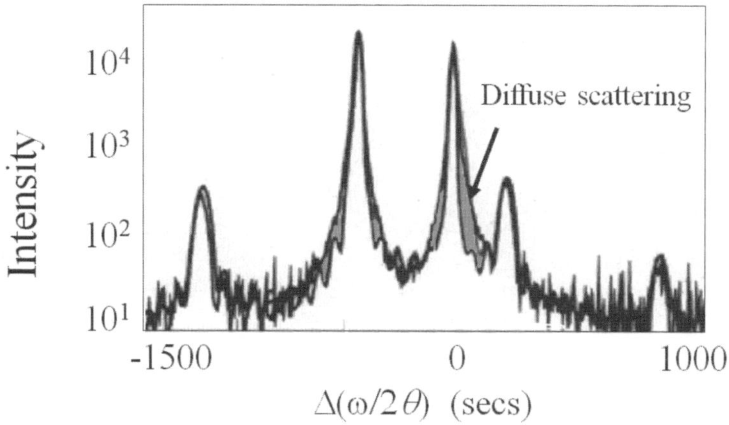

Figure 5.5.6. The profile from projecting a reciprocal space map of a 3μm layer of $In_{0.05}Ga_{0.95}As$ on GaAs with a $In_{0.1}Ga_{0.9}As$ / GaAs superlattice on top. The inner profile is the best fit based on a structure being perfect, and the outer profile is the best fit to the experimental profile, with all the diffuse scattering included. This, gave the extent of the interfacial distortion.

then fitted in an iterative manner until the best fit is found. The additional variables to model the diffuse scattering are purely the extent of the strain-fields above and below the interface, figure 5.5.6.

5.6. Analysis of laterally inhomogeneous multi-layers (textured polycrystalline):

In the example above we have taken a reciprocal space map from planes parallel to the interface and projected it onto a direction normal to the interface. Although we can achieve a good fit to the profile there is a considerable amount of information that is lost, for example the size of the mosaic blocks, the distance between defects and the distribution of tilts. This will be covered in this section. We shall firstly consider the analysis of lateral inhomogeneities that are too small to be observed by topography and then consider the larger aspects of inhomogeneity. In the first part we will consider a very direct analysis and show how simulation of reciprocal space maps gives a more complete picture before considering topographic methods.

5.6.1. *Direct analysis of laterally inhomogeneous multi-layers:*

From section 2.8.2 we derived some expressions for the influence of finite size regions on the scattering pattern. Also any region that is tilted with respect to the average will add to the broadening and consequently we have a combination of contributions. If we assume that the "mosaic blocks" are perfect and their bounds are primarily limited parallel to the surface plane, then we can consider the various components that contribute to the shape and position of the measured scattering, figure 5.6.1a.

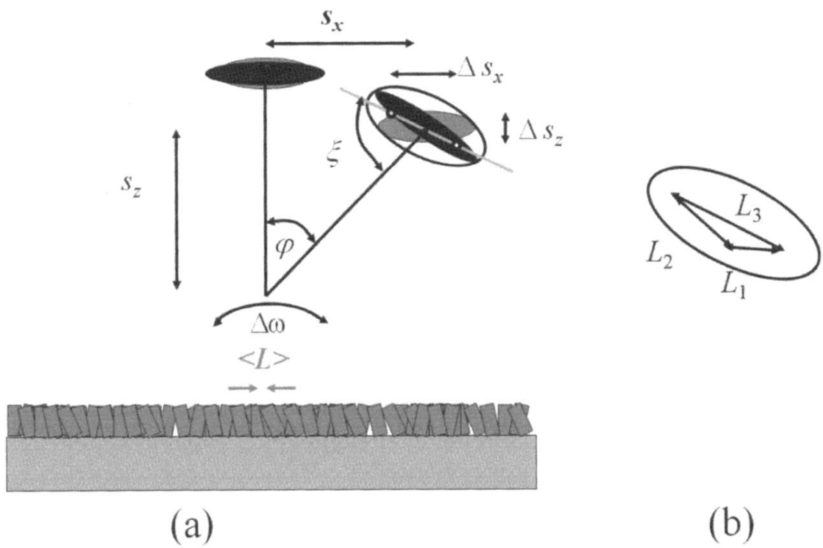

Figure 5.6.1. (a) The structural features that give rise to the shape of the scattering in reciprocal space are the microscopic tilt distribution (black ellipse) and lateral correlation length (grey ellipse). (b) gives the proportion of these components with respect to the whole general shape.

Scattering from planes approximately parallel to the surface will have overlapping contributions from lateral correlation lengths (lateral finite sizes) and the distribution of tilts (microscopic tilts) giving a characteristic broad elliptical scattering shape. For scattering from planes inclined to the surface plane these contributions will be inclined to each other and will rotate the diameter of the ellipse depending on the strength of the various contributions. Consider figure 5.6.1b, a schematic of a reciprocal space map, where the contribution from the lateral correlation length, L_1, is parallel to the surface plane and the microscopic tilt contribution, L_2, is normal to the radial direction (along an $\omega/2\theta$ scan direction). The full-width-at-half-maximum intensity of the overall ellipse can be characterised by two lengths in reciprocal space, i.e. Δs_x and Δs_z. Now the two contributions will be correlated and we can consider them as vector sums of the microscopic tilt perpendicular to the radial direction, L_2, and the lateral correlation length parallel to the surface plane, L_1.

Hence from trigonometry we obtain

$$\frac{L_2}{L_1} = -\frac{\cos\xi}{\cos(\varphi+\xi)} \tag{5.6.1}$$

Similarly we can write

$$\frac{L_3}{L_2} = \frac{\sin\varphi}{\cos\xi} \tag{5.6.2}$$

In reciprocal space units L_3 is simply given by

$$L_3 = \sqrt{\left(\{\Delta s_x\}^2 + \{\Delta s_z\}^2\right)} \tag{5.6.3}$$

and

$$\varphi = \tan^{-1}\left\{\frac{s_x}{s_z}\right\}$$
$$\xi = \tan^{-1}\left\{\frac{\Delta s_x}{\Delta s_z}\right\} \tag{5.6.4}$$

Hence

$$\text{Lateral correlation length} = \frac{1}{L_1} \tag{5.6.5}$$

and the microscopic tilt (in radians) is related to the dimension L_2 by

$$\text{Microscopic tilt} = \frac{L_2}{\sqrt{\{s_x^2 + s_z^2\}}} \qquad 5.6.6$$

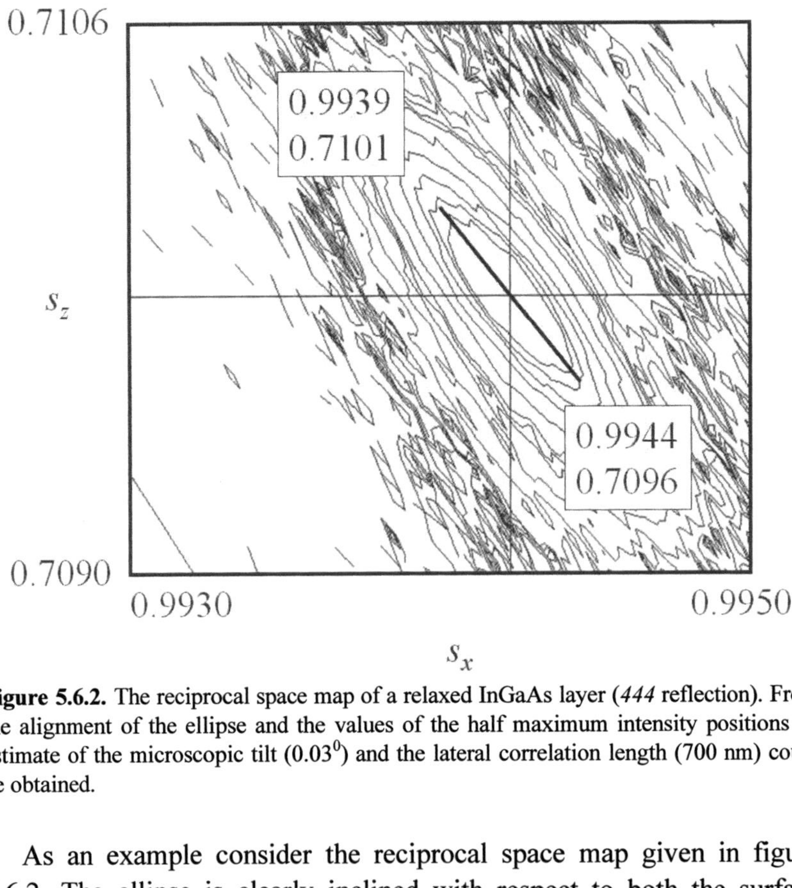

Figure 5.6.2. The reciprocal space map of a relaxed InGaAs layer (*444* reflection). From the alignment of the ellipse and the values of the half maximum intensity positions an estimate of the microscopic tilt (0.03^0) and the lateral correlation length (700 nm) could be obtained.

As an example consider the reciprocal space map given in figure 5.6.2. The ellipse is clearly inclined with respect to both the surface normal and radial direction and therefore includes both microscopic tilt and lateral correlation lengths. By measuring the positions of the half-height intensity along the ellipse we can determine the microscopic tilt as 0.03^0 and the lateral correlation length to be 700 nm. The location of the axis of the ellipse can be difficult to judge if the spreading is small, and is therefore most reliable when the microscopic tilting is large or the lateral dimension is small.

We can extend this method to two reflections from planes parallel to the surface and this can be more reliable than the single reflection method above. This could be applicable to materials that are textured with poor orientation preference in the surface plane. The full width at half maximum intensity (FWHM) of two reflections along the same direction in reciprocal space will be denoted by the coordinates as $(\Delta s_x)_1$ and $(\Delta s_x)_2$. The microscopic tilt, $\Delta\omega$, and the lateral correlation length, L, will influence both reflections, n; if we consider both these parameters and denote their contributions by $(\Delta_\omega s_x)_n$ and $(\Delta_L s_x)_n$ respectively then

$$\Delta\omega = 2\tan^{-1}\left\{\frac{\Delta_\omega s_x}{2s_z}\right\}_1 = 2\tan^{-1}\left\{\frac{\Delta_\omega s_x}{2s_z}\right\}_2 \qquad 5.6.7$$

and

$$L = \left\{\frac{1}{\Delta_L s_x}\right\}_1 = \left\{\frac{1}{\Delta_L s_x}\right\}_2 \qquad 5.6.8$$

As above we can consider the resultant broadening to be the correlation of the two contributions, and the profile are approximated to Gaussian, i.e.

$$\{\Delta s_x\}_n^2 = \{\Delta_\omega s_x\}_n^2 + \{\Delta_L s_x\}_n^2 \qquad 5.6.9$$

Then from equation 5.6.7, we have after substituting into equation 5.6.9

$$\{\Delta_\omega s_x\}_1 = \{\Delta_\omega s_x\}_2 \frac{\{s_z\}_1}{\{s_z\}_2} = \left[\{\Delta s_x\}_2^2 - \{\Delta_L s_x\}_2^2\right]^{1/2} \frac{\{s_z\}_1}{\{s_z\}_2} \qquad 5.6.10$$

If we now include the relationship given in equation 5.6.8, along with equation 5.6.9, we have

$$\{\Delta_\omega s_x\}_1^2 = \frac{\left[\{\Delta s_x\}_2^2 - \{\Delta s_x\}_1^2\right]}{\left[\{s_z\}_2^2 + \{s_z\}_1^2\right]}\{s_z\}_1^2 \qquad 5.6.11$$

Since we know s_z (compare with figure 5.6.1) we can derive the microscopic tilt $\Delta\omega$ from equation 5.6.7 and the lateral correlation length from rearranging equation 5.6.9 and substituting into equation 5.6.8

$$L = \left\{ \frac{1}{\Delta_L s_x} \right\}_1 = \left\{ \frac{1}{\{\Delta s_x\}_1^2 - \{\Delta_\omega s_x\}_1^2} \right\}^{1/2} \qquad 5.6.12$$

This can be a useful approach to the evaluation of these parameters and avoids the difficulty of determining the angle of the ellipse in the former method.

5.6.2. *Simulation of laterally inhomogeneous multi-layers:*

The analysis above makes several assumptions. The characteristic length scale is assumed to be uniform and really takes no real account of the detailed shape of the profile. The advantages of simulating the reciprocal space map will be become evident in the following example.

GaN is known to be full of defects when grown on sapphire and the spacing between these defects is an important parameter. The defect separation is analogous to the lateral correlation length. In the first instance we can analyse the reciprocal space map as above and extract some starting values for the microscopic tilt and lateral correlation length. In this example the lateral correlation length with a barely detectable microscopic tilt dominated the shape when analysing the *105* reflection. The derived lateral correlation length is 92 nm and this value was included in the simulation of the reciprocal space of the *002* reflection from the GaN. The GaN is in the hexagonal form with the *c*-axis normal to the surface and parallel to the *c*-axis of the sapphire substrate. The multiple-crystal scan along the radial direction ($\omega/2\theta$ scan) is shown in figure 5.6.3 along with the best-fit profile. The best-fit profile was found only by assuming that the In in the InGaN layer had segregated into the AlN cap region. The agreement was not perfect; in fact the quality of this particular sample was very poor, as we shall see later in section 5.6.3.3.

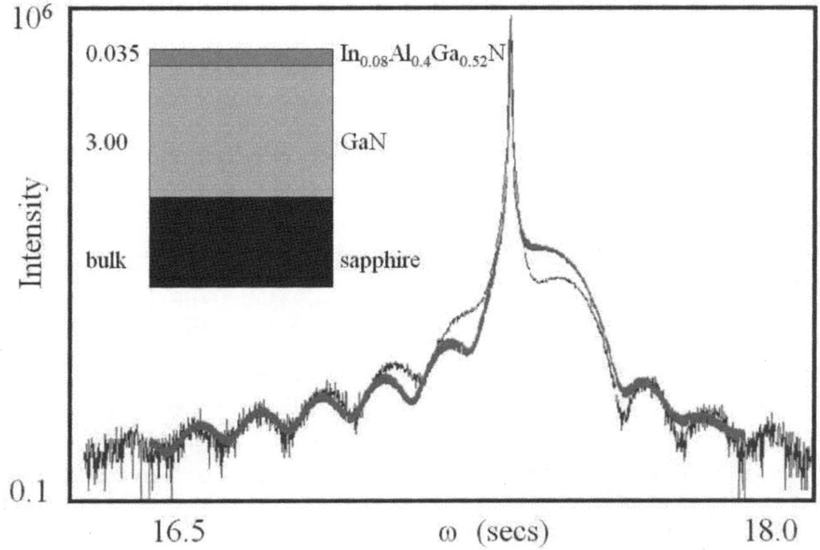

Figure 5.6.3. The best fit profile to a GaN structure on sapphire. The interface was heavily distorted and the InGaN had diffused into the AlGaN cap layer. The sapphire scattering peak occurs at a much higher incident angle. This *002* reflection was measured with the multiple-crystal diffractometer.

From section 2.8.2 we can simulate the reciprocal space map based on the determined lateral correlation length. The layer thicknesses and the composition determined from the radial scan. The measured reciprocal space map is given in figure 5.6.4a and the simulated profile in 5.6.4b. The general shape of the simulated profile is clearly not very close to the measured profile, although the full-width-at-half-maximum intensity for the "pseudo-substrate" and "layer" peaks are in agreement. The difference arises because there is a distribution of sizes that changes the distribution in the scattering parallel to the surface. When a distribution is included the appearance of the map changes quite dramatically. Another major contribution to the shape of the scattering arises from the distribution of tilts of the mosaic blocks, equation 2.8.16 and figure 2.8.5. So to obtain a good fit to the whole profile some average dimension should be fitted first, followed by a distribution in their sizes and finally the spread of tilts associated with the block wall edges. The best fit is given in figure 5.6.4c. It should be noted that the distribution does not encompass the apparent average derived from the direct analysis in this case, although the full-width-at-half-maximum intensity of both maps is the same. This could have important consequences in deriving quantitative information from these experiments.

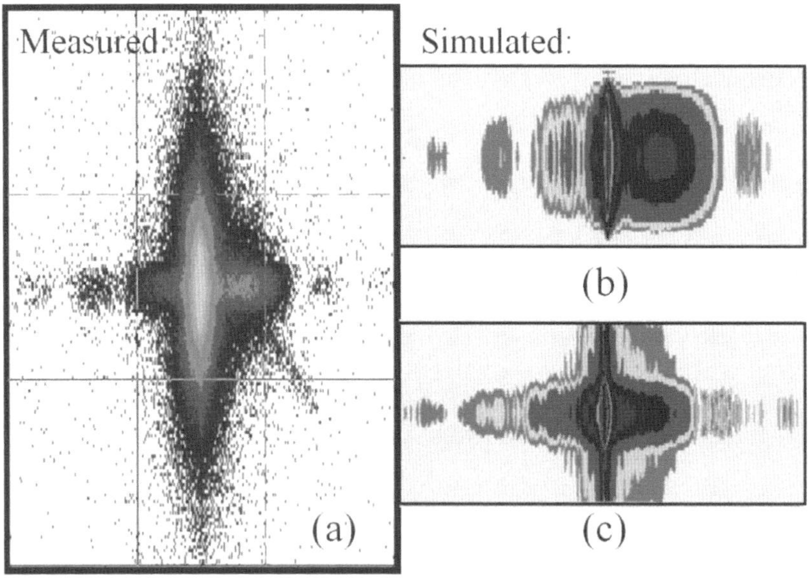

Figure 5.6.4. The *002* reflection from the GaN structure of figure 5.6.3; (a) measured reciprocal space map, (b) the simulated reciprocal space map assuming the separation between defects are all similar (92 nm) and (c) assuming a distribution of sizes (150±50 nm).

5.6.3. *Lateral inhomogeneities without large misfits:*

In the last few examples we have seen that large misfits have led to the breaking-up of the layers into mosaic blocks, however this situation can still exist without relaxation taking place. Examples include mosaic substrates; this is described in section 5.3.2 and illustrated in figures in that section. For example the layer can mimic the substrate indicating perfect registry between the substrate and the layer within each mosaic block and each block can be mis-orientated with respect to each other, figure 5.3.12. For detailed analysis of these structural types full three-dimensional mapping is ideal, section 4.2.2, or great care must be exercised in the interpretation of the reciprocal space maps since these represent projections of the scattering onto an arbitrary plane.

When lateral inhomogeneities are greater than about 5 μm then topography can prove to be a very powerful method of analysis. Again this is best performed in combination with the reciprocal space maps, since this assigns the scattering feature to the "real" space image. Examples of these have been given in previous sections but for completeness is mentioned here. Under some growth conditions twinning can exist and knowing the lateral sizes of the various components can be of great interest. Two examples will be given one for the case when the twins are too small to be observed by topography and the other when they can be observed.

5.6.3.1. Analysing epitaxial layers with very small twinned regions:

This is an example taken from Fewster and Andrew (1993b), who analysed a high critical temperature superconducting oxide on a strontium titanate substrate. The layer has orthorhombic symmetry with the a and b axes of very similar lengths (0.383 nm and 0.388 nm respectively), which are very close to the $SrTiO_3$ lattice parameter of 0.390 nm. This combination creates the possibility of the a and b parameters aligning along two directions at 90^0 to each other. This can be envisaged as different nucleation regions during deposition having different orientations and effectively appearing as a distribution of twin components. Using diffractometry the relative proportions can be obtained from the area under these peaks.

Clearly any analysis from planes inclined to the surface normal will have a well defined ω and 2θ that relates to the lattice parameters and orientation. Therefore scattering within the *0kl* and *h0l* planes for the two possible orientations will have slightly different angular settings when $h = k$ due to the small differences in lattice parameters along a and b. The chosen reflections for this example were *038* and *308*. The angular settings for Cu $K\alpha_1$ radiation for these reflections are $\omega = 4.360^0$, $2\theta = 105.5506^0$ and $\omega = 4.5873^0$, $2\theta = 106.891^0$ respectively. The angular separation is reasonable and material quality is poor resulting in weak scattering and therefore high intensity low angular resolution data is the most appropriate choice.

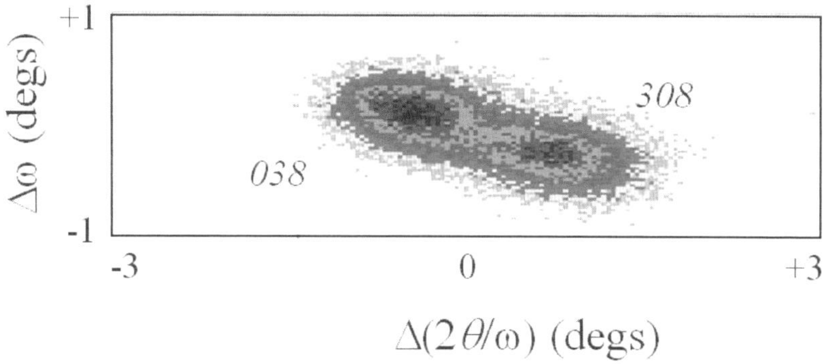

Figure 5.6.5. The scattering from the *038* planes of one orientation of YBaCuO with the *308* scattering planes of the 90^0 twin component.

The instrument used was a slit-based diffractometer with the configuration described in section 4.4.1. Since no single scan was able to scan through both peaks and create a direct comparison of their intensities a reciprocal space map was obtained, figure 5.6.5. From this the integrated areas and relative peak heights could be obtained, compared with the calculated intensities (based on kinematical theory) and the ratio of the two contributions obtained. In this particular case there are nearly equal proportions of the two twin components.

5.6.3.2. Analysing twin components larger than 5 microns:

The analysis of features greater than 5 microns can begin to be resolved spatially, e.g. their orientation and strain effects can be separated from the surrounding matrix and observed in diffraction, although judging their dimensions is best done by topography. It is interesting to note that the topography and diffractometry procedures for determining size of features are complementary and cover the full range of dimensions.

Topography can be applied very successfully in combination with diffractometry and examples have been given in previous sections in this chapter.

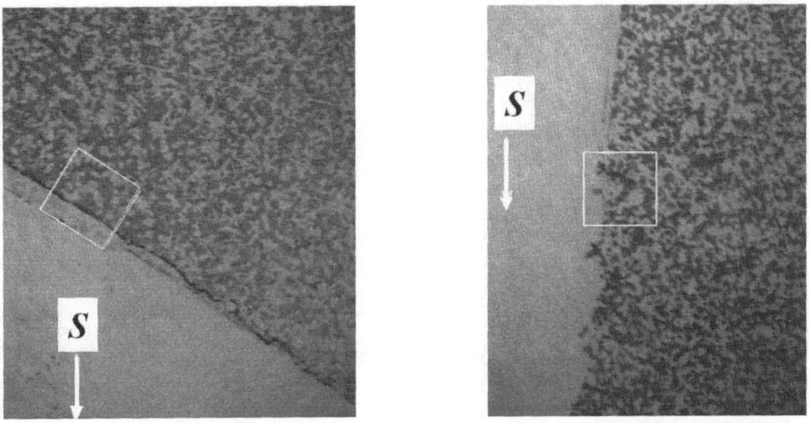

Figure 5.6.6. The evidence of twins in CdHgTe grown on CdTe (*111*). The two Berg-Barrett topographs were compared after a 60^0 rotation showing the reversal of contrast. The exposure times were a few minutes and the twin dimensions were approximately 20 μm.

This particular example is rather interesting in that a topograph from planes parallel to the surface will yield an even distribution of intensity. The sample is $Cd_{0.2}Hg_{0.8}Te$ grown on a (*111*) CdTe substrate. A Laue photograph indicates 6-fold symmetry, which in itself is suspicious, it should be 3-fold. This is the first indication of twinning but gives no real indication of the size of these twins. A simple Berg-Barrett topograph, figure 5.3.5, or with any reflection topography system, e.g. figure 4.3.1, where a photographic emulsion is held immediately after the sample, will yield images as in figure 5.6.6. The two images are both of the *115* type reflections, the second after a 60^0 rotation of the sample about the surface normal. The images are negatives of each other and clearly indicate that the two contributions are interlocking 180^0 twins. The exposure times are only a few minutes and clearly give a very quick analysis of this type of materials problem.

5.6.3.3. Analysis of textured polycrystalline semiconductors:

Working through these various examples we have ordered them in decreasing levels of "quality" from nearly perfect epitaxy, through partially relaxed structures to textured epitaxy. When the orientation dependence in the plane of the interface breaks down altogether the sample is essentially textured polycrystalline. When the scattering from planes parallel to the surface are predominately strong whereas those from planes inclined to the surface are weaker than expected the sample may well be of textured polycrystalline form. The GaN sample structure on a sapphire substrate described in section 5.6.2 has this characteristic; to confirm this possibility the scattering from planes normal to the surface plane were studied.

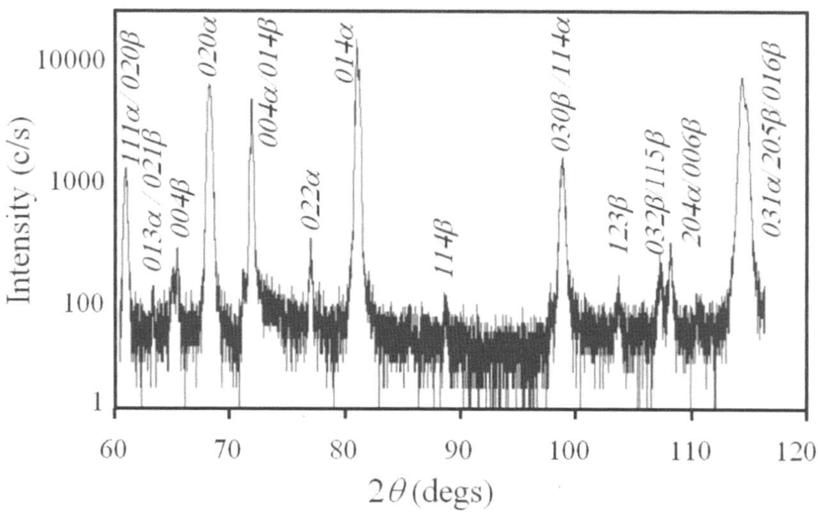

Figure 5.6.7. A low-resolution in-plane scan from a poor quality epitaxial GaN layer. The large number of reflections is indicative of many crystallites of different orientations.

The experimental set-up for in-plane scattering is described in section 4.5.2 and figure 4.5.2. Since this is slit based and low-angular resolution diffractometry the intensity is high and a full range of scattering peaks were found using this geometry, figure 5.6.7. If the material was nearly perfect epitaxy then only a few reflections should be observed, figure

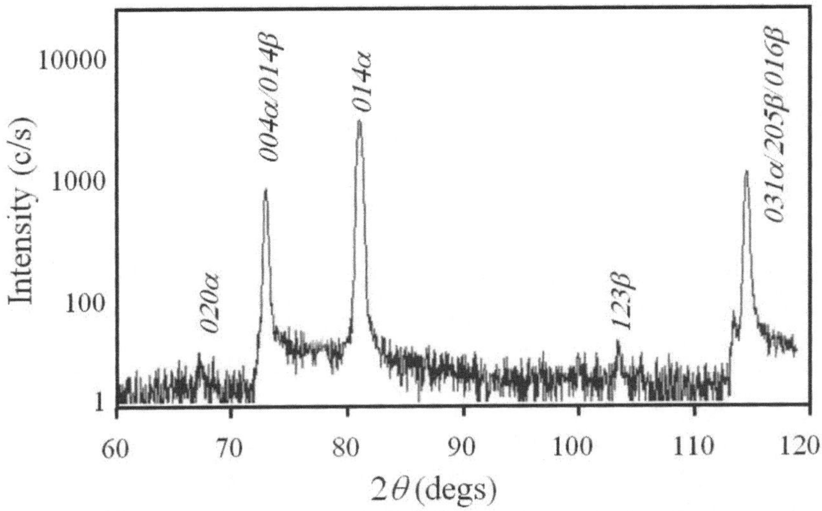

Figure 5.6.8. A low-resolution in-plane scan of a reasonable quality GaN layer with fewer in-plane reflections observed. This should be compared with figure 5.6.7.

5.6.8. From figure 5.6.7 the large range of reflections clearly indicates that the structure is composed of a large range of orientated crystallites with a very strong *00l* texture normal to the surface. This information including the simulation of the reciprocal space map now gives a reasonably complete picture of this sample.

5.7. Analysis of nearly perfect polycrystalline materials:

As the degree of texture declines and the crystallites in the sample become more randomly orientated the information that characterises the sample becomes less, Table 1.3.1. Working from Table 1.2.1, it can be seen that the relevant parameters are; shape (including layer thickness, etc., if relevant), composition (what phases are present), form (are there different size crystallites or amorphous content), orientation (is there any texture or preferred orientation), distortion (is there a strain distribution), homogeneity (variation over the sample laterally and in-depth), interfaces (if this is a multilayer structure then this is relevant) and density (how well are the crystallites packed). This may seem a fairly comprehensive list of useful information but we are dealing increasingly more with averages and attempting to estimate some distribution of values. There is a plethora of books on the analysis of polycrystalline materials that I shall not reproduce here but give some examples that are attempting to take the analyses further.

5.7.1. *Layer thickness analysis of nearly perfect polycrystalline materials:*

In the analysis of nearly perfect or textured single crystals where the surface normal dimension of each contribution (mosaic block for example) is sufficient to obtain fringing; then these can be related to the layer thickness, provided the single crystal occupies the full depth. For a polycrystalline material this is not the case, so we have to move to methods that are sensitive to density changes rather than finite crystal dimensions.

The obvious approach here is reflectometry, section 2.10. Because reflectometry is conducted at very small angles, for example out to 2^0 with a step resolution of 0.01^0 or better, then the spacings being probed (the separation between features, e.g. 0.01^0 as an example) will be ~ 0.44 μm, and the minimum length scale being probed is 22 nm. This is significantly greater than that for typical scattering planes and therefore an average refractive index can be assumed. As the resolution increases then larger length scales can be probed and this makes reflectometry very appropriate for typical layer thicknesses. The illustrative example in figure 5.4.24 for an AlAs / GaAs periodic structure is still appropriate, even if the components were polycrystalline or amorphous (although the macroscopic density may be less). The analysis of a single layer is considerably simpler as shown below.

The profiles in figure 5.7.1 are given for two polycrystalline layers samples, one is pure Cr and the other is CrO_x that have been deposited on amorphous glass substrates. The application of interest here was to know the layer thickness and the oxygen level, i.e. how accurately could x be determined. The assumption is that the porosity is identical in both samples, this is reasonable because the Cr layer gives a density matching

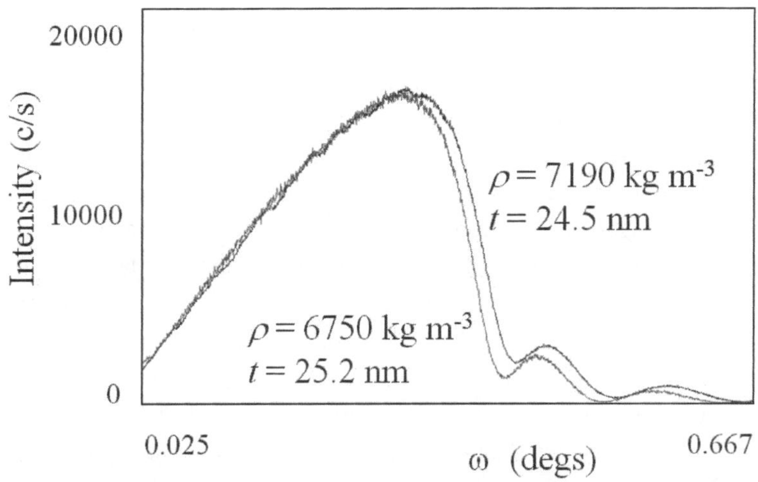

Figure 5.7.1. These two profiles indicate the sensitivity of a reflectometry profile to layer thickness and density. The data was collected with the multiple-crystal diffractometer to establish an absolute angular scale for a reliable measurement of the density.

the expected bulk value. Any curvature of the sample due to stresses within the sample or from mounting would smear the information making analysis more difficult. The instrument choice therefore becomes an important factor.

The best profile is achieved with the multiple-crystal diffractometer (with a line focus) because the instrument function is insignificant and all the contributions to the shape are related to the sample. This is the best configuration for complicated multi-layers requiring modeling of the intensity, or those that contain layers >300 nm, that require high angular resolution. This allowed the density to be determined to within <1%. To compare this with a slit based system, the best that can be achieved is closer to 5%. The intensity is compromised with the multiple-crystal diffractometer compared with a slit system, which could have been boosted with an X-ray mirror, but the incident beam size is increased and exceeded the sample projection length and was therefore of no positive benefit.

The geometry used in this example is as described above, i.e. without the X-ray mirror. This placed the profile on an absolute angular scale for obtaining the density (that is related to the critical angle): the higher the oxygen concentration the lower the density. Because the scattering angle can easily be placed on an absolute scale, section 5.3.4, the position of the whole profile can be placed on an absolute angle scale. Sample size effects and the relatively large incident beam size (despite not having an X-ray mirror) required the whole profile to be modeled to account for the shape of the scattering close to and below the critical edge. The accurate location of the critical edge is necessary for calculating the thickness as well as the composition as discussed in section 5.4.2.2 (Table 5.4.2). The derived values in figure 5.7.1 from fitting give an indication of the sensitivity of this approach. The density of pure Cr is 7190 kg m^{-3} and if we assume that O substitutes the Cr directly, with a negligible change in lattice parameter then the oxygen state is given as $x = 0.09$. The thickness levels do differ by a small amount, and to detect this difference the critical edge has to be known accurately. This is where the absolute angle measurement is crucial, otherwise this profiles would more-or-less overlay giving no useful comparative information.

5.7.2. *Composition analyses of polycrystalline materials:*

The main composition method in powder diffraction relies on the peak positions in 2θ and to a lesser extent on their intensity. This information is compared with one of the extensive data-bases of polycrystalline phases, to search for a match. This phase identification method is used and written about extensively and will not be covered further. The proportions of these phases rely much more heavily on the intensity measurement of all the peaks, once the components have been identified, and the basis of the approach was given in section 4.4.1.3. Figure 4.4.5b illustrates a profile of a multi-phase geological sample from a high-resolution powder diffractometer. This represents a typical powder diffraction profile, however in this section these methods are extended to very small proportions that are much more difficult to observe. Also we shall consider the case where the phase exists as a solid solution, i.e. the phase is not a fixed simple stoichiometric ratio. The composition can also be determined via the density as in the example in section 5.7.1.

5.7.2.1. Identification of trace phases:

Randomly orientated polycrystalline materials as thin layers will have very weak scattering and at times the scattering of interest is difficult to observe. With the help of reciprocal space mapping and a low angular resolution diffractometer (section 4.4.1 and mentioned in 4.4.1.3) exceedingly weak scattering can be observed. This can be important for identifying phases, determining the crystallite size or for analysing the state of strain in the plane of the interface for example. We shall just illustrate two examples using these two different geometries to illustrate what is possible with relatively simple apparatus.

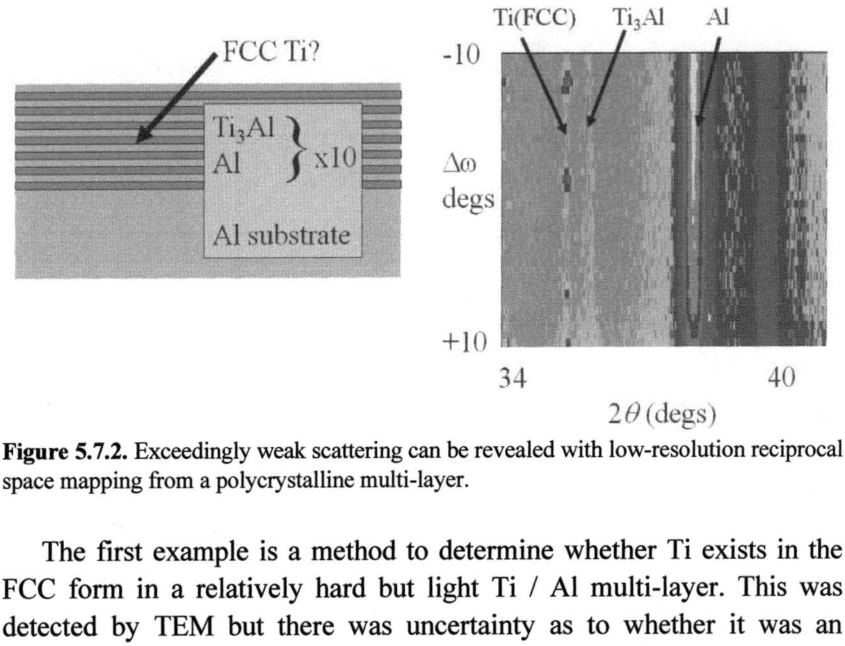

Figure 5.7.2. Exceedingly weak scattering can be revealed with low-resolution reciprocal space mapping from a polycrystalline multi-layer.

The first example is a method to determine whether Ti exists in the FCC form in a relatively hard but light Ti / Al multi-layer. This was detected by TEM but there was uncertainty as to whether it was an artifact of the TEM sample preparation. X-ray methods with no sample preparation should resolve this issue. The geometry of the apparatus was as described in section 4.4.1, essentially the Bragg-Brentano geometry. A very careful radial scan with $\omega = 2\theta / 2$ indicated no Ti FCC was present, however a limited area low resolution reciprocal space map using the same geometry revealed scattering associated with the Ti FCC phase, figure 5.7.2. The difficulty in seeing this detail using conventional methods also becomes rather obvious, the composition of this phase is very low (a few %) and also that which does exist would be a case of pure chance in finding this with a single scan. The visibility is very clear in the map.

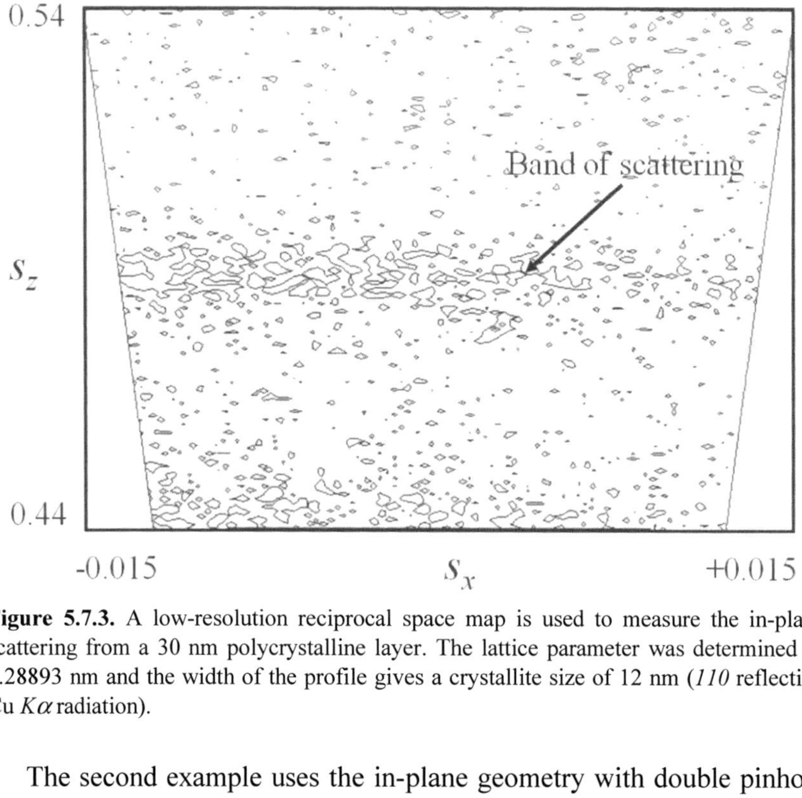

Figure 5.7.3. A low-resolution reciprocal space map is used to measure the in-plane scattering from a 30 nm polycrystalline layer. The lattice parameter was determined as 0.28893 nm and the width of the profile gives a crystallite size of 12 nm (*110* reflection Cu $K\alpha$ radiation).

The second example uses the in-plane geometry with double pinhole and parallel plate collimator, section 4.5.2. The sample is a 30 nm randomly orientated polycrystalline Cr layer on glass and this gives an indication how the very weak scattering from planes normal to the surface plane can be observed by low-resolution reciprocal space mapping. A single 2θ scan indicates no scattering observable, however the reciprocal space map produces a clear band of scattering, figure 5.7.3. Both these examples are using some of the simplest optics and a sealed 2 kW X-ray Cu source.

5.7.2.2. Determining phase composition in a solid solution:

Earlier in this chapter we discussed the measurement of alloy composition in semiconductors, this is important in polycrystalline materials as well. As an example, a 100 nm layer of In_2O_3 can easily be observed and identified with conventional Bragg-Brentano optics (section 4.4.1) or using a double-slit and parallel plate analyser (section 4.4.2, if the underlying layers need to suppressed); however in electronics and many other applications it is usually co-deposited with SnO_2 to form a conducting transparent layer. These exist in solid solution and not as separate phases, therefore the diffraction profile is similar to In_2O_3 but the peaks are shifted to lower angles with increasing Sn content. The space groups are different for the two phases and therefore there will be a limited solid-solubility, which means Vegard's rule cannot be used across the whole range of proportions, but it can be used as a guide to concentration of Sn after obtaining a calibration curve when combined with a chemical sensitive method. Reflectometry is very good at evaluating the thickness, but the density contrast from, say 5 % and 10 %, and 70 % SnO_2, is just 0.14 % and 2 %, which is too small for a reliable evaluation (e.g. compare with figure 5.7.1). An indication of the composition sensitivity with diffraction is much stronger, figure 5.7.4.

5.7.3. *Crystallite size analysis:*

The determination of crystallite size is fraught with problems, and in general the favoured approach is to use a standard reference material, e.g. LaB_6, to characterize the instrumental aberrations and use this to deconvolve from the measured profile. The resultant profile is then fitted with the Scherrer equation, equation 2.7.10. The problems arise from the fact that even a crystallite size of 1 μm has a diffraction width a factor of five below that of typical measured widths, and this assumes

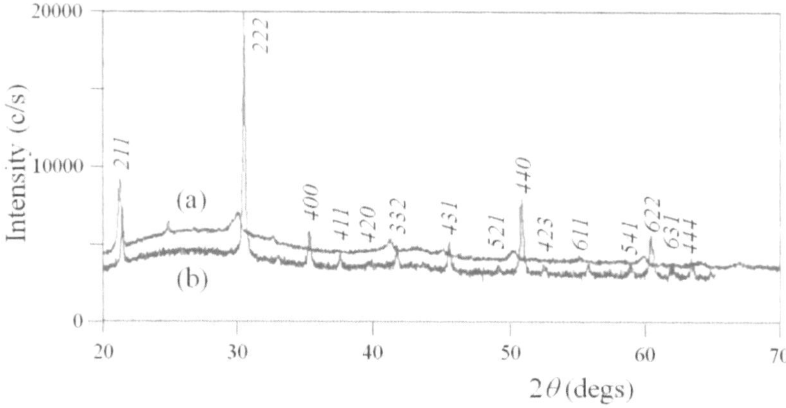

Figure 5.7.4. The peak search on the profile with the higher Sn concentration (a) will not be found in the data-base because it is a solid solution of In_2O_3 (b) and SnO_2. The consequence of this is the peaks are shifted, which can be used to reveal the proportion of In and Sn.

that a Bragg peak is captured (see section 2.9 on the new diffraction theory that is particularly relevant to powder diffraction). The whole scattering will create a profile broader than that just due to the Bragg condition. As the crystallite dimensions decrease then the instrumental aberrations become less significant and the reliability increases.

An analysis based on the Scherrer equation is given in figure 5.7.5, where there are two distinct peak widths. The most reliable approach is to simulate the profile, including the background, to extract this information. The advantage with the simulation approach is the noise level can also be included, which relates to the data collection time and signal strength making it easier to estimate the acceptable bounds for the fit and therefore the reliability of the method.

It is helpful at this stage to discuss in more detail and why the peak widths cannot be simply related. The data is collected by interaction with the enhanced scattering around the Debye-Scherrer ring, i.e. at the Bragg angle; however the Debye-Scherrer ring is very complex, Fewster and Andrew (1999). It is tempting to analyse the fluctuations in intensity as differences in crystallite size, but it is unlikely to be as simple as that. The complexity of the intensity of the Debye-Scherrer ring can be simulated with all the crystallites being the same size, figure 5.7.6. This

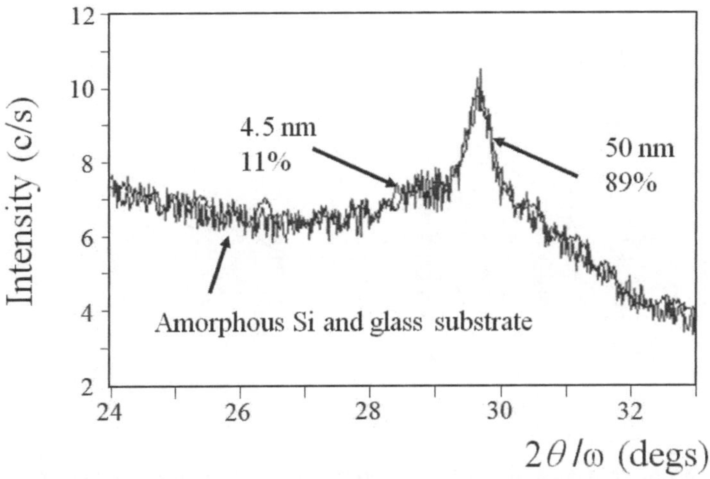

Figure 5.7.5. The scattering from small crystallites on glass indicating the two distinct size contributions to the scattering. The profiles have been fitted by overlapping Gaussian profiles with noise to aid the comparison.

is a consequence of the theory (section 2.9), that does not assume that all the scattering is associated with the Bragg condition. The accumulated peak shape is related to the crystallite size but not in a simple way. This indicates that the above conclusion is close as we can expect for the moment, i.e. use a very good reference sample that will give the narrowest peak possible to characterize all that is not known, and do not put too much reliance on the result above ~ 50 nm.

5.7.4. *Orientation distribution of polycrystals:*

Thin polycrystalline layers often show some orientation texture, which is a consequence of the different growth rates along different crystal directions. This is especially true in high symmetry space groups with significant anisotropy. If the crystals are grown on an amorphous surface, so that no preferred nucleation occurs, then the surface normal will be dominated by the slow growth planes. The faster growth directions fill the surface plane more rapidly to cause this bias in orientation texture.

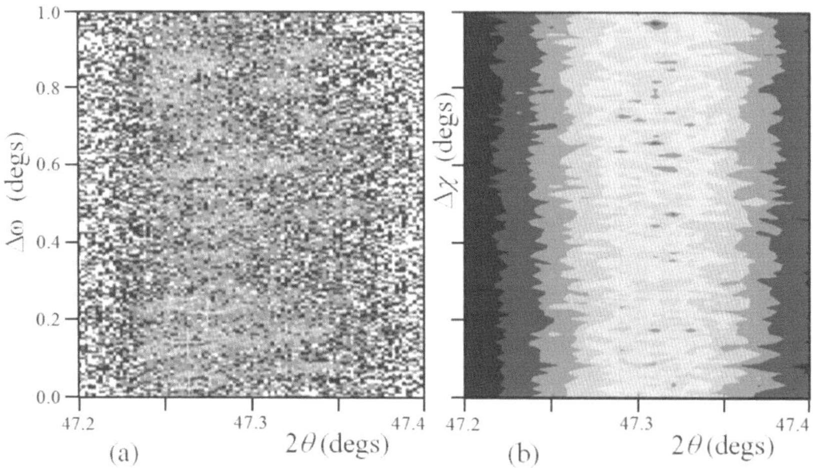

Figure 5.7.6. (a) Part of the Debye-Scherrer ring measured with a double pinhole (100 μm) and an analyser crystal chosen to remove wavelength dispersion, section 4.3.4. The sample is Si with a crystallite size range from 5 to 15 μm with an average of 10 μm, 80 % are thought to be very close to this average value (data collected by P Kidd). (b) the simulation of the Debye-Scherrer ring assuming all the crystallites are 10 μm, based on the new theory, section 2.9.

There are some well established methods for determining the orientation because it has importance in metallurgy. Any processing can cause orientation texture, which gives another variable for strength, thermal creep, etc. The traditional methods for collecting this data is to centre the detector at a specific 2θ angle for a reflection, and then the sample is manipulated in ω and ϕ or χ and the intensity monitored. If the intensity is plotted on a polar diagram and the intensity distribution is uniform (after accounting for any instrumental aberrations), then there is no dominant orientation preference. If the intensity distribution contains high and low regions then that is indicative of some orientation preference. If several of these data-sets are collected for different reflections, these can be brought together to establish an orientation distribution function giving an indication of the average three-dimensional alignment of grains, Bunge (1983). The calculation behind this last step is not trivial, especially for low symmetry space-groups.

The two sections that follow look at this in a different way. The first considered another experimental approach that we have found very useful, because it captures more and different information. The second section is more a cautionary note, which is consequence of the theory discussed in section 2.9.

5.7.4.1. Data collection to reveal orientation texture:

Collecting the full 3D scattering from a sample is a lengthy process, but the effort can be very revealing. The early attempts at a series of 2D maps with a modified Bragg-Brentano diffractometer (the main axes ω and 2θ were uncoupled and driven independently, and the diffracted beam monochromator was tuned to isolate and use the Cu $K\beta$ wavelength), was used by Fewster and Andrew (1993b, 1999). A single map contains all the data within the range presented in figure 4.2.1, e.g. figure 5.7.7a. In this example the data was collected whilst rotating in ϕ, and a projection onto 2θ produced a diffraction profile as if no texture existed, figure 5.7.7c. If the data was collected in steps of ϕ this would create the full 3D distribution of intensity. The resolution is not high and varies over reciprocal space. The normal Bragg-Brentano scan is given in figure 5.7.7.b. The data collection of this map is lengthy with this 0D detector; however with a further compromise in resolution this can be collected in ~ 3 h, by using a position sensitive detector, as described in section 4.3.7. A more recent reciprocal space map is presented in figure 5.7.8, which formed part of the analysis discussed in section 5.7.6.

5.7.4.2. Theoretical considerations in interpretation of orientation:

If the scattering was entirely associated with the Bragg condition then the analysis is relatively straightforward as given above, i.e. the distribution of orientations can be obtained in the conventional way. However if the intensity is distributed as suggested in section 2.9, shown in figure 2.9.7, then there is a bias in the orientation distributions towards higher ω. There is increased intensity capture closer to grazing incidence from the larger axial divergence and from the tails where $\Omega_X = \theta$. In the particular example at $2\theta = 60^0$, the centre of intensity in Ω is 0.002^0 above θ; clearly the tilt is symmetrical. This represents the position of the peak when tilting the sample in ω, which contains a random orientation of

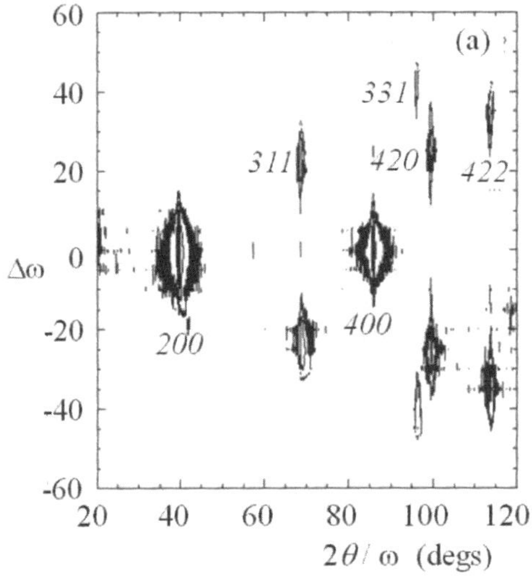

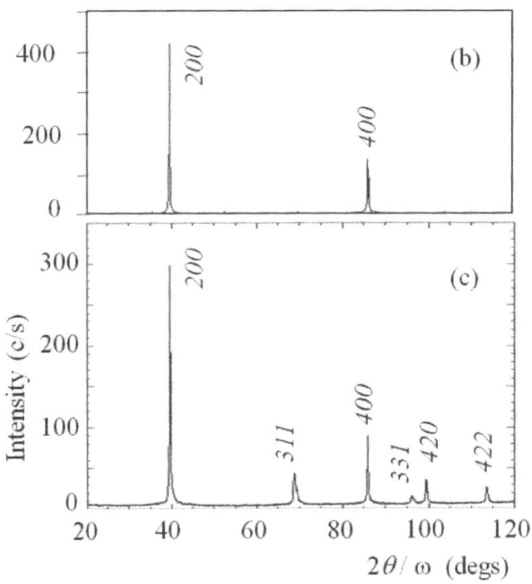

Figure 5.7.7. (a) the large area reciprocal space map from an Al textured sample, whilst rotating in ϕ to capture the whole of the available reciprocal space. (b) gives the profile with a conventional Bragg-Brentano scan and (c) gives the projection of this map to recover the whole pattern.

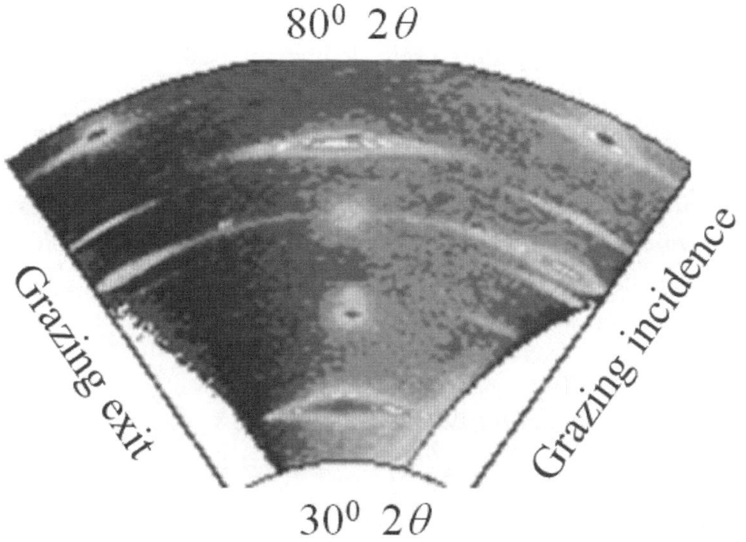

Figure 5.7.8. The reciprocal space map from a complex Nb/Al multilayer reveals the transition between textured polycrystalline and randomness in orientation. This shows numerous features for interpretation.

perfect 10 μm crystallites. In the conventional theory we would expect that $\omega = \Omega_0 = \theta$ exactly. As the crystallite quality deteriorates, the dominance of the Bragg peak is weakened and this bias can become larger, although could be lost within the resolution of the experiment; this quoted bias is equivalent to the intrinsic scattering width for this sample.

If we consider the scattering from an individual crystallite, the maximum enhancement at the Bragg angle will occur along Ω_0 and the symmetry of the diffraction pattern is conserved. That is a small crystallite will create a diffraction pattern that has the expected symmetry with the full array of reflections, although none of the crystal planes need to be satisfying the Bragg condition. In fact they could be tens of degrees away; although obviously their intensity contributions will become weaker.

5.7.5. *Distortion in polycrystals:*

The distortion relates to crystallite strain, which can be from grain to grain or within a crystallite grain. It is often assumed that the strain is within a grain (micro-strain) and the influence on the diffraction profile is representative of the average. However the same effects could arise from each crystallite having slightly different lattice parameters (distributed macro-strain). In general any difference in lattice parameter, either within a grain or from grain to grain, will create peak broadening. The Bragg angle is given by:

$$2\theta = 2\sin^{-1}\left(\frac{n\lambda}{2d}\right) \qquad 5.7.1$$

Thus the micro-strain ε (= $\Delta d/d$) will produce a spread in 2θ given by:

$$\Delta(2\theta) = -4\varepsilon \tan\theta \qquad 5.7.2$$

The factor 4 is taken as the spread in d values across the half-width of the profile on the 2θ scale. It must be remembered that another contribution to the width is that from crystallite size, equation 2.7.10 the Scherrer equation. The latter has a $1/\cos\theta$ dependence and so can be separated by measuring a series of profiles. For example the residual strain from the Al sample, figure 5.7.7, obtained by combining the *002* and *004* reflections is 0.0001.

With this analysis, we assume that the only contributions to the width are size and strain, i.e. the intrinsic diffraction contribution is negligible and that the instrumental aberrations have been accounted for. As we could see from figure 5.7.6 the shape of the profile is complex (the measured profile will be the integral of a section from the Debye-Scherrer ring) and that example indicates some spreading mainly due to the pinhole divergence giving a distribution of incident angles. If we analyse the Debye-Scherrer ring with the multiple-crystal diffractometer (section 4.3.3), employing 3D reciprocal space mapping (described in sections 4.2.2, 5.3.2.3) then it is possible to determine the exact location of the various contributions in absolute angles in reciprocal space, Fewster and Andrew (1999). An example for the textured Al sample is given in figure 5.7.9.

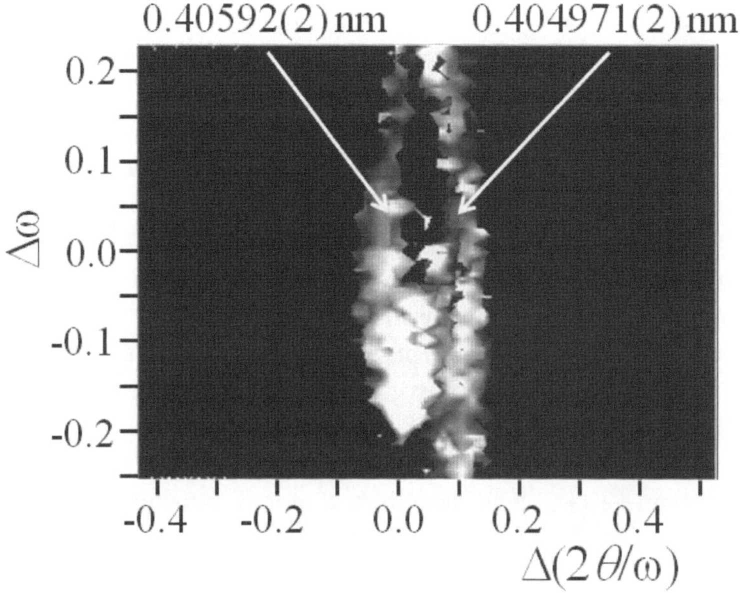

Figure 5.7.9. This measured 3D reciprocal space map of the Debye-Scherrer ring for the *002* reflection from a textured Al sample, which was collected with 2θ on an absolute scale, Cu $K\alpha_1$ radiation . The iso-intensity contours are at 20 % of the maximum level.

If the features in this 3D reciprocal space map are in some way related to the *d* spacing for the *002* reflection then there is spread of values, e.g. the two prominent regions of intensity in the figure correspond to 0.404971(2) and 0.405092(2) nm. These two regions give a strain difference of 0.0003; but this is not a weighted average, whereas the earlier figure of 0.0001 is the weighted average. But this spread comes from different contributions suggesting that this value is either 'distributed macro-strain' or is related to the diffraction effect shown in figure 5.7.6.

5.7.6. *Depth and interface analysis of polycrystalline multi-layers:*

The performance of a device based on a multilayer will depend on the interfaces and uniformity with depth. This section will consider the measurement of interfacial roughness, and the importance of choosing the right model, and a method to extract depth dependent information.

The theoretical methodologies for the interface and in-depth probing were covered in sections 2.11.1 and 2.11.4, respectively. The interface modeling posed several choices; the quality of the multi-layer (which scattering theory to use), the shape of the interface (fractal, castellated or stepped). The features in the scattering generally reveals the appropriate choice. The in-depth modeling relies on the variation of the X-ray field strength and its scatter to reveal the diffraction for analysis.

Until now we have been considering single layers or samples that are assumed to be uniform with depth. In-plane scattering appears to be a very attractive method for extracting information by varying the sampling depth as a function of incident angle. This variation is very rapid with a significant challenge for beam collimation and sample flatness, figure 2.11.9. If the sample is a periodic structure of layers then the interfaces can be studied, especially if their roughness is correlated, figure 2.11.5. Also depth information is easier to reveal because a standing-wave is generated from the periodicity of the multi-layer. The theory is presented in section 2.11.4. The example will be for a Nb/Al structure given in Fewster, Andrew, Holý and Barmak (2005). Since the standing-wave is set up based on density fluctuations the principle works with a mixture of amorphous and crystalline layers, this has been exploited by Bernstorff *et al.* (2013).

5.7.6.1. Lateral information from polycrystalline interfaces:

Figure 5.7.10 illustrates a reciprocal space map from a Nb/Al multilayer. The specular streak along s_z dominates and is punctuated with satellite reflections, which have strong lateral streaking (along s_x) associated with each: this is indicative of vertically correlated roughness. If we refer back to figure 2.11.3 that considers all the wave-vectors, then the strong scattering, spot-like features and additional streaking indicates that all the wave-vectors should be considered, i.e. this must be treated dynamically. A kinematic description would only include an incident and scatter beam (k_0 and k_H in figure 2.11.3). The reciprocal space map was collected with the geometry described in section 4.4.2, a double-slit incident conditioning and a parallel-plate collimator followed by a 40 μm slit (i.e. no analyser crystal).

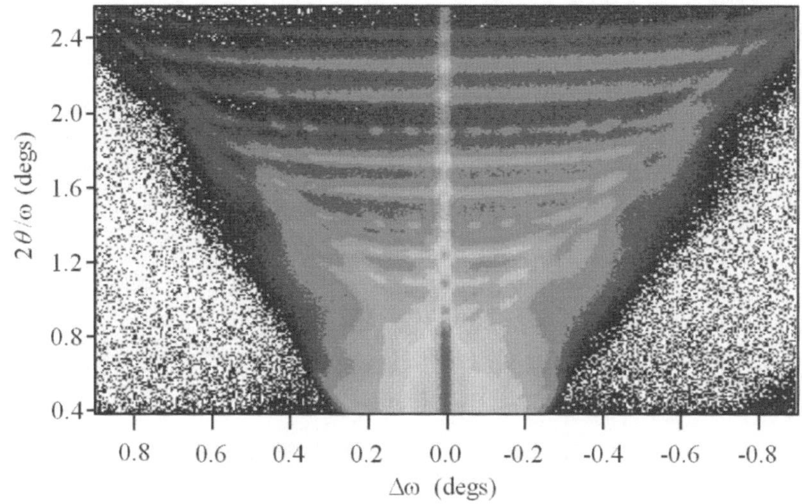

Figure 5.7.10. The reciprocal space map for the Nb/Al periodic structure obtained close to the strong specular *000* direction.

Although the full map can be simulated and fitted this is rather cumbersome, and so various ω scan were extracted and fitted. The combined fit of several profiles should give a good estimation of the lateral correlation lengths. Two ω scans are illustrated in figure 5.7.11a and b are fitted to a fractal model for the interface. The features can be reasonably well modeled with a lateral correlation length of 300 nm, a Hurst parameter of 0.2 and vertical interface spreading of 1.5 nm. But is this a correct characterization of the interfaces? Figure 5.7.11c and d shows the same data fitted to a castellated interface model to give a different characteristic lateral correlation length, which is composed of steps of height 2.0 nm separated by an upper and lower region both of length 20 nm, Lyford, Fewster and Holý (2004).

Because neither fit is perfect, it becomes a personal judgment on which best represents the description of the interface. As mentioned in section 5.4.6.2 a nearly perfect crystal is best described by a castellated, or if vicinal a staircase, interface. It may be reasonable to argue that a structure composed of small crystals of similar sizes cannot have

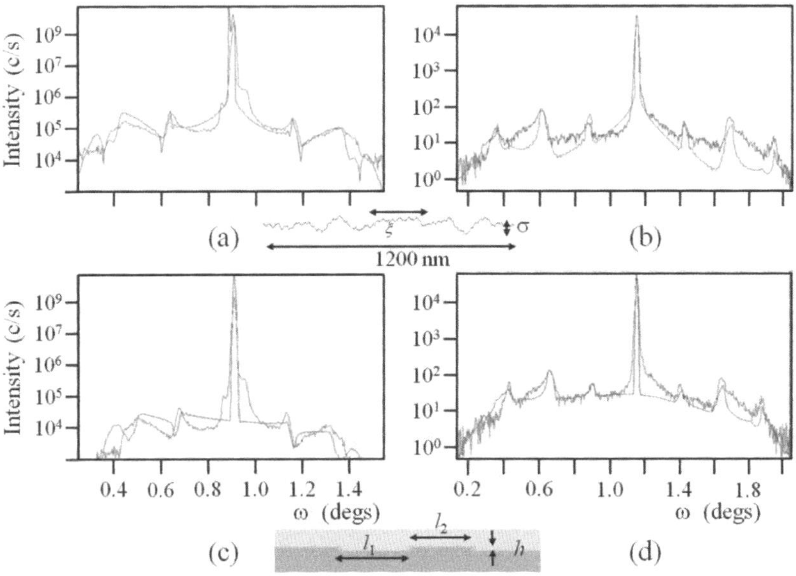

Figure 5.7.11. The ω scan profiles for the Nb/Al periodic structure obtained at 1.79^0 (a) and 2.34^0 (b) 2θ and fitted with a fractal model for the interface. (c) and (d) give the fits to a castellated interface with the same data.

interfaces described by a fractal. Therefore figure 5.7.11c and d and the derived parameters are more likely to be the most appropriate description; this is also fits within the same bounds observed in transmission electron micrographs, Barmak et al (1997).

5.7.6.2. Depth information from polycrystalline multi-layers:

The reciprocal space maps were obtained with the geometry described in section 4.5.3. The data was collected with little restriction in detector capture range normal to the sample surface. This helps in keeping the intensity at a good level, but this does require an integration in α_f for simulating the profile and the reciprocal space maps. Examples of two of these maps for the Nb/Al periodic structure mentioned in the last section is given in figure 5.7.12 showing the Nb *110* (with a small contribution from the Al *111* reflection) and the Al *220* reflection. The structure was composed of 14 periods of 35 nm with a Nb/Al molar ratio of 1:3.

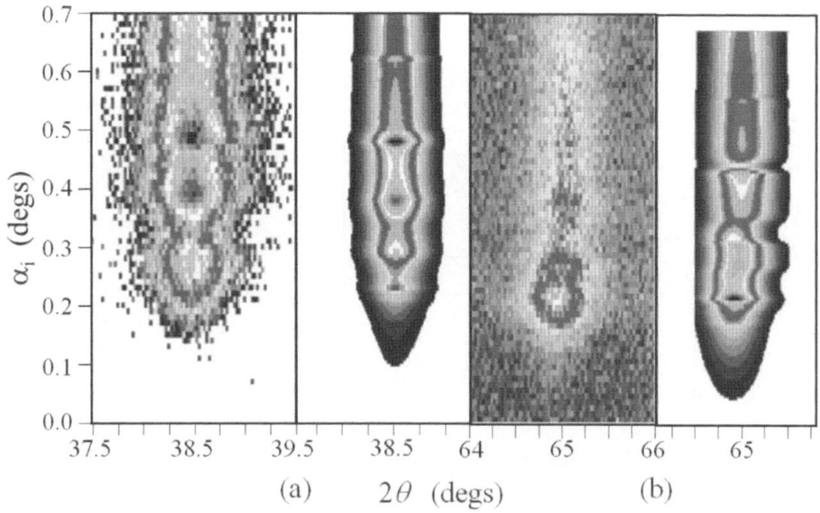

Figure 5.7.12. The measured reciprocal space map of 2θ versus α_i for the Nb/Al periodic structure centred on the Nb *110* reflection (a) and the Al *220* reflection (b) with the best fit simulated maps.

The first obvious observation of the reciprocal space maps and the α_i profiles are the peaks. These can be simulated based on the theory discussed in section 2.11.4: we can consider this as the period of the standing wave going in and out of phase with the period of the structure. So at certain period wavelengths the scattering from Al or Nb is enhanced. Also the Al *220* reflection is observed at lower α_i than Nb *110*, this is because the density of Nb is higher and pushes the critical angle to higher incidence values. The initial simulation assuming no interfacial roughness did not achieve a convincing fit (the first peak observed in the Al *220* profile was too low), the inclusion of the interfacial roughness changes this first peak dramatically, figure 5.7.13. This method is therefore very sensitive to interface quality.

Another feature of the maps is the good alignment in 2θ of the Nb *110* reflection, whereas the Al *220* alignment varies in α_i. The first peak associated with the Al layers is dominated by the uppermost layer in the structure, which also has the narrowest profile in 2θ. The growth conditions were unchanged during deposition and hence the crystallite size would be the same throughout. The lattice parameters of buried

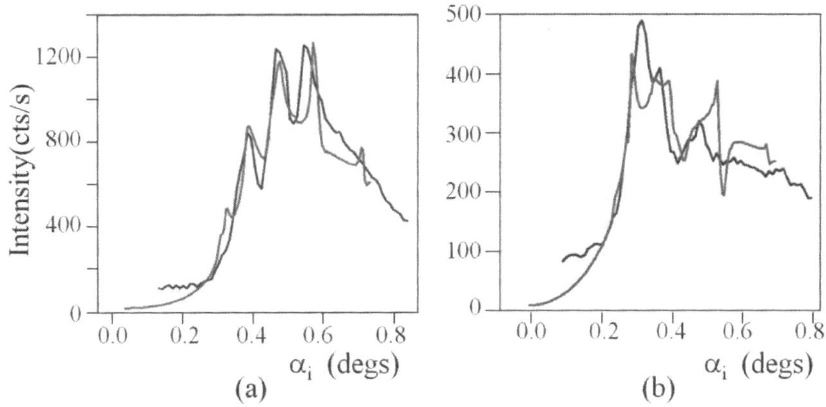

Figure 5.7.13. The α_i profile for the Nb/Al periodic structure centred on the Nb *110* reflection (a) and the Al *220* reflection (b) and the best fit simulated profiles.

layers match those of other samples and the expected bulk value. This leads to a reasoned conclusion that the buried layers have a micro-strain of 0.0013(3), the crystallite size is 12.43(60) nm, and the top layer has macro-strain of 0.00232(5). Considerably more detail is given in Fewster et al (2005), but this brief description should give an indication of what can be achieved.

5.8. Concluding remarks:

These examples should indicate the sensitivity to various parameters and the level of detail that can be achieved. X-ray scattering reveals deviations from perfection with the right method, and very quick analyses are achievable for all the basic parameters. The study of imperfect epitaxy offers the greatest challenge since more parameters are required to characterise the material, Table 1.3.1, chapter 1. Examples of these have been covered. There are still many challenging problems to be addressed as the diversity of materials and combinations keep appearing. This subject is still very much alive and dynamic!

Bibliography

Anderson, J C, Fewster, P F, Ozsvald, I, Tye, G A (2010) unpublished work.

Auvray, P, Baudet, M and Regreny, A (1989) J Cryst. Growth **95** 288

Barmak, K, Rickman, J M and Michaelsen, C (1997) J Electron. Mater. **26** 1009

Barrett, C S (1945) Transactions of the American Inst. Of Mining and Metallurgy Engineers **161** 15–64

Berg, W F (1931) Naturwiss **19** 391–396

Bernstoff, S, Holý, V, Endres, J, Vales, V, Sobota, J, Siketić, Z, Bognanović-Radović, I, Buljan, M and Dražić, G (2013) J. Appl. Cryst. **46** 1711–1721.

Birch, J, Sundgren, J-E and Fewster, P F (1995) J Appl. Phys. **78** 6582

Bond, W L (1960) Acta Cryst. **13** 814

Bunge, H J (1983) *Texture Analysis in Material Science: Mathematical Methods,* Butterworth-Heinemann

Cullity, B D (1978) *Elements of X-ray diffraction,* Addison Wesley, Reading

Dane, A D, Veldhuis, A, de Boer, D K G, Leenaers, A J G and Buydens, L M C (1998) Physica B **253** 254

Dismukes, J P, Ekstrom, L and Paff, R J (1964) J Phys. Chem. **68** 3021

Fewster, P F (1984) J Appl. Cryst. **17** 265

Fewster, P F (1986) Philips J Res. **41** 268

Fewster, P F (1987) *Thin Film Growth Techniques for Low Dimensional Structures* NATO ASI Series B: Physics **163** pp. 417–440, Ed: Farrow et al. New York: Plenum.

Fewster, P F (1988) J Appl. Cryst. **21** 524-529

Fewster, P F (1990) Proceedings Volume **90–15**, pp. 381–384, J Electrochem. Soc. *Superlattice Structures and Devices.*

Fewster, P F (1991a) *Analysis of Microelectronic Materials and* Devices Chapter 3.1.1. Ed: Grausserbauer and Werner, Wiley: New York

Fewster, P F (1991b) J Appl. Cryst. **24** 178

Fewster, P F (1993a) Semicond. Sci. Technol. **8** 1915

Fewster, P F (1993b) Patent Nos: EPO 0603943B, 5442676, B33826

Fewster, P F (1996a) Rep. Prog. Phys. **59** 1339

Fewster, P F (1996b) Critical Reviews in Solid State and Materials Sciences **22** 69

Fewster, P F (1998) *X-ray and Neutron Dynamical Diffraction: Theory and Applications*, pp. 289–299, Ed: Authier et al, Plenum Press: New York.

Fewster, P F (1999) Microscopy of Semicond. Mater. Conf. Oxford; Inst. Phys. Conf. Series **164** pp197–206.

Fewster, P F (2005) J Appl. Cryst. **38** 62–68

Fewster, P F and Andrew, N L (1993a) J Appl. Cryst. **26** 812

Fewster, P F and Andrew, N L (1993b) Materials Science Forum, **133** 221 Ed: Mittemeijer and Delhez, Trans Tech Pub: Switzerland

Fewster, P F and Andrew, N L (1993c) J Appl. Phys. **74** 3121

Fewster, P F and Andrew, N L (1995a) J Phys. D **28** A97

Fewster, P F and Andrew, N L (1995b) J Appl. Cryst. **28** 451

Fewster, P F and Andrew, N L (1998) Thin Solid Films **319** 1

Fewster, P F and Andrew, N L (1999) Reciprocal space mapping and ultra-high resolution diffraction of polycrystalline materials: *Defect and Microstructure Analysis by Diffraction*, Ed: Snyder, Fiala and Bunge, IUCr series 10: pp. 346–364. Oxford University Press.

Fewster, P F, Andrew, N L and Curling, C J (1991) Semicond. Sci. Technol. **6** 5

Fewster, P F and Curling, C J (1987) J Appl. Phys. **62** 4154

Fewster, P F, Andrew, N L, Holý, V and Barmak, K (2005) Physical Review B **72** 174105-1, 11

Fewster, P F, Holý, V and Andrew, N L (2001) Material Science in Semiconductor Processing **4** 475–481

Fewster, P F, Holý, V and Zhi, D (2003) J Phys. D: Appl. Phys. **36** A217–A221

Fewster, P F and Tye, G A (2001) Patent no: GB010114

Fewster, P F and Whiffin, P A C (1983) J Appl. Phys. **54** 4668

Fullerton, E E, Schuller, I K, Vanderstaeten, H and Bruynseraede, Y (1992) Phys. Rev. **B45** 9292.

Hart, L, Fahy, M R, Newman, R C and Fewster, P F (1993) Appl. Phys. Lett. **62** 2218–2220

Hart, L, Fewster, P F, Ashwin, M J, Fahy, M R, Newman, R C (1995) J Phys. D: Appl. Phys. **28** A154–A158

Heflen, L, Myagotin, A, Rack, A, Pernot, P, Mikulík, P, Michel, M D and Baumbach, T (2007) phys. Stat. solidi (a) **204** 2760–2765

Hornstra, J and Bartels, W J (1978) J Cryst. Growth **44** 513

Keissing, H (1931) Ann. Phys. Lpz. **10** 769

Kidd, P (2003) J Material Science: Materials in Electronics **14** 541–550

Kidd, P and Fewster, P F (1994) Mat. Res. Soc. Symp. Proc. **317** 291

Klappe, J G E and Fewster, P F (1993) J Appl. Cryst. **27** 103

Lang, A R (1959) Acta Cryst. **12** 249–250

Laugier, J and Filhol, A (1983) J Appl. Cryst. **16** 281

Lyford, T, Fewster, P F and Holý, V (2004) presented at XTOP 2004: Prague, Czech Republic.

Moram, M A and Vickers, M E (2009) Reports on Progress in Physics **72** 036502 1–40

Nagai, H (1974) J Appl.Phys. **45** 3789

Sevidori, M, Cembali, F and Milita, S (1996) *X-ray and Neutron Dynamical Diffraction: Theory and Applications*, pp.301–321, Ed: Authier et al, Plenum Press: New York.

Schiller, C (1988) Analysis **16** 402

Springholz, G, Holý, V, Pinczolits, M and Bauer, G (1998) Science **282** 734

Stojanoff, V and Siddons, D P (1996) Acta Cryst. A **52** 498–499

Tye, G A and Fewster, P F (2000) Unpublished work

Vegard, L (1921) Zeitschrift für Physikalishe Chemie **5** 17–26

Wilkins, S W, Gureyev, T E, Gao, D, Pogany, A & Stevenson, A W (1996). *Nature* **384**: 335–338.

Wormington, M, Panaccione, C, Matney, K M and Bowen, D K (1999) Phil. Trans. Roy. Soc. **A 357** No. 1761 p2827

Zeimer, U, Grenzer, J, Korn, D, Döring, S, Zorn, M, Pittroff, W, Pietsch, U, Saas, F and Weyers, M (2007) Physica Status Solidi (a) **204** 2753–2759

Zhang, J, Neave, J H, Li, X, Fewster, P F, El Mubarek, W A W, Ashburn, P, Mitrovic, I Z, Buiu, O, amd Hall, S (2005) J Crystal Growth **278** 505–511

Ziedses des Plantes, B G (1932) Acta Radiologica **13** 182–192

Epilogue

During my work in this subject area, there are numerous aspects that I think are still outstanding; many of which are associated with the latest work on "the New Theory of X-ray Diffraction". Although the latter helps explain many of the puzzling features in powder diffraction, as it is discussed in this book, there are numerous questions that are raised because of it and many unanswered questions may well be explained by it. This short epilogue is a list of these features, which is not exhaustive, but may be of interest to those who puzzle over such things!

- Since the non-Bragg scattering appears to be a significant proportion of the total scattering in small crystals, the scattering contribution to absorption may be more significant than assumed.
- The 'wiggles' associated with 'crystal truncation rods' are an interesting feature, and the assumption in the book is some interplay between the surface shape and reflecting planes; this new theory may help in giving a better understanding of the effect.
- Since, from this theory, scattering exists everywhere the diffraction pattern it is much richer than just that which exists at the Bragg condition. This could help reveal features that may help in determining molecular structures.
- There is no reason for path lengths of $n\lambda$, where $n > 1$, not to exist in diffraction. For example n=2 should exist in diamond and Si (*111*) orientated crystals, and intensity can be observed at this position corresponding to that of the 'systematically absent' *222* reflection. This has been interpreted in terms of bonding asymmetry, but perhaps there is a simpler alternative explanation or that this diffraction effect is a significant contributor to the intensity at this position?
- If there is intensity associated with the 'harmonics' where $n > 1$, then this will add to the second order reflection where $n = 1$. This could change the interpretation of the intensity for this second order reflection.

- The combination of 'harmonics' in the scattering should give an indication of the atomic vibrations, since each is a consequence of a different path length that has a different sensitivity. For example the *hkl* reflection for $n = 1$will be attenuated less than for $n = 2$, etc.
- Since there is intensity at the angle $2\theta_{Bragg}$ independent of the crystal orientation, it is difficult to see how the orientation of a crystallite can be inferred from a diffraction pattern without including intensity information, i.e. knowing how close the orientation is to a Bragg condition.
- Since a significant proportion of the intensity associated with a crystal plane is not in the vicinity of the Bragg condition, how much reliance can we put on the intensity measurements from small crystals?

Many of these comments can be answered rather easily with increased computing power. This would allow the calculation of the scattering from a full atomistic model of the crystal and would be the most exacting model. Moving to further interactions similar to the extended Born approximation will make it dynamical. The use of 'sinc' type functions does account for a distribution of scattering points, but is less transparent. At present most atomistic models are limited to very small crystals: the reason for this is simply for a 3D array of atoms 1.5 Å apart, a very small 1.5 μm crystal cube will contain 10^{12} atoms, creating a significant challenge.

APPENDIX 1

GENERAL CRYSTALLOGRAPHIC RELATIONS

A.1. Introduction

The crystal symmetry of any material is reflected in its physical properties. The sensitivity of certain crystallographic planes to these physical properties will be helpful in deciding the appropriate experiment to probe the structural details. To help choose a suitable scattering plane for an experiment, some stereographic projections for commonly used surface plane orientations are given for the cubic and hexagonal symmetries. The common surface plane projections are (*001*), (*110*) and (*111*) in the cubic case and (*0001*) in the hexagonal case.

The projections for cubic symmetries are independent of the value of the lattice parameters, however hexagonal projections are altered by the ratio of the *c* to *a* axis lengths. Only the one projection is given for the hexagonal case: this projection is not too sensitive to the *c/a* ratio and is therefore a reasonable representation of this system for this commonly used orientation. The additional index *i* in the hexagonal notation (*hkil*) can be derived from $i = -\{h + k\}$. The inclusion of this index is very useful to indicate similar planes, for example the plane index (*11-23*) has symmetry equivalent planes of (*1-213*) and (*-2113*). The similarity is clear in the (*hkil*) notation, whereas in the (*hkl*) notation this equivalence is less clear, i.e. (*113*), (*1-23*) and (*-213*).

The interplanar spacings for various sets of planes are considered first and given for the general case, which is then simplified for the cubic and hexagonal cases. The unit cell lengths are *a*, *b* and *c* and the angle between *a* and *b* is γ, *b* and *c* is α and *c* and *a* is β. The interplanar spacing is required to determine the stereographic projections for the non-cubic symmetries.

The relationship between the sample surface and the underlying crystal structure, e.g. crystal plane offsets, will change the nature of the distortion of deposited layers and may not correspond to the original space-group. To interpret the distortion and scattering from such structures the following more general equations (triclinic case) combined with the vector product will give the index for directions normal and parallel to the surface plane.

A.2. Interplanar spacings

The distance d between successive atomic planes of the type (hkl) is given by

$$d = \left[\frac{1}{V^2} \left\{ s_{11}h^2 + s_{22}k^2 + s_{33}l^2 + 2s_{12}hk + 2s_{13}hl + 2s_{23}kl \right\} \right]^{-1/2} \qquad \text{A1}$$

where

$$\begin{aligned} s_{11} &= \{bc\sin(\alpha)\}^2 \\ s_{22} &= \{ca\sin(\beta)\}^2 \\ s_{33} &= \{ab\sin(\gamma)\}^2 \\ s_{12} &= abc^2\{\cos(\alpha)\cos(\beta) - \cos(\gamma)\} \\ s_{13} &= ab^2c\{\cos(\gamma)\cos(\alpha) - \cos(\beta)\} \\ s_{23} &= a^2bc\{\cos(\beta)\cos(\gamma) - \cos(\alpha)\} \end{aligned} \qquad \text{A.2}$$

and the volume is given by

$$V = abc\{1 - \cos^2(\alpha) - \cos^2(\beta) - \cos^2(\gamma) + 2\cos(\alpha)\cos(\beta)\cos(\gamma)\}^{1/2} \qquad \text{A.3}$$

We can simplify these relationships for the cubic symmetry

<u>*Cubic:*</u>

$$a = b = c$$
$$\alpha = \beta = \gamma = 90^0$$

hence

$$d = \frac{a}{\left[h^2 + k^2 + l^2\right]^{1/2}} \qquad \text{A.4}$$

$$V = a^3 \qquad \text{A.5}$$

Hexagonal:

$$a = b \neq c$$

$$\alpha = \beta = 90^0, \gamma = 120^0$$

$$d = \left[\frac{3}{4\left\{\frac{h^2 + k^2 + hk}{a^2} + \frac{3l^2}{4c^2}\right\}}\right]^{1/2} \qquad \text{A.6}$$

$$V = \frac{3^{1/2} a^2 c}{2} \qquad \text{A.7}$$

A.3. Stereographic projections:

The normal to an atomic plane (*hkl*) will intersect a hemisphere at a position called a pole and some typical poles are given in figure A1 to A4. Thus for the (*001*) pole for the stereographic projection centred on the (*001*) plane will be directly above the centre. The distances along the surface of this hemisphere will represent the angles between any two planes. For example the angle between (*001*) and (*100*) is 90^0 in the cubic system and will therefore occur at a distance of one quarter of the circumference of the sphere. The angles between common crystal planes in the cubic system are given in Table A1. The lines in the projection are a guide to follow surface planes of a type, for example in figure A1 the line from the pole (*001*) to (*110*) will include (*115*), (*114*), (*113*), (*112*) and (*111*), remembering also that these are parallel to (*2210*), (*228*), (*226*), (*224*) and (*333*) respectively.

The angle between atomic planes ($h_1k_1l_1$) and ($h_2k_2l_2$) is determined from the following formulae.

$$\varphi = \cos^{-1}\left(\frac{d_1d_2}{V^2}\{s_{11}h_1h_2 + s_{22}k_1k_2 + s_{33}l_1l_2 + s_{12}(h_1k_2 + h_2k_1) + s_{13}(h_1l_2 + h_2l_1) + s_{23}(k_1l_2 + k_2l_1)\}\right) \qquad \text{A.8}$$

Again this expression can be simplified for cubic and hexagonal systems

Cubic:

$$\varphi = \cos^{-1}\left(\frac{h_1h_2 + k_1k_2 + l_1l_2}{\left[(h_1^2 + k_1^2 + l_1^2)(h_2^2 + k_2^2 + l_2^2)\right]^{1/2}}\right) \qquad \text{A.9}$$

Hexagonal:

$$\varphi = \cos^{-1}\left(\frac{h_1h_2 + k_1k_2 + \frac{1}{2}(h_1k_2 + h_2k_1) + \frac{3a^2}{4c^2}l_1l_2}{\left[\left(h_1^2 + k_1^2 + h_1k_1 + \frac{3a^2}{4c^2}l_1^2\right)\left(h_2^2 + k_2^2 + h_2k_2 + \frac{3a^2}{4c^2}l_2^2\right)\right]^{1/2}}\right) \qquad \text{A.10}$$

etc.

Table A.1: The angles between common crystallographic planes in the cubic system:

(hkl)	**Common reflections for FCC semiconductors**	*(1 0 0)*	*(11 0)*	*(1 1 1)*
(1 0 0)	*(002)*, *(004)*, *(006)*	0 90	45 90	54.7
(1 1 0)	*(022)*, *(044)*	45 90	0 60 90	35.3 90
(1 1 1)	*(111)*, *(222)*, *(333)*, *(444)*	54.7	35.3 90	0 70.5 109.5
(2 1 0)	*(024)*	26.6 63.4 90	18.4 50.8 71.6	39.2 75
(2 1 1)	*(224)*	35.3 65.9	30 54.7 73.2 90	19.5 61.9 90
(2 2 1)	*(244)*	48.2 70.5	19.5 45 76.4 90	19.5 61.9 90
(3 1 0)	*(026)*	18.4 71.6 90	26.6 47.9 63.4 77.1	43.1 68.6
(3 1 1)	*(311)*	25.2 72.5	31.5 64.8 90	29.5 58.5 80
(3 2 0)	*(046)*	33.7 56.3 90	31.5 64.8 90	29.5 58.5 80
(3 2 1)	*(246)*	36.7 57.7 74.6	19.1 40.9 55.5	22.2 51.9 72 90
(3 3 1)	*(331)*	46.5 76.7	13.1 49.5 71.1	22 48.5 82.4
(3 3 5)	*(335)*	62.8	49.7	14.4 84.9
(5 1 1)	*(115)*	15.6 78.9	74.2	38.9
(7 1 1)	*(117)*	11.3 82	78.6	43.3

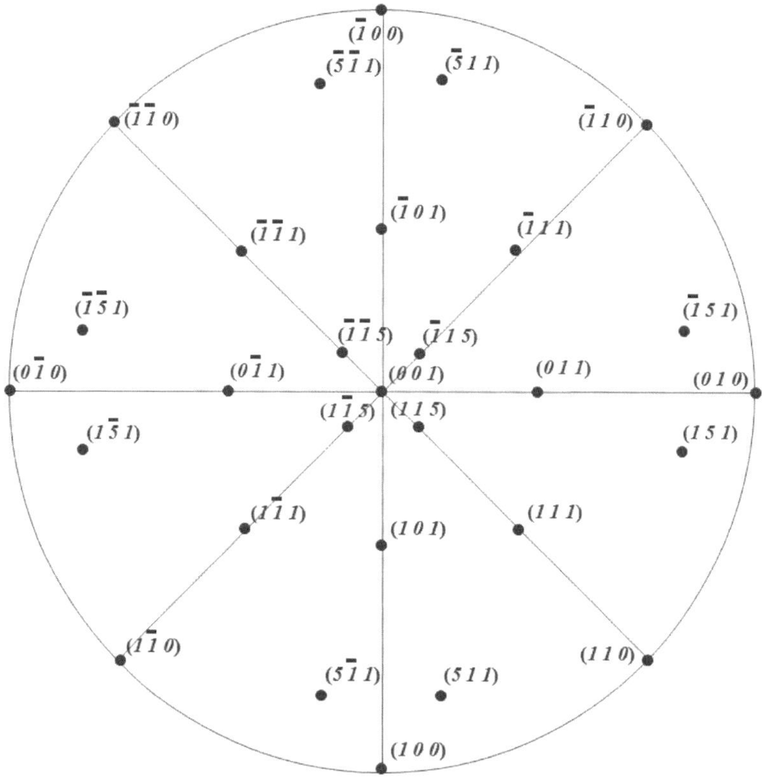

Figure A.1. This is the stereographic projection of the (*001*) plane for the cubic system. These indexes of the poles make it easier to visualise the angle between the (*hkl*) planes.

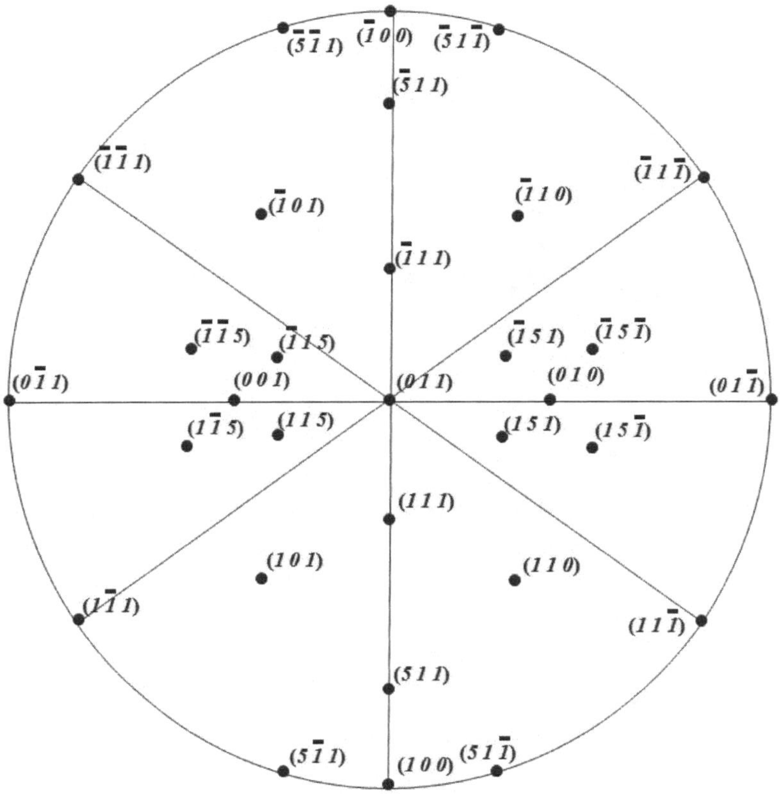

Figure A.2. This is the stereographic projection of the *(011)* plane for the cubic system.

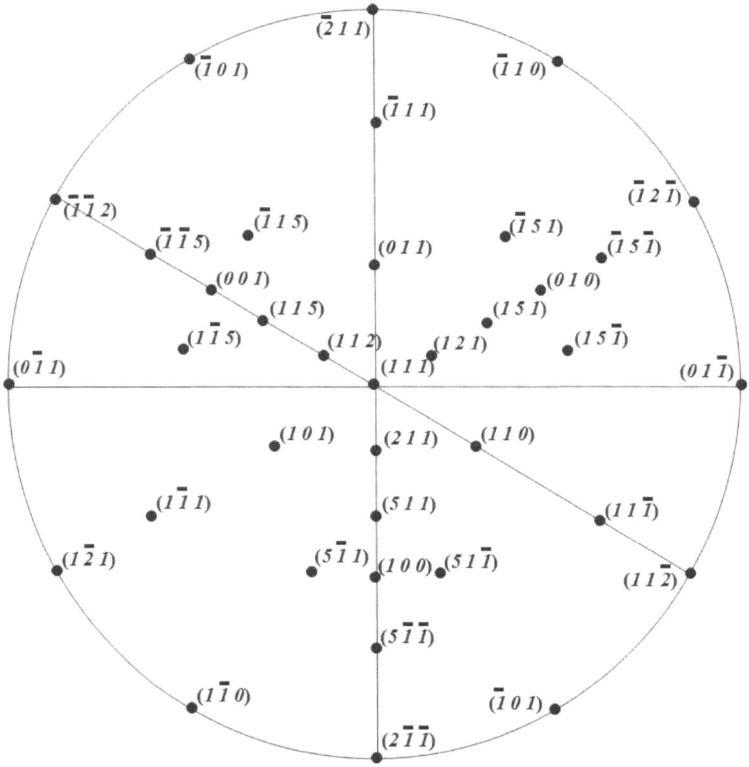

Figure A.3. This is the stereographic projection of the (*111*) plane for the cubic system.

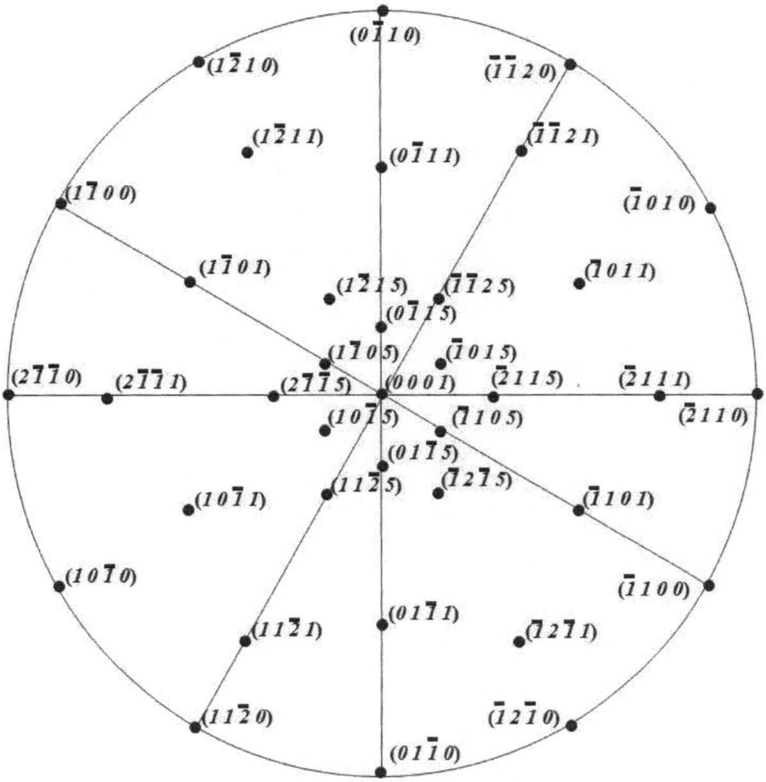

Figure A.4. This is the stereographic projection of the (*0001*) plane for the hexagonal system for a *c/a* ratio of 1.63 (similar to that for GaN).

INDEX